Advances in Solid Oxide Fuel Cells II

T0312124

Advances in Solid Oxide Fuel Cells II

A Collection of Papers Presented at the 30th International Conference on Advanced Ceramics and Composites January 22–27, 2006, Cocoa Beach, Florida

Editor

Narottam P. Bansal

General Editors

Andrew Wereszczak
Edgar Lara-Curzio

A JOHN WILEY & SONS, INC., PUBLICATION

For general information on our other products and services please contact our Customer Care Department within the U.S. at 877-762-2974, outside the U.S. at 317-572-3993 or fax 317-572-4002.

Wiley also publishes its books in a variety of electronic formats. Some content that appears in print, however, may not be available in electronic format.

Library of Congress Cataloging-in-Publication Data is available.

ISBN-13 978-0-470-08054-2
ISBN-10 0-470-08054-X

10 9 8 7 6 5 4 3 2 1

Contents

Interconnects and Protective Coatings

Seals

Mechanical Properties

Modeling

Preface

The third international symposium "Solid Oxide Fuel Cells: Materials and Technology" was held during the 30th International Conference on Advanced Ceramics and Composites in Cocoa Beach, FL, January 22–27, 2006. This symposium provided an international forum for scientists, engineers, and technologists to discuss and exchange state-of-the-art ideas, information, and technology on various aspects of solid oxide fuel cells. A total of 125 papers, including three plenary lectures and eleven invited talks, were presented in the form of oral and poster presentations indicating strong interest in the scientifically and technologically important field of solid oxide fuel cells. Authors from four continents and 14 countries (Brazil, Canada, Denmark, France, Germany, India, Iran, Italy, Japan, Sweden, Switzerland, Taiwan, Ukraine, and U.S.A.) participated. The speakers represented universities, industries, and government research laboratories.

These proceedings contain contributions on various aspects of solid oxide fuel cells that were discussed at the symposium. Forty one papers describing the current status of solid oxide fuel cells technology and the latest developments in the areas of fabrication, characterization, testing, performance analysis, long term stability, anodes, cathodes, electrolytes, interconnects and protective coatings, sealing materials and design, interface reactions, mechanical properties, cell and stack design, protonic conductors, modeling, etc. are included in this volume. Each manuscript was peer-reviewed using the American Ceramic Society review process.

The editor wishes to extend his gratitude and appreciation to all the authors for their contributions and cooperation, to all the participants and session chairs for their time and efforts, and to all the reviewers for their useful comments and suggestions. Financial support from the American Ceramic Society is gratefully acknowledged. Thanks are due to the staff of the meetings and publications departments of the American Ceramic Society for their invaluable assistance. Advice, help and cooperation of the members of the symposium's international organizing committee (Tatsumi Ishihara, Tatsuya Kawada, Nguyen Minh, Mogens Mogensen,

Nigel Sammes, Prabhakar Singh, Robert Steinberger-Wilkens, and Jeffry Stevenson) at various stages were instrumental in making this symposium a great success.

It is our earnest hope that this volume will serve as a valuable reference for the engineers, scientists, researchers and others interested in the materials, science and technology of solid oxide fuel cells.

NAROTTAM P. BANSAL

Introduction

This book is one of seven issues that comprise Volume 27 of the Ceramic Engineering & Science Proceedings (CESP). This volume contains manuscripts that were presented at the 30th International Conference on Advanced Ceramic and Composites (ICACC) held in Cocoa Beach, Florida January 22–27, 2006. This meeting, which has become the premier international forum for the dissemination of information pertaining to the processing, properties and behavior of structural and multifunctional ceramics and composites, emerging ceramic technologies and applications of engineering ceramics, was organized by the Engineering Ceramics Division (ECD) of The American Ceramic Society (ACerS) in collaboration with ACerS Nuclear and Environmental Technology Division (NETD).

The 30th ICACC attracted more than 900 scientists and engineers from 27 countries and was organized into the following seven symposia:

- Mechanical Properties and Performance of Engineering Ceramics and Composites
- Advanced Ceramic Coatings for Structural, Environmental and Functional Applications
- 3rd International Symposium for Solid Oxide Fuel Cells
- Ceramics in Nuclear and Alternative Energy Applications
- Bioceramics and Biocomposites
- Topics in Ceramic Armor
- Synthesis and Processing of Nanostructured Materials

The organization of the Cocoa Beach meeting and the publication of these proceedings were possible thanks to the tireless dedication of many ECD and NETD volunteers and the professional staff of The American Ceramic Society.

ANDREW A. WERESZCZAK
EDGAR LARA-CURZIO
General Editors

Oak Ridge, TN (July 2006)

Overview and Current Status

DEVELOPMENT OF TWO TYPES OF TUBULAR SOFCS AT TOTO

Akira Kawakami, Satoshi Matsuoka, Naoki Watanabe, Takeshi Saito, Akira Ueno
TOTO Ltd.
Chigasaki, Kanagawa 253-8577 Japan

Tatsumi Ishihara
Kyushu University
Higashi, Nishi 819-0395 Japan

Natsuko Sakai, Harumi Yokokawa
National Institute of Advanced Industrial Science and Technology (AIST)
Tsukuba, Ibaraki 305-8565 Japan

ABSTRACT

The current status of two types of SOFC R & D at TOTO is summarized. We have developed 10kW class tubular SOFC modules for stationary power generation using Japanese town gas (13A) as fuel. A small module which consisted of 5 stacks (a stack consisted of 2×6 cells) generated 1.5kW at $0.2A/cm^2$ and achieved 55%-LHV efficiency at an average temperature of 900°C. A thermally self-sustaining module consisting of 20 stacks achieved 6.5kW at $0.2A/cm^2$ and 50%-LHV.

We have also developed micro tubular SOFCs for portable application, which operate at relatively lower temperatures. The single cell generated 0.85, 0.70, and $0.24W/cm^2$ at 700°C, 600°C, and 500°C, respectively. We built and evaluated a stack consisting of 14 micro tubular cells, and it successfully demonstrated 43W, 37W and 28W at a temperature of 700°C, 600°C, and 500°C, respectively.

INTRODUCTION

TOTO is the top sanitary ware manufacturer in Japan and highly experienced in traditional and advanced ceramic products. Solid Oxide Fuel Cells (SOFCs) are mainly composed of ceramics, and our fabrication technology has been utilized to produce high performance SOFCs at a low cost. TOTO started the research and development of tubular type SOFCs in 1989. From 2001 to 2004, we successfully completed a 10kW class thermally self-sustaining module test in a New Energy and Industrial Technology Development Organization (NEDO) project[1]. Since 2004, TOTO started a new collaboration with Kyushu Electric Power Co., Inc. and Hitachi, Ltd. in a new NEDO project. We are developing a co-generation system by integration with the TOTO stack. On the other hand, TOTO also started the development of micro SOFCs using micro tubular cells (diameter is less than 5mm) from 2002 under another NEDO project. In this paper, we summarized the current status of two types of SOFCs R & D at TOTO, i.e. tubular SOFC for stationary power generation and micro tubular SOFC for portable power application.

TOTO TUBULAR SOFC

Cell development

The schematic viewgraphs of the TOTO tubular cell are shown in Figure 1. A perovskite cathode tube is formed by extrusion molding. A zirconia electrolyte and a nickel/zirconia cermet

3

anode are coated onto the tube by the TOTO wet process[2]. A vertical interconnector is coated along the tube in a strip. The diameter of the cell is 16.5mm and the active length is 660mm. Fuel gas is supplied to the outside of the cell, and air is supplied to the inside by a thinner air supply tube. Recently, the materials used for cells are changed as shown in Table 1. The new material configuration resulted in the improvement of cell performance as shown in Figure 2, especially in lower operating temperatures. Our latest cell worked best at temperatures over 850°C.

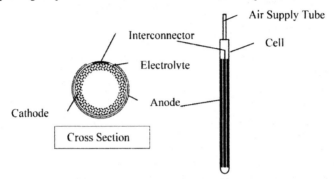

Figure 1. Schematic view of TOTO tubular cell.

Table 1. Materials used in TOTO tubular cell.

Components	Previous Type Cell	New Type Cell
Cathode Tube	$(La,Sr)MnO_3$	$(La,Sr)MnO_3$
Electrolyte	YSZ	ScSZ
Anode	Ni/YSZ	Ni/YSZ
Interconnector	$(La,Ca)CrO_3$	$(La,Ca)CrO_3$

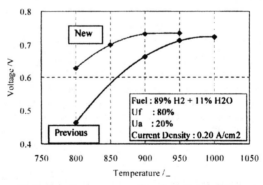

Figure 2. Cell performances as a function of operating temperature.

Stack development

Twelve tubes are bundled in a 2×6 stack with nickel materials connecting an interconnector of one cell and the anode of the next (Figure 3). A stack was installed in a metal casing and heated by an electric furnace. Simulated fuel of partially steam reformed town gas was supplied to the stack. The town gas was assumed to be 50% steam reformed under S/C (steam carbon ratio) =3.0. A stack generated 0.34kW at a current density of 0.2A/ cm^2 at a temperature of 940°C. The maximum efficiency for DC output of the stack was 57%-Lower Heating Value (LHV) calculated on the basis of equivalent town gas (Figure 4).

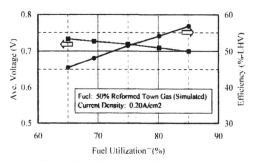

Figure 3. Appearance of TOTO stack.   Figure. 4 Performance of 2×6 stack.

Small module (quarter size module)

The thermally self-sustaining operation indicates that the modules generate power without any external heat supply. A quarter-size small size modules, which consisted of 5 stacks were made and tested for basic evaluations to realize the thermally self-sustaining operation. The small module and its metal casing were covered with a ceramic insulator. In this test, the desulfurized town gas was partially steam reformed through a reactor. The reformer was installed outside of the module with an electric furnace as shown in Figure 5. The conversion of steam

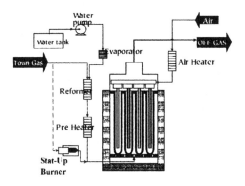

Figure 5. Flow diagram of small SOFC module.

reforming can be controlled independently with reformer temperature. However the higher hydrocarbons such as ethane (C_2H_6), propane (C_3H_8) and butane (C_4H_{10}) were completely converted. The residual CH_4 was internally reformed on the anode, and the endothermic effect was utilized to maintain the homogeneous temperature distribution of the module. The outlet fuel and air were mixed and burned above the module, and this combustion heat was used for air pre-heating. Those improvements in temperature distribution and fuel gas distribution strongly affected on the module performance. We succeeded to operate a module with 1.6kW at 0.2A/cm^2 at an average temperature of 900°C which corresponds to the efficiency of 40%-LHV for 3000 hours. The maximum efficiency was 55 %- LHV, which obtained for different module.

Thermally self-sustaining module

A ten kW class module consisting of four quarter-size modules is fabricated for thermally self-sustaining operation (Figure 6). An integrated heat exchanging steam reformer was mounted above the module. It consisted of an evaporator, pre-heater and reformer. However, the stability of the evaporator was not clearly demonstrated, therefore steam was supplied by another evaporator with an electric furnace. The gas flow was simple, and it does not include any gas recycles as shown in Figure 7. The module, the after-burning zone, and the reformer were covered with a ceramic insulator. A commercial steam reforming catalyst was embedded in the reforming section. The desulfurized town gas was supplied to the modules after being partially steam reformed. In the test, the heat of exhaust gases was used for steam reforming through the reformer. The module was heated by a partially oxidation burner and air heaters from room temperature.

The voltages of each stack, and of the 2-cells in a quarter module, were monitored, and the variation of the voltages was quite small. The module generated 6.5kW at 0.19A/cm^2 and 46%-LHV (Table 2 Test 1). The improved module generated 6.5kW at 0.2A/cm^2 and 50%-LHV (Table 2 Test 2). The thermally self-sustainability of both operating condition were confirmed. The test was shifted to evaluation of long-term stability under the condition of Test 1. No degradation was observed in the module or the integrated heat exchanging steam reformer during the operating time of 1000 hours (Figure 8).

Stack

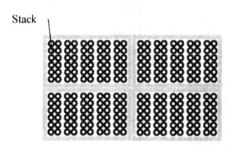 

Figure 6. Stack layout and appearance of thermally self-sustaining SOFC module.

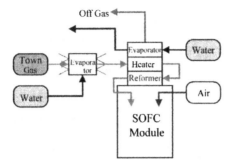

Figure 7. Flow diagram of thermally self-sustaining SOFC module system.

Table 2. Performance of thermally self-sustaining SOFC module.

Items	Test 1	Test 2
Power(kW)	6.4	6.5
Uf (%)	70	75
Current Density (A/ cm^2)	0.19	0.20
Ave. Cell Voltage (V)	0.69	0.70
Efficiency (%-LHV)	46	50

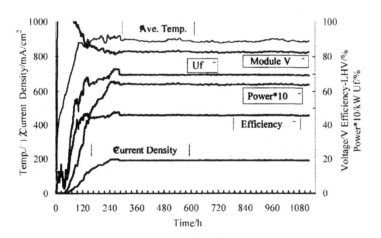

Figure 8. Long term stability of thermally self-sustaining SOFC module.

TOTO MICRO TUBULAR SOFC

Background

A micro tubular SOFC has potential for portable and transportation applications because of its advantages, including high volumetric power density and high thermal shock resistance which enables rapid start up[3]. We started the development of micro SOFCs using micro tubular cells (diameter is less than 5mm) from 2002 under a NEDO project[4]. The objectives of this project are to lower operation temperature to 500-700°C and to develop a compact stack that will achieve downsizing and rapid start up (in minutes or less) for the SOFC system. One of the significant advantages of SOFCs is its fuel flexibility compared to other kinds of fuel cells. Liquid petroleum gas (LPG) or dimethyl ether (DME) was selected as fuels because they can be easily handled and stored in portable cartridges.

To achieve high power density at lower temperatures, the anode-supported design with thin lanthanum gallate with strontium and magnesium doping was selected. The single cells and cell stacks were tested with various fuels, i.e. hydrogen, simulated reformate gas of LPG, and direct fueling of DME.

Micro tubular cell development

Table 3 shows materials and fabrication processes of the TOTO micro tubular cell. The anode substrate tube made of $NiO/(ZrO_2)_{0.9}(Y_2O_3)_{0.1}$ (NiO/YSZ) was formed by extrusion molding. The anode interlayer and the electrolyte were subsequently coated onto the anode substrate by slurry coating, and co-fired. These techniques are suitable for mass production and cost reduction. $NiO/(Ce_{0.9}Gd_{0.1})O_{1.95}$ (NiO/GDC10) anode interlayer was inserted between the substrate and the electrolyte to enhance the performance at lower temperatures. $La_{0.8}Sr_{0.2}Ga_{0.8}Mg_{0.2}O_{2.8}$ (LSGM) was selected for the electrolyte material, which is considered as promising material for the intermediate temperature operating SOFCs [5],[6]. An important point in cell fabrication is the insertion of a dense $(Ce_{0.6}La_{0.4})O_{1.8}$ (LDC40) layer between LSGM and NiO/GDC10 layer to prevent the undesirable nickel diffusion from the anode to LSGM during the co-firing procedure[7]. However, there was a problem that sintering temperature required for the preparation of dense LDC40 layer was much higher than that required for co-firing temperature. Therefore, we investigated the various additives for the sintering promotion of LDC40. As a result, it was found that the addition of a small amount of Ga_2O_3 to LDC40 was effective in obtaining a fully dense LDC40 layer at the co-firing temperature and it also improved electrical conductivity of LDC40 itself. From energy dispersed X-ray analysis (EDX) and X-ray diffraction analysis (XRD), it was seen that the reaction between LSGM and Ni was avoided by the introduction of LDC40 layer, and there was no obvious element diffusion between LSGM and LDC40. For the cathode material, $(La_{0.6}Sr_{0.4})(Co_{0.8}Fe_{0.2})O_3$ (LSCF) was coated by slurry coating and fired.

Table 3. Materials and fabrication process of micro tubular cell.

Component	Material	Fabrication	Firing
Anode Tube	NiO/YSZ	Extrude Molding	
Anode Interlayer	NiO/GDC10		Co-Firing
Electrolyte	LDC40(Ga_2O_3)-LSGM (Double Layered)	Slurry coating	
Cathode	LSCF		Firing

Figure 9 is a picture of a micro tubular single cell. The diameter of cell is 5mm, and the active length is 50mm. The single cell was jointed to the current collector cap with silver braze metal and its performance was tested in a furnace. Figure 10 shows the evaluation method for single cell performance. A fuel gas was supplied inside the cell, and air was supplied to the outside of the cell. The current voltage and impedance of the single cells were measured using a potentiostat and a frequency response analyzer in the 500 to 700°C temperature range.

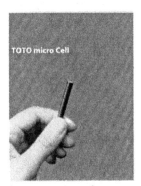

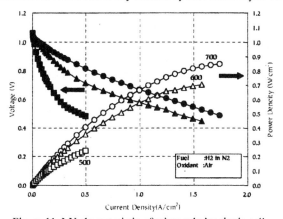

Figure 9. TOTO micro tubular cell.

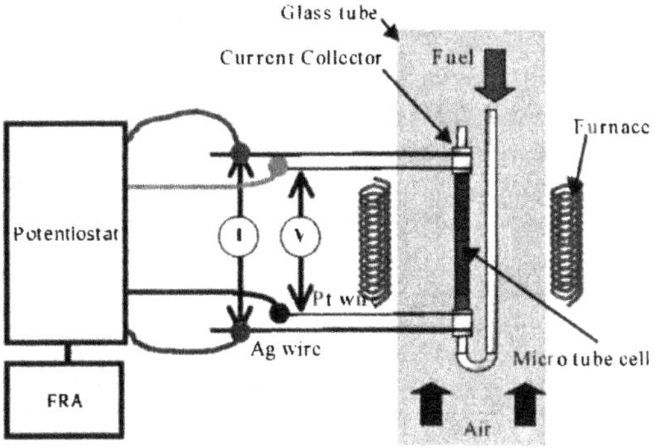

Figure 10. Evaluation method for single cell performance.

Figure 11 shows the typical I-V curves of a micro tubular single cell using dry H_2 in N_2 as fuel. H_2 flow was fixed at 0.12L/min. The open circuit voltage (OCV) was close to the theoretical value. It indicated that the electrolyte has a good gas tightness, and the chemical reaction between LSGM and Ni are effectively avoided by the LDC40 layer. The maximum

Figure 11. I-V characteristic of micro tubular single cell.

power densities were 0.85, 0.70, and 0.24W/cm^2 at 700°C, 600°C, and 500°C, respectively. Figure 12 shows the impedance spectra of a micro tubular cell measured under 0.125A/cm^2 at various temperatures. It has been generally assumed that the intercept with the real-axis at the highest frequency represents the ohmic resistance, and the width of low frequency arc represents the electrode resistance. The electrode resistance increased significantly with decreasing operation temperature, and ohmic resistance at 500°C was very high. The most likely cause of the high resistance is the low ionic conductivity of LDC40. Therefore, it is expected that the cell performance can be improved by optimizing the anode electrode and the thickness of LDC40 layer.

Figure 13 shows the fuel utilization effects on micro tubular cell performance measured under 0.125A/cm^2 at 700°C and 600°C. The observed cell voltage was close to the theoretical value calculated by the Nernst equation. It indicates that the micro tubular cell can be operated at a high efficiency.

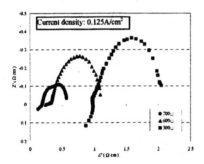

Figure 12. Impedance of micro tubular cell.   Figure 13. Fuel utilization effect on micro tubular cell performance.

The cell performances using H_2 in N_2 (1:1) gas mixture or simulated reformate gas were compared in Figure 14. The composition of simulated reformate was 32%H_2, 13%CO, 5%CO_2, and 50%N_2 based on the preliminary experiment of the catalytic partial oxidation (CPOX) reforming of LPG. As shown in the figure, the difference in cell performances was small at lower current densities. However, the performance using reformate gas was lower at higher current densities at temperatures of 600°C and 700°C. The differences became significant with increasing operating temperatures. To identify the differences, the impedance spectra were measured under a current density of 0.8A/cm^2 at 700°C (Figure 15). The electrode resistance on simulated reformate was higher than that on H_2 in N_2, and it was thought that this difference was caused by the CO transport resistance from the anode in the high current density area. Therefore, we are now trying to improve the anode performance.

Figure16 shows the cell performance using DME + air mixture as fuel at 550°C. The DME flow rate was fixed at 85ml/min and the excess air ratios (air-fuel ratio/ theoretical air-fuel ratio) were 0.1, 0.2, 0.3, and 0.4 respectively. Direct use of fuel without a reformer in SOFCs will simplify the system greatly, and this is important for SOFCs, especially in portable and transportation applications. DME is an attractive fuel because it is highly active and easily

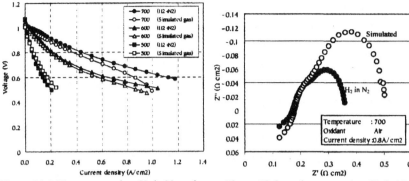

Figure 14. I-V curve tested on H_2 in N_2 and simulated reformate gas.

Figure 15. Impedance tested on H_2 in N_2 and simulated reformate gas.

liquefied and stored. As shown in the figure, the use of DME + air as fuel resulted in higher performance than that of H_2 in N_2 and no carbon deposition was observed during operation. Figure 17 shows the DME conversion rate and the exhaust gas composition analyzed by a gas chromatography. (The water content was not measured.) The DME conversion rate and CO_2 content increased with excess air ratio. Therefore, the increased performance achieved by using DME + air mixture is probably due to the raising cell surface temperatures caused by the decomposition and the combustion of DME. (The furnace temperature was kept at 550°C). These results demonstrated the high possibility of micro tubular cells being used for direct fueled operations.

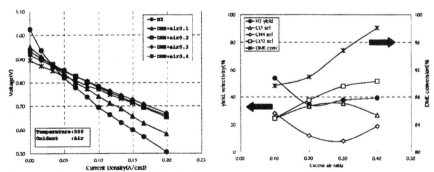

Figure 16. I-V curve tested on DME+air mixture as fuel at 550°C.

Figure 17. DME conversion rate and exhaust gas composition tested on DME+air mixture.

Micro tubular stack development

In order to evaluate the performance of cells in a bundle, we built the stack consisting of 14 micro tubular cells as shown in Figure18. This stack was evaluated in a furnace using hydrogen as a fuel, and successfully demonstrated 43W, 37W and 28W power generation at a temperature of 700°C, 600°C, and 500°C, respectively (Figure19). Table 4 summarizes the results we have obtained from the stack evaluation. The maximum stack power densities were 478W/L and 239W/kg at 700°C. These results demonstrated that micro tubular SOFCs have a high potential for portable and transportation applications.

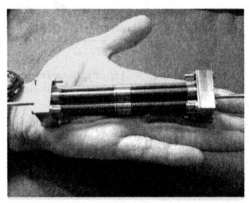

Figure 18. Appearance of micro tubular SOFC stack.

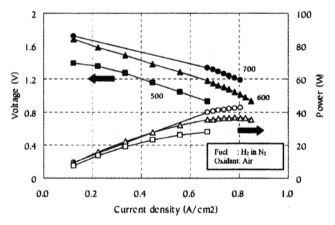

Figure 19. Performance of micro tubular SOFC stack.

Table 4. Performance of micro tubular SOFC stack at 700°C.

Item	Result
Maximum power (W)	43
Stack volume (L)	0.09
Stack weight (kg)	0.18
Stack power density (W/L)	478
Stack power density (W/kg)	239

SUMMARY

TOTO tubular SOFC

The 2×6 stack, small module and thermally self-sustaining module of the TOTO tubular SOFC were designed and made. They were evaluated using town gas or simulated fuel and showed excellent performance. We continually improve the cell performance and the durability to advance the module performance. TOTO started the small-scale production of SOFC and trial delivery in 2004. The SOFC can be supplied as a stack for the development of stationary power generation systems. In 2004, TOTO started a new collaboration with Kyushu Electric Power Co., Inc. and Hitachi, Ltd. in a new NEDO project. We are developing a co-generation system by integration with the TOTO stack.

TOTO micro tubular SOFC

The anode-supported micro tubular cells with thin lanthanum gallate with strontium and magnesium doping were developed. The single cells and the cell stack were tested using various fuels, i.e. hydrogen, simulated reformate gas of LPG, direct fueling of DME, and they showed excellent performance at lower temperatures from 500-700°C. These results demonstrated that micro tubular SOFCs have a high potential for portable and transportation applications. Further development on durability, quick start up, and compactness of the stack is being undertaken.

ACKNOWLEDGMENT

The development was supported by NEDO in Japan.

REFERENCES

[1]T. Saito, T. Abe, K.Fujinaga, M. Miyao, M. Kuroishi, K. Hiwatashi, A. Ueno, "Development of Tubular SOFC at TOTO", SOFC-IX, Electrochemical Society Proceedings Volume 2005-07, 133-140(2005)

[2]K. Hiwatashi, S.Furuya, T. Nakamura, H. Murakami, M. Shiono, M. Kuroishi, "Development Status of Tubular type SOFC by Wet Process", Proc. of the 12th Symposium on Solid Oxide Fuel Cells in Japan, 12-15(2003)

[3]J. Van herle, J. Sfeir, R. Ihringer, N. M. Sammes, G. Tompsett, K. Kendall, K. Yamada, C. Wen, M. Ihara, T. Kawada and J. Mizusaki, " Improved Tubular SOFC for Quick Thermal Cycling", Proc. of the Fourth European Solid Oxide Fuel Cell Forum, 251-260 (2000)

[4]A. Kawakami, S. Matsuoka, N. Watanabe, A. Ueno, T. Ishihara, N. Sakai, K. Yamaji, H. Yokokawa, "Development of low-temperature micro tubular type SOFC", Proc. of the 13th Symposium on Solid Oxide Fuel Cells in Japan, 54-57(2004)

[5]T. Ishihara, H. Matsuda, Y. Takita, "Doped LaGaO₃ perovskite type oxide as a new oxide ionic conductor", J.Am.Chem.Soc, 116, 3801-3803 (1994)

[6]M.Feng, J.B.Goodenough, "A superior oxide-ion electrolyte", Eur.J.Solid State Inorg.Chem.,31,663-672 (1994)

[7]K. Huang, J.H Wan, J.B.Goodenough, "Increasing Power Density of LSGM-Based Solid Oxide Fuel Cells Using New Anode Materials", J.Electrochem.Soc, 148(7), A788-A794 (2001)

Cell and
Stack Development

DEVELOPMENT OF SOLID OXIDE FUEL CELL STACK USING LANTHANUM GALLATE-BASED OXIDE AS AN ELECTROLYTE

T. Yamada, N. Chitose, H. Etou, M. Yamada, K. Hosoi and N. Komada
Mitsubishi Materials Corporation, Fuel Cell Group, Business Incubation Department
1002-14 Mukohyama, Naka, Ibaraki, 311-0102, Japan

T. Inagaki, F. Nishiwaki, K. Hashino, H. Yoshida, M. Kawano and S. Yamasaki
The Kansai Electric Power Company, Inc., Energy Use R&D Center
11-22 Nakoji, 3-chome, Amagasaki, Hyogo, 661-0974, Japan

T. Ishihara
Department of Applied Chemistry, Faculty of Engineering, Kyushu University
744 Motooka, Nishi-ku, Fukuoka, 819-0395, Japan

ABSTRACT

One of the important trends in recent years is to reduce the operating temperature of solid oxide fuel cell (SOFC). Since FY2001, Mitsubishi Materials Corporation (MMC) and The Kansai Electric Power Co., Inc. (KEPCO) have been collaborating to develop intermediate temperature SOFC modules, which use lanthanum gallate based electrolyte, for stationary power generation.

Our recent study has been focused on the durability of the stack repeat unit which is composed of a disk-type electrolyte-supported cell, an anode-side current collector, a cathode-side current collector and two interconnects, and on the improvement of the output power density of the cell. A long-term test of a stack repeat unit has been performed at 750 °C under constant current density of 0.3 A/cm^2 with hydrogen flow rate of 3 ml/min/cm^2 and air flow rate of 15 ml/min/cm^2 for over 10,000 hrs. The decrease in terminal voltage was not observed the initial 2,000 hrs, but was 1~2 %/1,000 hrs after then. The maximum electrical efficiency attained was 54 %(LHV) at 750 °C and 0.292W/cm^2 with 90 % hydrogen utilization.

The third-generation 1-kW class module was operated as CHP demonstration system for 2,000 hrs without significant degradation.

The fourth-generation 1-kW class module successfully provided the output power of 1 kW with thermally self-sustained operation below 800 °C. The average electrical efficiency calculated from the experimental data for 21 hrs stable operation was 60 %(LHV).

INTRODUCTION

A solid oxide fuel cell (SOFC) is an energy conversion device receiving great deal of interest because of its high electrical efficiency, environmental compatibility, and ability to utilize variety of fuels. However, there are a number of problems before realizing a low-cost high-performance SOFC, arising from the high operating temperature as high as 1,000 °C. A reduction in operating temperature (<800 °C) in SOFC can decrease materials degradation and facilitate the use of low-cost metallic components for interconnect, etc. instead of more expensive ceramic materials.

Recently Ishihara et al. reported a reduced-temperature SOFC using the perovskite-type oxide of doped LaGaO₃ as electrolyte [1]. In particular, LaGaO₃ where Sr was substituted for the

La-site and Mg and Co for the Ga-site ($La_{0.8}Sr_{0.2}Ga_{0.8}Mg_{0.2-x}Co_xO_{3-\delta}$) [2,3] was highly interesting from the viewpoint of decreasing the operating temperature.

Mitsubishi Materials Corporation (MMC) and The Kansai Electric Power Co., Inc. (KEPCO) have been collaborating to develop intermediate-temperature SOFC (IT-SOFC) modules, which use the above-mentioned lanthanum gallate-based electrolyte, for stationary power generation units since FY2001. The target of the development for the forthcoming several years is to commercialize 10-kW class high-efficiency IT-SOFC module. The module of this particular size is being designed and constructed by MMC and will be integrated into the combined heat and power (CHP) generating system which is subsequently being developed by KEPCO under New Energy and Industrial Technology Development Organization (NEDO) program.

In this paper, the performances of the SOFC stack repeat unit and the recent generations of 1-kW class modules operated at intermediate temperatures are reported.

EXPERIMENTAL

The electrolyte powder mixture is prepared by using the conventional solid state reaction technique with commercially available starting powders of La_2O_3 (99.99 %), $SrCO_3$ (99.9 %), Ga_2O_3 (99.99 %), MgO (99.99 %) and CoO (99 %). In particular, these commercial powders were mixed proportionally to the composition of $La_{0.8}Sr_{0.2}Ga_{0.8}Mg_{0.15}Co_{0.05}O_{3-\delta}$(LSGMC) by ball-milling and then calcined at 1,200 °C in air. The calcined powders were ground again and mixed with a binder and an organic solvent for 1 day, and the prepared slurry was tape-cast to make green sheet. After drying, disks were cut out from the sheets, and sintered at 1,400~1,500 °C in air after removing organic additives at temperatures lower than 1,000 °C. The thickness of the sintered specimens were 200~250μm, and their densities were larger than 98 % of the theoretical value.

Ni/$Ce_{0.8}Sm_{0.2}O_{2-\delta}$(Ni/SDC) and $Sm_{0.5}Sr_{0.5}CoO_{3-\delta}$(SSC) were used for the anode and the cathode, respectively. The area of 113.1 cm^2 (φ120 mm) was first coated with a slurry of a mixture of the anode powders and an organic binder on a surface of the electrolyte, followed by calcination at 1,200~1,300 °C in air. Then, the cathode was coated similarly on the opposite surface of the electrolyte, followed by calcination at 1,100~1,200 °C. Disk-type electrolyte-supported cells are shown in Figure 1.

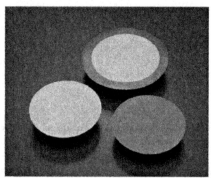

Figure 1. Disk-type electrolyte-supported cells.

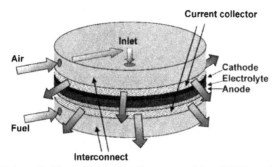

Figure 2. Schematic illustration for the disc-type seal-less SOFC stack repeat unit.

For the power-generation experiment, the seal-less stack repeat unit shown in Figure 2 was used. As current collectors, porous metallic sheets were attached to the electrodes. The current collectors were sandwiched between separators made of ferritic stainless steel with special surface treatment to maintain contact resistances within acceptable levels and the cell. The electric current was taken out directly from the interconnects. The stack repeat unit set with a test cell was placed in a uniform temperature field generated by electric heater plates. Dry hydrogen was preheated and supplied to the anode regulated by a mass flow controller. Air was preheated and supplied to the cathode regulated by a floater-type flow meter at a volumetric rate at five times of hydrogen flow. All power generation tests were carried out at 650, 700, 750 and 800 °C with typical voltage-current measurements starting from the open circuit voltage (OCV) down to 0.5 V.

PERFORMANCES OF THE STACK REPEAT UNITS

Figure 3 shows the temperature dependence of the electrical conductivities of doped lanthanum gallate and stabilized ZrO_2. It is noted that the electrical conductivity of $La_{0.8}Sr_{0.2}Ga_{0.8}Mg_{0.1}Co_{0.1}O_{3-\delta}$ is much higher than that of the conventional electrolytes.

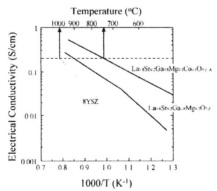

Figure 3. Electrical conductivity of various oxides.

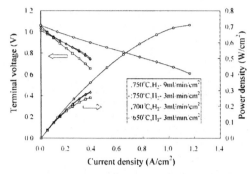

Figure 4. Voltage-current characteristics of the stack repeat unit.

For a given thickness, the same level of electrical conductivity with other materials can be achieved at much lower temperatures when $La_{0.8}Sr_{0.2}Ga_{0.8}Mg_{0.1}Co_{0.1}O_{3-\delta}$ is used as an electrolyte.

Figure 4 shows the typical power generating characteristics of the stack repeat unit at 650~750 °C. It is observed that the output power density was 0.248 W/cm² at 0.3A/cm² and 750°C with 70 % hydrogen utilization under air flow rate of 15 ml/min/cm² and dry hydrogen flow rate of 3 ml/min/cm². And thus, the electrical efficiency was 45.8 %(LHV). The maximum output power density of 0.71 W/cm² was attained at 1.2 A/cm² and 750 °C with 90 % hydrogen utilization under air flow rate of 45 ml/min/cm² and dry hydrogen flow rate of 9 ml/min/cm². The corresponding electrical efficiency was 43 %(LHV). Table 1 summarizes the typical power generating characteristics of the stack repeat unit as a function of temperature. The data correspond to measurements performed at a specified current density and fuel utilization (Uf) for temperatures between 650 °C and 800 °C (see columns 2,3,4,5). On the other hand, the column number 6 is for values obtained for the highest conversion efficiency and the column number 7 is for the maximum power density at 750 °C. Although the peak output power was achieved at 750°C, the variation in terminal voltage at 0.3A/cm² and Uf 70 % was only a few percent in the temperature range between 700 and 800 °C. Therefore, it is confirmed from these experiments that the operating temperature of the SOFC may be decreased to ca. 700 °C without compromising the performance of cells developed in the current program.

Table 1. Typical performances of the stack repeat unit.

Temperature (°C)	650	700	750	800	750	750
H₂ flow rate (ml/min/cm²)	3	3	3	3	3	9
Fuel utilization (%)	70	70	70	70	90	84
Terminal voltage (V)	0.748	0.815	0.825	0.816	0.751	0.668
Output power (W)*	25.4	27.7	28.0	27.7	33.0	78.2
Current density (A/cm²)	0.30	0.30	0.30	0.30	0.39	1.04
Power density (W/cm²)	0.225	0.245	0.248	0.245	0.292	0.692
Efficiency (%LHV)	41.5	45.2	45.8	45.3	53.9	42.7

*Cell active area : 113.1 cm².

Despite the fact that the cells are capable of generating much higher power densities, we decided to operate them at lower power densities and lower fuel utilizations for eliminating complexities such as cell degradation, heat management etc.

A long-term test of a stack repeat unit has been done at 750 °C and 0.3 A/cm^2 with hydrogen flow rate of 3 ml/min/cm^2 and air flow rate of 15 ml/min/cm^2 for over 10,000 hrs. The variation in terminal voltage is shown in Figure 5. It was found that no decrease in terminal voltage occurred during the initial 2,000 hrs. However, during the rest of the test period the decrease in terminal voltage was 1~2 %/1,000 hrs.

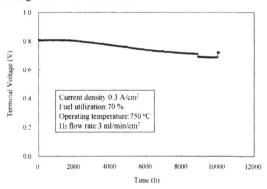

Figure 5. Variation of the terminal voltage during the stability test of the stack repeat unit.

PERFORMANCES OF THE 1-kW CLASS MODULE

Following the success of earlier 1-kW class modules, the third-generation 1-kW class module has been manufactured for CHP generation system demonstration. The module shown in Figure 6 contains a single-stack of 46 cells, a pre-reformer, a steam generator, and heat exchangers for fuel and air. Desulfurized town gas, deionized water, and air are supplied to the module at room temperature. Electric heaters are used for the initial start-up then turned off once the stack reached the temperature high enough for the steam-reforming and for the electrochemical reactions. During the thermally self-sustained operation, the stack temperature is controlled by adjusting the air flow rate.

Figure 6. Demonstration system with the third-generation 1-kW class module.

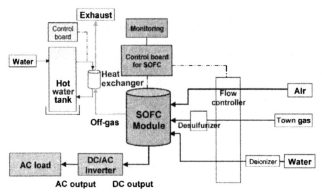

Figure 7. Schematic diagram of the 1-kW class demonstration system.

In addition to the above-mentioned 1-kW class module the demonstration system is composed of, a DC-AC inverter, a desulfurizer, a heat recovery unit, and an automatic control unit. All processes of start-up, power generation, hot-standby and shut-down are carried out automatically. During the steady-state power generation, flow rates of gases and water are regulated to keep the specified AC output power and the stack temperature lower than a specified value. The heat recovery unit, which produces hot water by exchanging heat with the exhaust gas, is controlled independently from the module. Figure 7 illustrates the schematic diagram of the system.

Table 2. Performance of the 1-kW class demonstration system.

Fuel	Town gas*1
DC power output	1095 W
AC power output	992 W
Inverter efficiency	92%
DC conversion efficiency	56%(LHV)
AC conversion efficiency	51%(LHV)*2
Average stack temperature	766°C
S/C	3.5

*1 CH$_4$:89%, C$_2$H$_6$:7%, C$_3$H$_8$:3%, C$_4$H$_{10}$:1%
*2 Parasitic loss excluded.

Typical performance data of the 1-kW class demonstration system are summarized in Table 2. The targeted 1-kW AC power output was obtained under thermally self-sustained operation below 800 °C utilizing commercial town gas with internal steam reforming. The electrical efficiency was 51 %(LHV) based on AC output, and 56 %(LHV) on DC output. The parasitic loss is not included in the AC efficiency given above, because the system components have not been optimized in terms of power consumption yet. For the heat recovery unit test, the hot water of 90 °C was produced utilizing the module exhaust gas during the operation at AC output power of 1 kW.

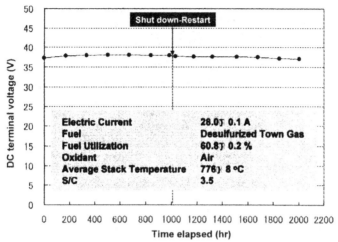

Figure 8. Long-term stability of the third-generation 1-kW class module.

A long-term stability test of the demonstration system was also performed. During the test, the system was controlled to provide 1 kW-AC power output, then the DC terminal voltage of the module was periodically measured at prescribed current and fuel flow rate. As shown in Figure 8, durability of the module over 2,000 hrs was demonstrated successfully.

The fourth and the latest generation module confirmed the integrity of the design concept of the cell stack and the hot BoP components such as pre-reformer, steam generator, and heat exchanger of both air and fuel built within the module. The external view of the module is shown in Figure 9.

One of the major design improvements from the earlier generations to the fourth generation was the optimization of both temperature profile and heat-flux distribution within the module to attain higher efficiency. In order to complete this task for higher efficiency, in the fourth generation stack, a novel interconnect plate with internal gas manifolds was designed. The new design requires reduced number of stack components. Air and fuel inlets are situated at the opposite corners of the square interconnect plate. The interconnect plate is designed in such a way that the gas manifolds are constructed by stacking the interconnect plates and the ceramic rings alternately as shown in Figure 10. Stud bolts are used to fixate the entire stack between the end plates while providing the necessary compression for preventing gas leakage between ceramic rings and interconnects. Axial compressions needed for the gas manifolds and for the electrochemically active region are separated by introducing cantilevered arms to interconnect plates. With this elegant design, the axial compression applied to the electrochemically active region is optimized to minimize the contact resistances and at the some time to eliminate the possibility of mechanical failure of the stack components.

The DC power output of 1,160 W was recorded at the output terminals under thermally self-sustained and stable operation of the forth-generation module against the design target of 1,120 W. Desulfurized town gas was supplied to the module as fuel and air was utilized as oxidant at room temperature.

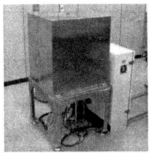

Figure 9. External view of the fourth-generation 1-kW class module.

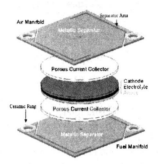

Figure 10. The fourth-generation stack unit design with internal gas manifolds.

Deionized water was pumped into the module again at room temperature for steam reforming of the town gas, which was carried out within the module. The observed stack average temperature was 758 °C. The average conversion efficiency carefully calculated from the experimental data for the 21 hrs stable operation was 60 %(LHV). The typical performance of the fourth generation module is shown in Table 3.

Table 3. Typical performance of the fourth-generation 1-kW class module.

Fuel	Town gas*1
DC power output	1145W
DC current	31.9A
DC terminal voltage ·	35.9V
DC conversion efficiency	60%LHV
Fuel utilization	81%
Average stack temperature	758 °C
S/C	3.4

*1 CH₄:89%, C₂H₆:7%, C₃Hₓ:3%, C₄H₁₀:1%

CONCLUSIONS

We have characterized the power-generation of the stack repeat unit using LSGMC electrolyte in the temperature range between 650 and 800 °C. The maximum output power density of 0.71 W/cm^2 is attained at 1.2 A/cm^2 and 750 °C with 90 % hydrogen utilization under air flow rate of 45 ml/min/cm^2 and hydrogen flow rate of 9 ml/min/cm^2. The voltage drop was 1~2 %/1,000 hrs for 2,000~10,000 hrs at 750 °C and 0.3 A/cm^2 with 70 % hydrogen utilization under hydrogen and air.

The third-generation 1-kW class module was operated as CHP demonstration system for 2,000 hrs without significant degradation. The fourth-generation 1-kW class module was successfully constructed using interconnect plates with internal manifolds. Output power of 1-kW and the electrical efficiency of 60 %(LHV) with thermally self-sustained and stable operation was achieved below 800 °C.

ACKNOWLEDGEMENTS

Part of this work has been supported by the New Energy and Industrial Technology Development Organization (NEDO) under the "Development of Solid Oxide Fuel Cell System Technology" P04004 project.

REFERENCES

[1] T. Ishihara et al., *J. Am. Chem. Soc.*, **116**, 3801 (1994).
[2] T. Yamada et al., *Sol. State Ionics*, **113-115**, 253 (1998).
[3] T. Ishihara et al., *Chem. Eng. Sci.*, **54**, 1535 (1999).
[4] T. Ishihara et al., *Sol. State Ionics*, **113-115**, 585 (1998).
[5] K. Kuroda et al., *Sol. State Ionics*, **132**, 199 (2000).
[6] N. Chitose et al., *Proceedings International Hydrogen Energy Congress and Exhibition IHEC 2005 Istanbul, Turkey*, (2005).
[7] T. Kotani et al., *Abstracts for 2005 Fuel Cell Seminar, California, USA*, 81 (1998).

ANODE SUPPORTED LSCM-LSGM-LSM SOLID OXIDE FUEL CELL

Alidad Mohammadi[1], Nigel M. Sammes[2], Jakub Pusz[1], Alevtina L. Smirnova[1]

[1]Department of Materials Science and Engineering; University of Connecticut; 97 Eagleville Road; Storrs, CT 06269, USA
[2]Department of Mechanical Engineering; University of Connecticut; 191 Auditorium Road; Storrs, CT 06269, USA

ABSTRACT:
This paper describes an intermediate temperature solid oxide fuel cell (ITSOFC), based on porous $La_{0.75}Sr_{0.25}Cr_{0.5}Mn_{0.5}O_3$ (LSCM) anode, $La_{0.8}Sr_{0.2}Ga_{0.8}Mg_{0.2}O_{2.8}$ (LSGM) electrolyte, and porous $La_{0.6}Sr_{0.4}MnO_3$ (LSM) cathode. Using different amounts of poreformers, binders and firing temperatures, the porosity of the anode was optimized while still retaining good mechanical integrity. The effect of cell operation condition under saturated hydrogen fuel on the SOFC open circuit voltage (OCV) was also investigated. It is shown that 20 mL/min flow rate of saturated hydrogen results in an initial OCV up to about 1.0V for a single cell. The cell was tested for more than 500 hours maintaining high values of OCV (0.9V). Increasing the hydrogen flow rate up to 200 mL/min, results in enhanced OCV values up to 0.99V.

1. INTRODUCTION:

In conventional SOFCs Ni-zirconia cermet (NiO-YSZ) is used as the anode material while the electrolyte is yttria stabilized zirconia (YSZ) and the cathode is a strontium doped lanthanum manganite (LSM) [1,2,3]. Such a cell operates in the temperature range of 850-1000°C. However, the intermediate temperature operation conditions (600-800°C) not only reduce the cost, but also improve performance and the SOFC stack long-term endurance. In order to optimize the SOFC performance in the intermediate temperature range, the corresponding alternative materials can be used [4,5], such as ceria or lanthanum oxides that are cost effective alternatives for YSZ. The Gd_2O_3 doped CeO_2 (GCO) has higher ionic conductivity than YSZ [6,7] and lower thermal expansion coefficient than Co-based perovskites [8], which make it attractive for intermediate temperature SOFC electrolyte applications. However, the reduction of Ce^{4+} to Ce^{3+} can cause some electronic conduction, which results in a power loss [9]. Another alternative for YSZ electrolyte is Mg doped lanthanum gallite (LSGM) that is known to possess lower resistivity and higher ionic conductivity compared to that of YSZ [10,11]. High catalytic activity for oxygen dissociation and good chemical stability over wide range of oxygen partial pressures are other advantages of LSGM [10,12,13]. LSM can be used as a proper cathode material for LSGM electrolyte since there is no interfacial reaction reported between these two materials applying firing temperatures up to 1470°C [14], however the fabrication process is important to achieve proper cathode microstructure [15].

The goal of this work was to optimize the anode and cell structure considering both microstructural changes during sintering process and cell performance under OCV conditions.

2. EXPERIMENTAL PROCEDURE:

2.1. Manufacturing of anode pellets

Different amounts of poreformer were used for making pellets for anode support. Mixing of 5-10wt% potato fiber as poreformer with LSCM powder resulted in 40-50% porosity of anode supported pellets. Organic binder composition [16] was used in some samples in addition to the pore-former. In this case the LSCM powder was mixed with 5wt% of poreformer and 5wt% of binder. The mixture dried overnight, was heated for couple of hours at 100±7°C, and then kept overnight again before grinding. The anode supported pellets were uniaxially pressed under 3500psi load and fired at 1200°C for 2 hrs.

2.2. Electrolyte supported cells preparation

Pellets of electrolyte were made using LSGM powder under applied pressure of 3500psi. These pellets were fired for 2 hrs at 1400°C. Anode slurry was made adding methanol to the mixture of LSCM powder and 5% potato fiber poreformer. This slurry was brushed on a surface of sintered electrolyte pellet and after drying at room temperature fired for 2 hrs at 1200°C. Then LSM cathode ink was painted on the other side of the pellet and after drying at room temperature fired at 1200°C for 2hrs. Compared to the anode supported cells, the electrolyte supported cells have the advantage of having higher porosity in anode layer since the electrolyte is fired first. Thus, lower anode sintering temperature results in higher mass transfer rate in anode.

2.3. Fuel cell testing procedure

The spiral silver wires used as current collectors were attached to the anode and cathode of the cell using Alfa Aesar silver conductive adhesive paste. The paste was cured at 150°C for 0.5-1 hour to obtain strong bonding and good contacts and then (Figure 1) the cell was attached to an alumina tube using sealant to prevent reaction between air and fuel mixtures. Ceramabond 552 from Aremco products, Inc. was used as the sealant. The sealant was dried at room temperature for 2 hrs and then cured for 2 hrs at 93°C and 2 hrs at 260°C. Steel tubes were sealed on the other end of the alumina tube to let the fuel in and out. Saturated hydrogen passing inside the tube was blown to the anode side, while air was blown to the cathode side using a small blower. The hydrogen flow rate was controlled using FCTS GMET gas box by Lynntech Industries, Ltd. and then hydrogen was passed through a humidifier at room temperature. The entire system, including a stainless steel tube to blow air to the cathode, was fitted in the furnace. The furnace was fired up to 800°C and kept at that temperature during the entire experiment. Both air and saturated hydrogen were blown initially while heating up the furnace and during the entire experiment. This experiment is repeated several times and the final results are the average of individual experiment's results.

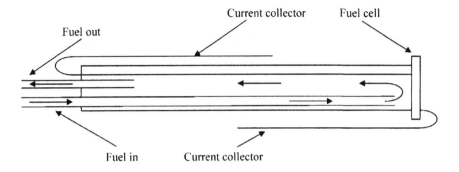

Fig.1. Button SOFC testing rig design

3. RESULTS AND DISCUSSION:

3.1. Evaluation of interfacial anode-electrolyte reactions

Possible interfacial reactions between anode and electrolyte were studied using XRD technique. For this purpose LSCM and LSGM powders were mixed in 50:50 ratio and the pellets made out of this mixture were fired for 2 hrs at 1200°C and 5 hrs at 1400°C. The XRD results obtained for these pellets (Figure 2) do not indicate any interfacial reaction since the corresponding peaks exactly correlate with XRD database. In both case, the major peaks indicating an orthorhombic structure with a = 5.487, b = 5.520 and c = 7.752 Å lattice parameters, which represents lanthanum gallium oxide. The standard lanthanum gallium oxide $LaGaO_3$ / 0.5 ($La_2O_3.Ga_2O_3$) pattern is shown in Figure 2 to compare with the obtained results. The other peaks about $2\theta=30°$ are related to the anode as well as some other peaks which overlap with those of the standard lanthanum gallium oxide.

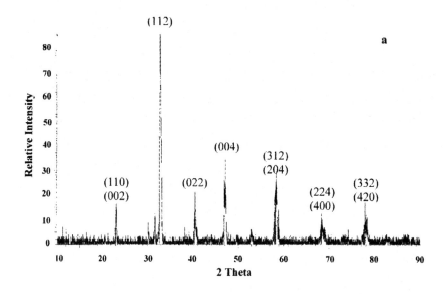

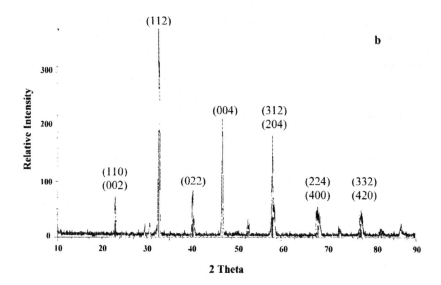

Figure 2. XRD results of LSCM:LSGM mixed pellets. Pellets were fired at
(a) 1200°C for 2 hrs   (b) 1400°C for 5 hrs.

3.2. Scanning electron microscopy

Microstructure of the anode supported pellets was investigated using SEM. Proper anode microstructure improves the ionic conductivity and has an important influence on anode performance [17]. Presence of potato fiber poreformer with or without organic binder in the anode SOFC layer leads to excellent porosity (more than 40%) while retaining good mechanical properties. Adding 5% potato fiber poreformer showed better mechanical integrity compared to the samples containing 10% poreformer, while still more than 40% porosity was observed. Addition of 5% organic binder to the LSCM powder at the same sintering temperatures showed similar porosity and mechanical integrity as anode supports with 5% potato fiber poreformer. In addition, the less number of micro cracks that are typical for pellets containing only poreformers was obtained.

Dense electrolyte layer and relatively porous cathode layers are also observed (Figure 3). It has been reported that increase in cathode firing temperature leads to coarser cathode microstructure [18], thus the anode structure in the electrolyte supported cell indicates high porosity sufficient for minor anode mass transport limitations and higher surface area of the catalysts.

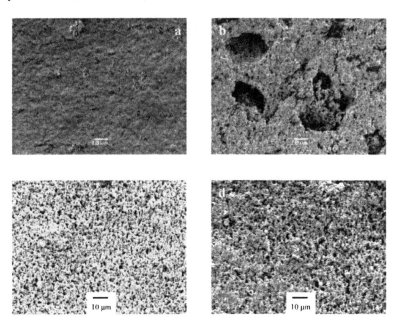

Figure 3. Microstructures of: (a) Electrolyte (b) Cathode
(c) Anode using 5wt% of potato fiber poreformer
(d) Anode using 5wt% of organic binder performer.
The indicated bars correspond to 10μm scale

3.3. SOFC performance at the OCV testing conditions
    The OCV of the electrolyte supported SOFC was measured at 800°C and 20 mL/min hydrogen flow rate. Starting from 400°C the increase in OCV was observed and at 800°C the cell voltage of about 1.0V was obtained in the first hours of cell testing. However, using only 20 mL/min hydrogen, the OCV reduces about 6.5% after more than 20 days or first 500 hrs (Figure 4). The rate of the reducing of the OCV decreases drastically and it almost reaches a constant value about 0.9V after 500 hrs.

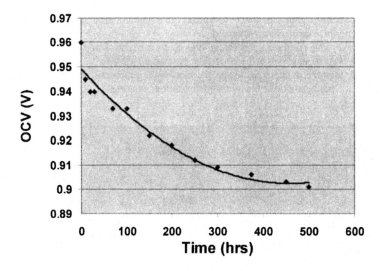

Figure 4. The OCV values vs. time at 800°C using only 20 mL/min hydrogen

Using different hydrogen flow rates after 500 hrs running the cell results are shown in Table 2. It is also shown in Figure 5 that increasing hydrogen flow rate gives higher OCV readings. However, the slope of OCV increasing after 100 mL/min reduces compare to the initial increase of hydrogen flow rate from 20 to 100 mL/min. It is also observed that not only increasing of hydrogen flow rate from 20 to 100 mL/min leads higher OCV readings but also after reducing the hydrogen flow rate back to 20 mL/min higher OCV values obtained compare to the readings before increasing the hydrogen flow rate. This increasing in OCV applying temporary higher hydrogen flow rate was about 5% and remained for couple of hours.

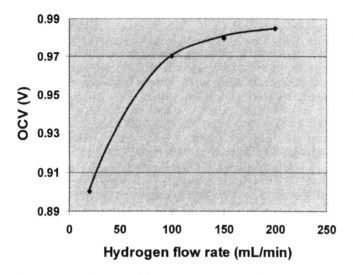

Figure 5. Change in OCV of the SOFC operating at different flow-rates

4. CONCLUSIONS:

The XRD results indicate that there are no chemical reactions between anode and electrolyte at sintering temperatures up to 1400°C. The microstructural characterization data demonstrate that addition of 5wt% of poreformer with or without binder gives up to 40% of porosity in anode layer, which is favorable for minimizing mass transport limitations. Utilization of organic binder in addition to the poreformer increases integrity while retaining good porosity and therefore fuel transportation through anode porous structure. The obtained values of OCVs for the SOFC tested in saturated hydrogen for 500 hours indicate degradation effects that can be due to microstructural changes in electrode layers. At low hydrogen flow rates the OCV values are around 0.9V after running the cell for 500 hrs. As expected, increase in hydrogen flow results in higher OCVs, due to better mass transport and availability of triple phase boundary.

ACKNOWLEDGEMENTS:

We gratefully acknowledge the financial support from US Army (Advanced Technology for Portable Miniature and Micro Fuel Cells, US ARMY CECOM. Agreement No. DAAB07-03030K-415).

RFERENCES:

[1] N.M. Sammes, Y. Du, R. Bove, *J. Power Source*, 145 (2005) 428–434.
[2] J. Pusz, A. Mohammadi, N.M. Sammes, R. Bove, The Potential for Running Landfill Gas as a fuel in NiO-8YSZ and NiO-3YSZ Based Anode Supported Micro-Tubular Solid Oxide Fuel Cells, in: Conference Proceedings of *Materials Science and Technology 2005*, September 25-28, 2005, Pittsburgh, PA, U.S.A., 2005.
[3] D.Rotureau, J.-P. Viricelle, C. Pijolat, N. Caillol, M. Pijolat, *J. Eur. Ceram. Soc.*, 25 (2005) 2633-2636.
[4] J.P.P. Huijsmans, F.P.F. van Berkel, G.M. Christie, *J. Power Sources*, 71 (1998) 107-110.
[5] J.Y. Yi, G. M. Choi, Solid State Ionics, 175 (2004) 145-149.
[6] M. Mogensen, T.Lindegaard, U.R. Hansen, *J. Electrochem. Soc.*, 141 (1994) 2122.
[7] L.M. Navarro, F.M.B. Marques, J.R. Frade, *J. Electrochem. Soc.*, 144 (1997) 267.
[8] S. Carter, A. Selcuk, R.J. Chater, J. Kajda, J.A. Kilner, B.C.H. Steele, *Solid State Inonics*, 53-56 (1992) 597.
[9] A. Mineshige, T. Yasui, N. Ohamura, M. Kobure, S. Fujii, M. Inava, Z. Ogumi, *Solid State Ionics*,152-153 (2002) 493.
[10] V.V. Kharton, F.M.B. Marques, A. Atkinson, *Solid State Ionics*, 174 (2004) 135-149.
[11] R. Pelosato, I. Natali Sora, V. Ferrari, G. Dotelli, C.M. Mari, *Solid State Ionics*, 175 (2005) 87-92.
[12] P.N. Huang, A. Petric, *J. Electrochem.ical Soc.*, 143 (1996)1644.
[13] M.Feng, J.B. Goodenough, *Eur. J. Solid State Inorg. Chem.*, 31(1994)663.
[14] K. Huang, M. Feng, J.B. Goodenough, M. Schmerling, *J. Electrochem. Soc.*, 143 (1996) 3630.
[15] S.P. Jing, *J. Power Sources*, 124 (2003) 390-402.
[16] A. Smirnova, G. M. Crosbie, R.A. Pett, *United States Patent*, US 6827892 B2, December 7, 2004.
[17] C.Lu, S. An, W.L. Worrell, J.M. Vohs, R.J. Gorte, *Solid State Ionics*, 175 (2005) 47-50.
[18] A. Mai, V.A.C. Haanappel, S. Uhlenbruck, F. Tietz, D. Stöver, *Solid State Ionics*, 176 (2005) 1341 – 1350.

Characterization/Testing

INFLUENCE OF ANODE THICKNESS ON THE ELECTROCHEMICAL PERFORMANCE OF SINGLE CHAMBER SOLID OXIDE FUEL CELLS

B. E. Buergler, Y. Santschi, M. Felberbaum, and L. J. Gauckler
Department Materials, ETH Zurich
8093 Zurich, Switzerland
e-mail: brandon.buergler@mat.ethz.ch

ABSTRACT

The influence of the anode thickness (9-60 µm) on the behaviour of SC-SOFCs was investigated in different total flows of a methane-air mixture to the anode. In order to eliminate the effects of the cathode with varying flow, a double chamber setup with separated fuel and air streams to the anode and cathode was used. The open circuit voltage (OCV) decreased with increasing gas flow and at the same time the power density increased. Oscillations of the OCV at high gas flows were observed for cells with thin anodes. Cells with thick anodes showed the highest OCV and the highest maximum power density. These anodes have a higher catalytic activity for the partial oxidation of methane and create a lower oxygen partial pressure at the anode/electrolyte interface. They also provide more hydrogen than thin anodes and can give a higher power output.

INTRODUCTION

Single Chamber Solid Oxide Fuel Cells (SC-SOFCs) are fuel cells with only one gas compartment operating in a non-equilibrium gas mixture of fuel and oxygen[1-7]. The Single Chamber configuration of Solid Oxide Fuel Cells promises a substantial simplification of fuel cell systems and this has led in the past to an increased interest in these fuel cells[8]. The operation principle is based on the different catalytic activities of the electrode materials for the partial oxidation of methane[3]. A very recent publication showed that SC-SOFCs can be operated in a self-sustained (no external heating) manner by using a novel design. By the interconnection of two individual cells[9] Shao et al. obtained approximately 246 mW/cm^2 at 1.0 V from two anode supported SC-SOFCs operating in propane-oxygen-helium mixtures.

The anode of a SC-SOFC has two functions: In the upper part of the anode, the partial oxidation of the fuel-air mixture is promoted by the highly active Nickel. At the anode/electrolyte interface the hydrogen and carbon monoxide formed by the preceding chemical reaction are electrochemically oxidised. In previous studies on electrolyte supported cells with anode thicknesses in the range of 8-17 µm[4,6] the electrode thicknesses were kept constant. In our recent work with much thicker anodes of 150 µm this was also the case [10]. Thus, there is a lack of understanding of the influence of anode thickness in SC-SOFCs. The aim of this study was to elucidate the effect. In Single Chamber SOFCs both electrodes are exposed to the same stream of fuel and air. If the gas flow or composition is changed both electrodes are affected. In order to solely investigate the influence of the gas flow and gas composition on anodes with different thicknesses, the experiments were carried out with a double chamber setup, with separate cathode and anode compartments. Platinum was used as the current collector material as in previous studies about SC-SOFCs with methane-air mixtures[4]. The reason why Pt was used is that it can be co sintered together with the electrodes, whereas gold or silver would have melted. Nickel could not be used because it would have oxidised during the sintering. The current collector meshes were always covered with a layer of anode material, thus the Pt was not directly exposed to the reactive gas mixture. Furthermore it was found that at 700°C in CH_4-air mixtures a porous Pt-layer did not promote the reaction between CH_4 and O_2[11]. The catalytic surface of the Ni-cermet is by orders of magnitude higher than the one of the Pt-current collector.

EXPERIMENTAL

Cells based on the system Air /$Sm_{0.5}Sr_{0.5}CoO_{3-\delta}$/$Ce_{0.9}Gd_{0.1}O_{1.95}$/Ni-$Ce_{0.8}Gd_{0.2}O_{1.90}$/ CH_4-Air were investigated. All the fabricated cells were supported by a 1 mm thick electrolyte disc having a diameter of 23 mm. The electrolyte discs were fabricated by uniaxially pressing CGO10 powder ($Ce_{0.9}Gd_{0.1}O_{1.95}$ Lot# 03-P5049BM.1, Praxair, Woodinville, USA) in a 30 mm diameter dye at 40 MPa. Then these discs were isostatically pressed at 300 MPa and sintered at 1500°C for four hours in air. The density of the discs always exceeded 98% of the theoretical density of CGO10 (ρ_{th} = 7.29 g/cm³). Grinding and subsequent lapping with a Staehli FLM 500 machine using B_4C-suspension (BC-800, Staehli, Switzerland) was then carried out until the desired thickness had been reached.

Both anodes and cathodes (1×1 cm²) were similarly prepared by screen printing. Three different anode thicknesses were fabricated by using 20 µm and 100 µm masks. For obtaining even thicker anodes than achievable with a single print through the 100 µm mask, double printing with this mask was carried out. This was done by drying the first layer and then printing a second layer on top. After the screen printing step the current collector was gently placed on the freshly printed layer and dried at 50°C for a few hours in a drying oven. 60 wt% NiO-containing $Ce_{0.8}Gd_{0.2}O_{1.9}$ (Lot # 03-P4832DM, Praxair, Woodinville, USA) was used as the anode material. Current Collectors consisted of a Pt-wire that was point-welded onto a completely flat Pt-mesh 0.9×0.9 cm² in size (52 mesh woven from 0.1 mm wire, Alfa Aesar). The cathode fabrication was always the same, using the same paste and the same screen printing mask (50 µm) for all the fabricated cells. $Sm_{0.5}Sr_{0.5}CoO_{3-\delta}$ powder (Lot# 91-82, Nextech Materials, Worthington, USA) was used as the cathode material. The anode was fabricated before the cathode and the sintering of both electrodes was done in two steps with 1300°C peak temperature for the anode and 1100° for the cathode. The heating rate was 1 K/min from room temperature to 500°C with a dwell time of 1 hour at this temperature and 3 K/min from there to the peak temperature for 2 hours. The samples were cooled down to room temperature at 5 K/min. The Pt-current collectors were fixed to the electrolyte disc with a small amount of ceramic adhesive. A photograph of the cathode side of one of the fabricated cells is shown in Figure 1.

In order to evaluate the printed anode thickness, reference samples were fabricated in the same way as the real cells. These specimens were broken in two and the cross section was analysed in a Scanning Electron Microscope (SEM, LEO 1530, Oberkochen Germany).

The cells were mounted in the test rig shown in Figure 2 with the anode facing down. The test rig consisted of two glass tubes with cylindrical attachments that were placed in a horizontal tubular furnace. Between those attachments the cell was inserted by the use of two alumina rings.

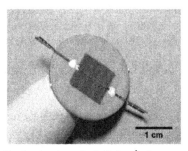

Figure 1 One of the fabricated cells, showing the cathode (1x1 cm^2), the current collector and the Pt-wires attached by ceramic adhesive.

Between the alumina rings and the glass tubes some ceramic paper (Insulfrax, 1 mm) was used for allowing a soft contact and also the Pt-wires to exit. The setup is unsealed but gas can only leak out of the system and not into it. The cathode was exposed to a constant flow of 500 ml/min of dry air, while the anode was fed with a varying total flow of CH$_4$-air mixture with CH$_4$/O$_2$=2. A thermocouple that had been covered with ceramic adhesive was placed very close to the anode. All the measurements were done after slow heating (2 K/min) to an anode temperature of 733°C. The test rig furnace was always adjusted in such a way as to obtain a temperature of 733°C measured with the anode thermocouple placed very closely to the anode.

When the CH$_4$-air flow was increased usually the furnace temperature had to be lowered in order to maintain the same anode temperature. When the flow was decreased the furnace temperature had to be raised. The anode gas flow was varied from 70-1000 ml/min.

Two Pt-wires were connected to each electrode allowing to carry out four point measurements of voltage-current characteristics by using a potentiostat (IM6, Zahner, Kronach, Germany) scanning from the OCV to 0 V with a slew rate of 5 mV/s.

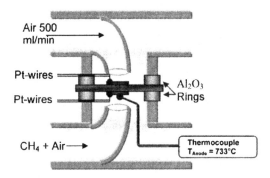

Figure 2 Test rig for the characterisation of SC-SOFCs in a double chamber configuration. In all the experiments the cathode (facing upwards) was fed with a constant airflow of 500 ml/min. The anodes of different thicknesses were exposed to a varying flow of CH$_4$-Air mixtures (CH$_4$:O$_2$=2).

RESULTS AND DISCUSSION

In Figure 3 the cross-sectional views of the reference samples for the evaluation of the anode thickness are shown. The three obtained thicknesses were 9, 21 and 60 μm.

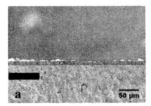

Figure 3   SEM micrographs of reference samples for the evaluation of the anode thickness after screen printing and sintering. a) 20 μm mask giving a final thickness of 9 μm b) 100 μm mask giving a final thickness of 21 μm and c) 100 μm mask with two screen printed layers on top of each other giving a final thickness of 60 μm.

In Figure 4 the OCV of a cell with a very thin anode (9 μm) is plotted as a function of time for three different CH_4-air flows. At 1000 ml/min the OCV fluctuated strongly around a mean value of 0.76V with a period between 20 and 50 seconds. At 500 ml/min the period increased to 60-100 seconds, the fluctuations had a greater magnitude but the mean OCV increased to 0.81V. From the maximum voltage of a period there was a continuous decrease of the cell voltage and at a certain point it dropped massively but increased immediately again to a maximum. At 250 ml/min gas flow to the anode the cell voltage seemed to be stable at 0.85V, at least within the time of the measurement, showing only a slight decrease in the range of some ten millivolts.

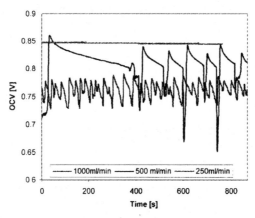

Figure 4   OCV of a cell with a 9 μm thick anode as a function of time for three different CH_4-air gas flows.

In Figure 5 to Figure 7 the measured voltage-current characteristics of the same cell are shown at the same gas flows (1000, 500 and 250 ml/min). The oscillatory behaviour was also observed during the discharge of the cell. At high gas flows the cell current broke down more frequently during the measurement. The reason for the unstable cell behaviour were temperature oscillations during the partial oxidation caused by the periodic oxidation/reduction of the NiO-CGO catalyst. Such temperature fluctuations in catalyst beds of silica supported Ni-catalyst have already been reported by Hu et al.[12]. When the gas phase temperature was increased, the oscillation time decreased and so did the magnitude of the oscillations. The authors stated that the oscillatory behaviour does not occur for every support material, e.g. Ni-Al$_2$O$_3$ did not show any oscillations. At 250 ml/min (Figure 7) a regular voltage-current relation was obtained and at 150 ml/min too (not shown). At lower gas flows the situation changed somewhat. The OCV was 0.80 V at 70 ml/min but at high current densities a severe voltage drop could be observed (Figure 8). Obviously the used gas flow was too low and led to a too small amount of CO and H$_2$ being available for the reaction of the O^{2-}-ions at the anode side. The results indicated that the Ni-anode was partially electrochemically reoxidised to NiO.

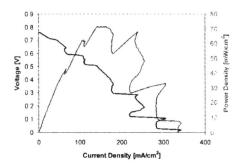

Figure 5   Voltage-current characteristic of the same cell as in Figure 4 at a flow of 1000 ml/min to the anode.

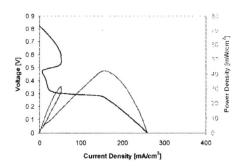

Figure 6   Voltage-current characteristic of the same cell as in Figure 4 at a flow of 500 ml/min to the anode.

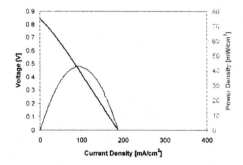

Figure 7   Voltage-current characteristic of the same cell as in **Figure 4** at a flow of 250 ml/min to the anode.

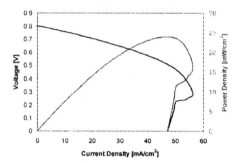

Figure 8   Voltage-current characteristic of the same cell as in **Figure 4** at a flow of 70 ml/min to the anode. Above 50 mA/cm^2 the voltage drops. Note the different scale as compared to the previous figures.

The dependence of the OCV and maximum power density on the anode gas flow is summarized in Figure 9 for a cell with a 60 μm thick anode. The OCV reaches a maximum at around 150 ml/min while the power density increases monotonically with increasing gas flow. Below 70 ml/min there is not enough gas to obtain a low oxygen partial pressure at the anode/electrolyte interface. This is the reason why the OCV drops at low flows. At high flows more and more unreacted oxygen reaches the anode-electrolyte interface causing the oxygen partial pressure to increase. However, there is a second effect that also plays a role. With increasing gas flow the effective anode temperature increases. It is known that the OCV decreases with increasing cell temperature for similar SC-SOFCs[4,10].The increased cell temperature will also cause an enhanced catalytic activity as well as lower polarization resistance of the anode.

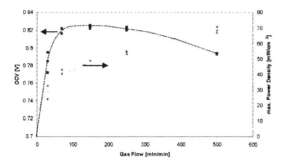

Figure 9 OCV and maximum power density of a cell with a 60 μm thick anode as a function of the gas flow to the anode. The lines are for guiding the eye only.

In Figure 10 the voltage-current characteristics of three cells are shown at 500 ml/min gas flow to the anode and an airflow of 500 ml/min to the cathode. There seemed to be an ideal gas flow to the anode, where regularly shaped voltage-current characteristics could be obtained. This range was 150-250 ml/min for the 9 μm thick anode, 150-500 ml/min for the 21 μm thick anode and 150-1000 ml/min for the 60 μm thick electrodes. The cell with the 60 μm thick anode showed the highest OCV and the highest power output, whereas the cell with the 9 μm anode showed the lowest power output and a large voltage drop at around 150 mA/cm². The reason for this drop was the unstable behaviour (Figure 4) of the thin anode, which caused irregular voltage-current characteristics of this cell, as shown already in Figure 6.

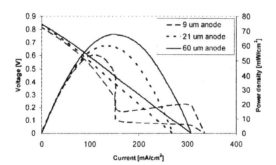

Figure 10 Voltage-current characteristic of three individual cells with 9, 21 and 60 μm thick anode at a flow of 500 ml/min CH₄-air mixture.

The influence of the gas flow on the OCV is shown in Figure 11 for the three different anode thicknesses. As seen already in Figure 9 the OCV reaches a maximum at around 150-250 ml/min. It seems that the maximum becomes broader as the anode thickness increases. At low gas flows the thick anode maintains a higher OCV as compared to the other cells. The cell with

21 μm thick anode had the lowest OCV, which could be explained by the fact that there were some cracks between the current collector and the anode material. These cracks were due to the fabrication process of the anode and were only observed for this electrode thickness. It is possible that the gas can reach directly the electrolyte/anode interface, locally increasing the oxygen partial pressure and consequently lowering the OCV.

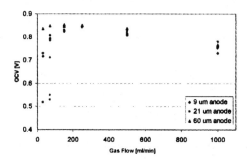

Figure 11   Open Circuit voltage of three individual cells with 9, 21 and 60 μm thick anode at a flow of 500 ml/min CH$_4$-air mixture to the anode and 500 ml/min air to the cathode.

SUMMARY

Single Chamber SOFCs with different anode thickness ranging from 9-60 μm were fabricated and measured in a double chamber setup, in which the anode was separated from the cathode compartment. The results for the cell with a 9 μm thick anode showed that under certain conditions there were large oscillations of the measured OCVs as well as voltage-current characteristics. The reasons for this are reduction/oxidation cycles of the Ni/NiO during the partial oxidation of CH$_4$ causing periodic temperature changes, which manifested themselves as oscillations of the OCV. There seemed to be an ideal gas flow to the anode, where regularly shaped voltage-current characteristics could be obtained. Thick anodes allowed more stable operation at higher gas flows. It was observed that even for the unstable conditions the power density was higher at large gas flows. However, high gas flows are inappropriate because they favour large fluctuations of the power output.

REFERENCES
[1]W. van Gool, "The Possible Use of Surface Migration In Fuel Cells And Heterogeneous Catalysis". *Philips Research Reports*, **20**, 81 (1965).
[2]C. K. Dyer, "A novel thin-film electrochemical device for energy conversion", *Nature*, **343**, 547 (1990).
[3]T. Hibino and H. Iwahara, "Simplification of Solid Oxide Fuel Cell System Using Partial Oxidation of Methane", *Chem. Lett.*, **7**, 1131 (1993).
[4]T. Hibino, A. Hashimoto, M. Yano, M. Suzuki, S.-i. Yoshida and S. Mitsuru, "High Performance Anodes for SOFCs Operating in Methane-Air Mixture at Reduced Temperatures", *J. Electrochem. Soc.*, **149**, A133 (2002).
[5]P. Jasinski, T. Suzuki, F. Dogan and H. U. Anderson, "Impedance spectroscopy of single chamber SOFC", *Solid State Ionics*, **175**, 35 (2004).

[6]I. C. Stefan, C. P. Jacobson, S. J. Visco and L. C. De Jonghe, "Single Chamber Fuel Cells: Flow Geometry, Rate, and Composition Considerations", *Electrochem. Solid St.*, **7**, A198 (2004).

[7]T. Suzuki, P. Jasinski, H. U. Anderson and F. Dogan, "Single Chamber Electrolyte Supported SOFC Module", *Electrochem. Solid St.*, **7**, A391 (2004).

[8]M. A. Priestnall, V. P. Kotzeva, D. J. Fish and E. M. Nilsson, "Compact mixed-reactant fuel cells", *J. Power Sources*, **106**, 21 (2002).

[9]Z. Shao, S. M. Haile, H. Ahn, P. D. Ronney, Z. Zhan and S. A. Barnett, "A thermally self-sustained micro solid-oxide fuel-cell stack with high power density", *Nature*, **435**, 795 (2005).

[10]B. E. Buergler, M. E. Siegrist and L. J. Gauckler, "Single chamber solid oxide fuel cells with integrated current-collectors", *Solid State Ionics*, **176**, 1717 (2005).

[11]T. Hibino, A. Hashimoto, T. Inoue, J.-i. Tokuno, S.-i. Yoshida and S. Mitsuru, "Single-Chamber Solid Oxide Fuel Cells at Intermediate Temperatures with Various Hydrocarbon-Air Mixtures", *J. Electrochem. Soc.*, **147**, 2888 (2000).

[12]Y. H. Hu and E. Ruckenstein, "Catalyst Temperature Oscillations during Partial Oxidation of Methane", *Ind. Eng. Chem. Res.*, **37**, 2333 (1998).

INVESTIGATION OF PERFORMANCE DEGRADATION OF SOFC USING CHROMIUM-CONTAINING ALLOY INTERCONNECTS.

D.R. Beeaff, A. Dinesen, P.V. Hendriksen
Risø National Laboratory
Frederiksborgvej 399, P.O. 49,
DK-4000 Roskilde, Denmark

ABSTRACT

The long-term aging of a stack element (fuel cell, current collectors, and interconnect materials) was studied. A pair of tests were made in which one sample contained an interconnect, a high-temperature stainless steel (Crofer 22 APU), treated with an LSMC coating applied to the cathode-side interconnect plate and the other sample containing a similar fuel cell but no steel interconnect. The interconnect-bearing sample was evaluated for thermochemical compatibility of cell components, including interconnect materials, under conditions typical of and/or expected during the lifetime of an installed multi-cell stack. This was done in an attempt to gain an understanding of reactions between sealant and cell materials, the oxidation of interconnect steels, and the formation of non-conducting species at the electrode interfaces. Additionally, it has been proposed by several researchers that the use of chromia-forming stainless steels can lead to the deposition of material at the triple phase boundary (TPB), a process known as chromium poisoning, which leads to a reduction in the electrochemical performance of the cell over time.

Sealing was accomplished using aluminosilicate glass with 30wt. % MgO filler. Long-term degradation of each sample was determined using a current density of 250 mA•cm-2 using humidified hydrogen as the fuel and air as the oxygen source. Additionally, the stack element was cycled and an investigation into the effect of cathode atmosphere was undertaken to elucidate the aging mechanism.

PROCEDURE

Table I summarizes the materials used in this test. The cell used was a standard anode-supported cell with a 16 cm^2 active area. Ceramic plates consisting of, on the anode side, Ni/YSZ presintered at 1500°C and, on the cathode side, LSM were used as gas distribution layers and contact components between the electrodes and interconnects. The architecture of both plates consisted of parallel gas flow channels. The interconnect (IC) plates were fabricated from Crofer 22 APU steel [ThyssenKrupp VDM GmbH, Werdohl, Germany] and were cleaned and etched with HF prior to the application of contact layers. The IC plates were sprayed with NiO or LSMC layers, dependent on which side of the cell they were used. Sealing of the stack assembly was achieved through the used of an alkali aluminosilicate glass-ceramic (designated YS2B) with 30 wt. % MgO filler to control the thermal expansion.

Table I. Summary of materials used for test.

Component	Material
Interconnect steel	Crofer 22APU
Cathode IC coat	15-μm LSMC
Cell	Anode-supported YSZ electrolyte with LSM cathode

Plantinum voltage probes were inserted into the cathode and anode gas distribution layers to measure the electrical potential across the cell and the potential across the interconnect – current collector interfaces. Within the cathode chamber, a Pt/Pt•Rd thermocouple was embedded to measure the internal temperature of the cell. Unless otherwise stated, all temperatures listed in this report are in reference to this internal thermocouple rather than the furnace or external cell temperature, although both were also monitored.

Subsequent to the initial assembly of the stack, as described previously, the operating temperature was briefly increased to 850 °C and 1000 °C to reduce the NiO anode support and anode interconnect coating. During the majority of the test, the cell temperature was maintained at 750°C. In addition to investigating the effects of cathode atmosphere, the stack was thermally cycled once during the course of the stack element test at t = 2000 h and five times just prior to termination. The thermal cycle consisted of cooling at a rate of 3°C·min^{-1} to room temperature and back to operating conditions. At the conclusion of the thermal cycles, the leak rate was measured to determine if any catastrophic fracturing had occurred within the glass seals or electrolyte.

Variation of Cathode Atmosphere

During the course of the test, the atmosphere of the cathode was varied to elucidate any effects this may have on the electrochemical performance of the cell. The purpose was primarily to determine the origin of the long-term cell degradation. The use of chromium bearing steel has been shown to cause degradation of cell performance [1-5]. The water content of the cathode atmosphere should have an increasing effect on the evolution of volatile chromium species and, thereby, loss of cell performance.

For the majority of the test, air was used as an oxygen source. At t = 1200 hours, the cell was briefly switched to pure oxygen, but a problem in the gas supply required the test to return to utilizing air. When the source of the problem was corrected, the cell was run on pure oxygen from t = 1700 h to t = 2000 h. Starting at t = 3300 h, the test was cycled between dry air and "moist" air, containing 3% water vapor as the cathode gas.

During the course of the test, impedance spectra was recorded using a Solartron impedance analyzer. The objective was to determine any particular process that was affected by the addition of water. At the conclusion of the test, the data was compared to a similar test run on a cell containing no chromium bearing components to determine if a method could be used *in situ* to determine the efficacy of different interconnect surface treatments.

RESULTS

Figure 2 gives the performance versus time of the cell during the life of the test in terms of area specific resistance (ASR) measured by performing *I-V* measurements. The horizontal line indicates the temperature of the cell (750°C) and show points at which a thermal cycle was performed. The cathode gas is indicated by the shading of data points. Two points are marked with asterisks, the first being the point at which oxygen flow was suddenly lost from the cathode chamber; the second indicating a change in ASR following the first thermal cycle. The plot is characterized by having two degradation rates. The cell performance initially deteriorates at a rate of ~100 mΩ·cm^2 until *t* = 1200 h at which time the rate decreases to ~15 mΩ·cm^2, continuing at the reduced rate while the cathode atmosphere is changed to pure oxygen. As expected in the increase oxygen partial pressure improves the cell performance.

The subsequent thermal cycle causes an instantaneous loss of performance but the long term rate remains the same until *t* = 3300 h. At this time the cathode atmosphere was cycled between dry and humidified air. Over this period the degradation rate increase back to 100 mΩ·cm^2 until the cell was allowed to stabilize in dry air for the remaining duration of the test.

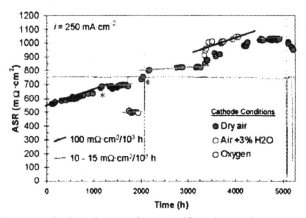

Figure 2. Performance of cell test in term of area specific resistance during the course of the test highlighting the effect of cathode atmosphere on degradation rate.

Figure 3 presents the same processes in terms of stack potential. Again, an initial rapid degradation of ~60 mV is observed during the first 1000 hours of operation after which, during operation using either dry air or dry oxygen, the stack shows a significant decrease in performance. The thermal cycle/oxygen loss at *t* = 2000 hours results in a sudden decrease of

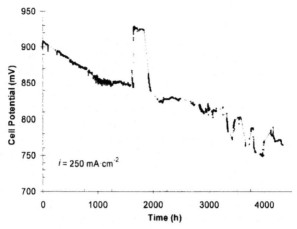

Figure 3. Performance of cell test in terms of cell potential during the course of the test.

A closer view of the dry/humid air cycling portion of the test is given in Figure 4. This figure shows the final 2000 hours of the test in terms of stack potential at a current density of 250 mA·cm^{-2}. The atmosphere was humidified four time during this period: once at 3350 hours for a brief time and again at 3400, 3600, and 3900 hours, each for a duration of 168 hours. During each of these times, there was an immediate decrease in cell performance during the introduction of water vapor of about 10 mV, which was recoverable upon returning to dry conditions. After the initial decrease, the cell performance was reduced at a rate of about 0.14 mV·h^{-1}, a decrease which was not recoverable upon returning to dry conditions.

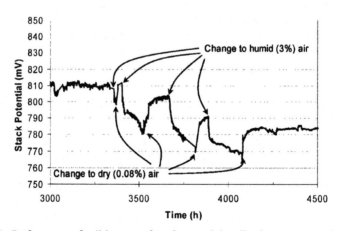

Figure 4. Performance of cell in term of stack potential (cell + interconnects) during the final 2000 hours of the test highlighting the effect of cathode atmosphere on degradation rate.

The brief loss of oxygen to the cathode at $t = 1200$ hours seem to have had an impact on the interface resistance of the cathode interconnect and, therefore, one would expect on the microstructure as well. Figure 5 shows the ASR of the interconnect interfaces during the course of the test. From the time of the loss of oxygen, the cathode interconnect interface (IC-CCC) showed a continuous parabolic increase in resistance that is unaffected by the humidification that occurs in the latter stages of the test. The anode interconnect interface (IC-ACC) generally shows good aging behavior, with a degradation rate of < 1 m$\Omega \cdot$cm$^2 \cdot$kh^{-1}; however, the thermal cycle that occurred at $t = 2000$ h results in an immediate increase in ASR. This is most likely due to a slight variation in contact area.

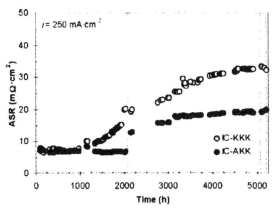

Figure 5. Performance of interconnect interfaces in term of area specific resistance during the course of the test. The lighter shade data points indicate the cathode interconnect and the darker indicate the anode interconnect interface. Vertical dashed lines are placed during a thermal cycle.

DISCUSSION

Variation of Cathode Atmosphere

The data taken from the impedance analyzer is compared to a test of a single cell containing no chromium bearing interconnect components in Figure 6. The single cell was also exposed to humidified air at the same rate of 140 l·h^{-1} at a temperature of 750°C and using hydrogen as a fuel. Every effort was made to replicate the condition of the interconnect bearing test. The figure is a standard Bode plot showing real versus imaginary impedances. The top portion of the graph is that of the cell/interconnect test while the lower portion is that of the single cell test. Note that the time scale is not equivalent.

There are some features of note that can be observed within the plots. First looking at the stack element portion of the figure, the initial effect of cathode gas humidification can be discerned. A spectrum was obtained prior to (t = 3338 h) and immediately after (t = 3362 h) humidification. No change is observed between the serial resistances of these two measurements, but there is a increase in the polarization resistance of 3 mΩ. This increase could be due in part to increased diffusion polarization within the porous electrode structure or surface adsorption of water. Both groups of spectra show an increase in both serial and polarization resistance during operation. For the single cell, the rate of change in serial resistance is approximately 14.5 mΩ·kh⁻¹. The cathode conditions during the time from which the stack element data was taken was alternately humidified and dry, somewhat convoluting the interpretation of the spectrum, but the change in serial resistance can be estimated to be around 10 mΩ·kh⁻¹. This correlates to the slightly greater decrease in cell performance during humidification of the single cell versus that observed in the stack element test, although both are equivalent within experimental error.

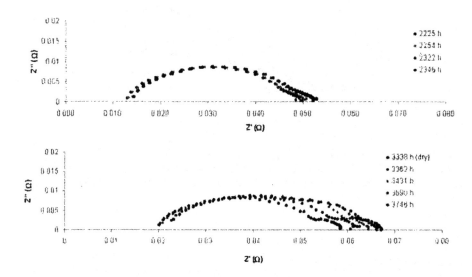

Figure 6. Bode plot of cell/interconnect (bottom) and comparable single cell (top) taken during humidification of cathode gas stream.

The equivalency in the loss of cell performance during operation in humidified air means that it is impossible to deduce the influence of water content on cell degradation in the presence of the interconnect. However, there are slight variations observed between the impedance spectra. Particularly if the variation in the imaginary component of impedance versus with frequency is examined, a notable difference is seen between the cell with an interconnect and one without. Figure 7 illustrates this. Both cells show an increase in the imaginary component in the higher frequency ranges (> 1000 Hz), but the cell with an interconnect shows an additional feature developing around $f = 100$ Hz. We are looking into ways to increase the resolution of the process activity versus frequency, most likely using an approach similar to Bessler's computational approach [6]; however, this analysis will not be completed at the time of paper submission. It has been show that the impedance spectrum can be deconvoluted into component electrode processes [7,8]. The increase in polarization resistance centered around $f = 120$ Hz has been identified by Barfod, *et al* as particular to cathode reaction processes, specifically the dissociative adsorption of oxygen and transfer across the triple-phase boundary [9], indicating that the presence of the chromium-bearing interconnect has an effect on the electrochemical processes occurring within the cathode.

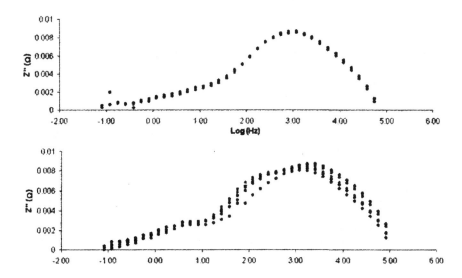

Figure7. Plot of frequency versus the imaginary component of impedance of cell/interconnect (bottom) and comparable single cell (top) taken during humidification of cathode gas stream. Note the development with time of a process occurring at a frequency of ~100 Hz in the case of the stack element test.

CONCLUSIONS

A test consisting of an anode-support fuel cell using chromium-bearing interconnect material was tested for a long (>5000 hour) duration. During this time, the performance of the cell was measured while the cathode air flow was humidified with 3% water vapor. This results in an immediate, recoverable loss in performance most likely due to increase concentration polarization or water adsorption. A longer duration effect is non-recoverable and leads to permanent cell degradation. However, this feature is also observed in a cell with no chromium-bearing components. Impedance spectra taken from both tests indicates a process inherent to the system containing an interconnect in a frequency range consistent with cathodic processes although the resolution of the spectrum is not sufficient to fully realize this. If the resolution is improved, the use of impedance spectroscopy could significantly assist in assessing the efficacy of different interconnect treatments on the evolution of chromium and subsequent poisoning of the cathode.

REFERENCES

[1]C. Gindorf, L. Singheiser, and K. Hilpert, *Steel Res.*, **72**, 528 (2001)
[2]C. Gindorf, L. Singheiser, and K. Hilpert, M. Schroeder, M. Marin, H. Greiner, and F. Richter, in *Solid Oxide Fuel Cells*, S.C. Singal and M. Dokiya, Eds., The Electrochemical Society Proceedings Series, Pennington NJ, (1999)
[3]K. Hilpert, D. Das, M. Miller, D.H. Peck and R. Weiss, *J. Electrochemical Soc.*, **143** 3642 (1996)
[4]S.C. Paulson and V.,I. Birss, *J. ElectrochemicalSoc.*, **151** A1961 (2004)
[5]Y. Matsuzaki and I. Yasuda, *J. Electrochemical Soc.*, **148** A126 (2001)
[6]W.G. Bessler, *Solid State Ionics*, **176** 997-1011 (2005).
[7]Hendriksen, P. V., Koch, S., Mogensen, M., Liu, Y.-L., and Larsen, P. H., in SOFC-VIII,inghal, S. C. and Dokiya, M., Editors, **PV2003-07**, p.1147, The Electrochemical Society Proceedings Series, Pennington, NJ, (2003)
[8]Barfod, R., Hagen, A., Ramousse, S., Hendriksen, P. V., and Mogensen, M., in Sixth European Solid Oxide Fuel Cell Forum Proceedings, Mogensen. M., Editors, **Volume 2**, p.960, European Fuel Cell Forum, Oberrohrdorff. Switzerland, (2004)

DEGRADATION MECHANISM OF METAL SUPPORTED ATMOSPHERIC PLASMA SPRAYED SOLID OXIDE FUEL CELLS

D. Hathiramani, R. Vaßen, J. Mertens, D. Sebold, V.A.C. Haanappel, D. Stöver
Forschungszentrum Jülich GmbH
Institute for materials and processes in energy systems
52425 Jülich, Germany

ABSTRACT
        Sufficient thermal shock resistance, good thermal conductivity, as well as its solderability makes metallic supports highly attractive for SOFCs used in auxiliary power units (APUs). It is known that metallic supported SOFCs based on ferritic steels show high degradation rates during operation. Life times of such SOFCs are in the range of 200 h instead of up to 5000 h required for typical mobile applications. To solve this problem it is important to identify the mechanism responsible for the degradation.
        SOFCs were produced by subsequently atmospheric plasma spraying of the anode (NiO/YSZ) and the electrolyte layer (YSZ) on top of porous metallic substrates. The cathode (LSFC) was applied by screen printing. Two different types of alloys were used as metallic substrates, Crofer22APU first (ThyssenKrupp) and the ODS alloy ITM-14 (Plansee). Electrochemical tests were performed for 200 h at 800°C under a constant current load.
        Possible degradation mechanisms were listed and systematic experiments were designed to study the individual contribution of each degradation process. From both the characterization of the single cell tests as well as the individual experiments for each degradation mechanism, it can be concluded that chromium diffusion from the metallic support to the anode layer seems to be responsible for the high degradation rates. Other mechanism like degradation of the metallic substrates due to oxidation or the formation of $La_2Z_2rO_7$ or $SrZrO_3$ along the electrolyte-cathode interface does only play a minor role to the total degradation of the SOFC during the first 200 h.

INTRODUCTION
        For solid oxide fuel cell (SOFC) based systems, the most important issue to become a commercial product is the reduction of the manufacturing costs. Atmospheric plasma-spraying (APS) seems to be an attractive and cost effective technique for a commercial based production of SOFCs[1]. Several thermal spray techniques are already investigated to produce SOFC layer systems, such as flame spraying[2], vacuum plasma spraying[3, 4] and APS[5-7]. To realize the production of SOFCs by plasma spraying a porous substrate is required which is able to withstand the thermal spray process. The typical wet chemical production routes for SOFC components are applied on ceramic substrates[8, 9]. These types of substrates are normally not useful as substrates for thermal spraying. Porous metallic substrates from ferritic steels are here an option[10]. Due to oxidation at high temperatures, a metallic support is limited to intermediate SOFC operation temperatures. On the other hand, the metallic properties like high electronic and thermal conductivity, good thermal shock resistance, and the possibility of conventional metallic joining (solderability) makes metallic supports highly attractive for SOFCs used in auxiliary power units (APUs)[11, 12].

EXPERIMENTAL

Two different types of porous metallic tape-casted and sintered plates were used as the substrate: the ferritic alloy CroFer22APU[13, 14] frequently used as interconnector material, and the other based on the powder metallurgical, rare-earth oxide dispersion strengthened (ODS) alloy ITM-14[10, 15]. The geometry of this substrate was planar with dimensions of 5x5 cm² and a thickness of about 1.5 mm. All plasma-sprayed layers are produced using the multi-cathode Triplex APS torch (Sulzer Metco AG) mounted on a six-axes robot. To deposit APS coatings the powder was injected perpendicular into the plume carried by an Ar gas stream. The anode layer was produced by injecting of NiO (OGM Harjavalta) and fully-stabilized YSZ (ZRO-292, Praxair) powder separately into the plasma plume[16]. Fused and crushed powder of 8mol% Y_2O_3 fully-stabilized ZrO_2 was used to deposit a gastight electrolyte layer by APS[5]. Cathode layers were produced by screen printing a $(La_{0.58}Sr_{0.4})(Fe_{0.8}Co_{0.2})O_{3-\delta}$ (LSFC) slurry on top of the electrolyte layer. The cathode layer was sintered during the start-up of the SOFC at operation temperature (800 °C). A polished cross section of a so produced SOFC (as-received) is shown in Figure 1. The metallic substrate is CroFer22APU and the anode layer is still in the oxidized state.

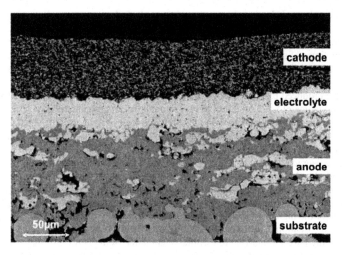

Figure 1: SEM micrograph (BSE) of a cross section of an SOFC layer system on top of a metallic substrate. Anode as well as electrolyte layer are produced by APS. The cathode layer is applied by screen printing.

The microstructure of the coatings was inspected on cross-sections by light microscopy, scanning electron microscopy (SEM) and energy-dispersive microanalysis (EDX). Images of the microstructure were taken using the Ultra55 (Carl Zeiss NLS AG) at 15 kV. In order to distinguish the different phases, backscatter electron (BSE) images were taken. The coating as well as the substrate porosity is determined by digital image analysis. Qualitative XRD phase analysis was performed using a D500 (Siemens AG) diffractometer equipped with diffracted-beam monochromator for Cu-K_α. Porosities of coatings and substrates are determined by digital image analysis. Long time measurements of the substrate conductivities are determined by the four point method in an Ar/4%H$_2$ atmosphere at 910°C. The conductivity was determined every 2 minutes at 1, 10, 50, and 100 mA. During the measurement intervals the current density was kept constant at 0.63 A/cm².

Single fuel cell tests were carried out to determine the electrochemical performance of the SOFC layer system. The cell tests are carried out using alumina housing. For the current collection on the anode, a Ni-mesh has been used, whereas a Pt-mesh has been used for the current collection on the cathode. No additional layers were applied between electrodes and current collecting meshes. The mechanical load on the meshes was always 20 MPa or higher (kgf per area of current collecting mesh). All current leads and potential wires are made from Pt-wire.

Sealing of the single cells was done using either gold or a glass-ceramic seal. The thermocouple, used to measure and taken as the operating temperature of the cell is located in the cell housing itself, closer than 10 mm from the cell.

SOFC start-up procedure: Heating the single cells (anode in its oxidized state) to 800°C is carried out by flushing the cathode side (compartment) with 0.5 l/min of air, and the anode side (compartment) with 0.5 l/min argon. All single cells were heated up to 800°C with a rate of 1°C/min. After reaching this temperature, the cells were conditioned for another 4 hours in argon – air atmosphere. This was followed by reduction of the anode layer carried out at 800°C by step wise increasing the amount of hydrogen on the anode side.

The recording of current-voltage (I-V) curves is conducted under galvanostatic control starting at open circuit voltage (OCV) using a current step size of 62.5 mA/cm². All electrochemical measurements were performed with 1 slpm hydrogen humidified with 0.03 slpm H$_2$O and 1 slpm air. Long-term stability tests were performed at 800°C under a constant load of 0.3 A/cm².

RESULTS AND DISCUSSION

The electrochemical performances of similar SOFCs produced by atmospheric plasma-spraying are published elsewhere[5]. Since the ferritic steel CroFer22APU are normally used as interconnector material, first metallic substrates are based on this kind of steel. A cross section of an APS SOFC layer system on top of a CroFer22APU substrate is shown in figure 1. The typical layer thickness of the anode is 80 μm, of the electrolyte 30 μm, and of the cathode 60 μm.

Figure 2 shows the cell voltage versus current density of such a SOFC layer system directly after the start-up of the cell and after 190 h of operation at 800°C under a constant electrical load of 0.3 A/cm². The initial open circuit voltage (OCV) of the tested cell was 910 mV, obviously lower than expected from the theoretical values. Probably gas diffusion due to pores, micro-cracks, or even between the gold seal and the cell, can play a role, and also electronic conductivity cannot be excluded. After about 190 h of exposure at 800 °C including a constant electrical load, the OCV was significantly higher than measured at the initial stage. Probably, the gas tightness of the gold seal was improved, or the microstructure was changed in such a way that less gas diffusion took place. At the initial stage of the experiment, the area specific resistance (ASR), which is the absolute slope of the curve at 700 mV was 672 mΩ.cm² at 800°C. After 190 h of exposure at 800 °C including a constant electrical load, the ASR significantly increased to a value of 1256 mΩ.cm². As a consequence, the current density at 700 mV obviously decreased; from 0.42 down to 0.24 A/cm². Since an increase of the OCV results in a gain of performance, the degradation is clearly linked to the increase of the ASR. From similar experiments on APS SOFCs it is known that the increase of the OCV happens during the first phase of operation (10-20 h).

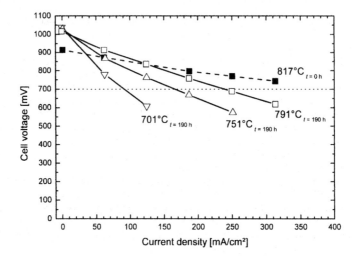

Figure 2: Current-voltage curves for a 16-cm² single cell with a Crofer22APU substrate as a function of the temperature. Solid symbol: after start-up; open symbols: after 190 h at 800°C under a constant electrical load (0.3 A/cm²). Fuel gas: H_2 (3% H_2O) = 1 l/min, oxidant: air = 1 l/min.

Figure 3 shows the cell voltage of the tested cell versus the time of exposure at 800°C and under a constant current load of 0.3 A/cm². After the first 24 h of exposure the cell voltage reaches a maximum followed by a relatively high degradation rate of about 130%/1000 h. In comparison to the demands for APU applications, which require typically 5000 h of operation, such an observed degradation rate is way beyond the scope. Therefore, it is necessary to identify the mechanism responsible for the degradation of the APS SOFC. In the following possible degradation mechanisms are discussed, which could be responsible for the described behavior.

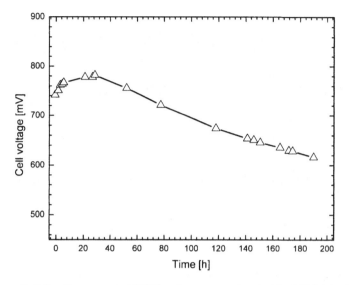

Figure 3: Cell voltage output at 800°C under constant electrical load (0.3 A/cm²) for a 16-cm² single cell with a CroFer22APU metallic substrate. Fuel gas: H_2 (3% H_2O) = 1 l/min, oxidant: air = 1 l/min).

The SOFC layer system was tested for an exposure time of 190 h at 800°C under a constant current load of 0.3 A/cm². Figure 4 shows the BSE image of the metallographically prepared cross section of the tested cell taken by SEM. By EDX analysis diffusion of chromium and iron was observed from the ferritic, metallic substrate into the anode layer. Additionally, diffusion of nickel into the first 10 μm of the metallic substrate was detected. Due to Ni diffusion the substrate can become austenitic and a diffusion of Cr into the anode layer will cause the formation of a Cr_2O_3 layer on top of the Ni particles. Such a Cr_2O_3 scale could be responsible for an electrochemical performance loss of the anode layer. In the bottom part of figure 4 iron domains between the spherical CroFer22APU particles of the substrate are present. The iron domains are frequently not directly attached to the CroFer22APU particles (arrow in the bottom right micrograph of figure 4). This microstructure of the iron indicates that probably iron oxide was reduced to metallic iron during the switch-off of the SOFC under OCV conditions, and therefore the oxygen partial pressure is decreased at the anode. A catastrophic oxidation of the ferritic steel CroFer22APU takes place once the chromium content of the steel falls below a certain level (between 16 and 18 wt%)[14]. Therefore, from the presence of iron oxide it can be concluded that a catastrophic oxidation of the metallic substrate has already started in the bottom part of the substrate.

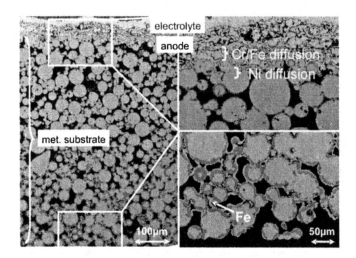

Figure 4: Cross section of a single SOFC with a CroFer22APU substrate after electrochemical test of 190 h at 800°C under constant electrical load (0.3 A/cm²). Right side: magnifications of the white marked rectangles shown on the left side.

To identify the main mechanism responsible for the aging SOFC specific experiments were designed to separate individual degradation processes.

(i) Two types of SOFCs were prepared to check if interactions between the metallic substrate and the anode layer are responsible for the high degradation rate. In a first experiment the CroFer22APU based substrate was replaced by an optional ferritic ODS steel ITM-14 (Plansee). Additionally, by varying the anode thickness of the SOFC layer system conclusions could be drawn to which extent diffusion processes into the anode are responsible for the cell aging. Results of the electrochemical performance of these special prepared SOFCs are shown in Figure 5. To better compare the individual degradation rates the plotted curves are normalized to 1 at an exposure time of 190 h. Degradation rates are determined from the maximum cell voltage to the cell voltage at 190 h. Degradation rates, anode thicknesses, and the OCVs of each tested cell are given in table I. The experiments clearly show that both the steel selected for the substrate as well as the thickness of the anode layer have an influence on the aging of the SOFC.

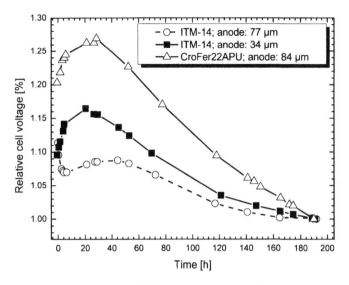

Figure 5: Cell voltage output at 800°C under constant electrical load (0.3 A/cm²) for 16-cm² single cells on different metallic substrates. To compare degradation rates, cell voltages are plotted relative to their cell voltage at 190 h. Fuel gas: H_2 (3% H_2O) = 1 l/min, oxidant: air = 1 l/min).

Table I: Properties of the tested SOFCs with metallic support.

Substrate	CroFer22APU	ITM-14	ITM-14
Symbol. in figure 5	Δ	■	O
Anode thickness [μm]	84	34	77
Degradation [%/1000 h]	134	85	58
OCV after start-up [mV]	910	957	955

(ii) Electronic conducting paths of the substrate are provided by the sinter necks of the individual metallic powder particles. Due to oxidation of the substrate during operation at 800°C in humid fuel atmosphere a rapid decrease of the conducting property of the substrate might occur. Therefore the electronic conductivity of both kinds of substrates was measured as a function of time by the four point method. The experiments were performed at elevated temperatures of 910°C to accelerate the oxidation process[14]. As atmosphere, similar to anodic condition, streaming Ar/4% H_2 humidified with 3% H_2O was used. During the whole exposure time a constant current density of 0.63A/cm² was applied on the samples. Figure 6 shows the measured conductivity of the CroFer22APU and the ITM-14 substrates versus time of exposure. For comparison the plotted conductivities are corrected by the porosity of the respective substrate. Whereas the CroFer22APU substrate shows a slight decrease of the conductivity on a long time range, the degradation of the ITM-14 substrate is more pronounced. After about 50 h the ITM-14 shows a strong, exponential drop of the conductivity. Since the conductivity of the metallic support is in the order of 100 S/cm, the conductivity of the substrate only contributes to a minor extent to the total ASR of the SOFC. Considering additionally that the conductivity was measured at accelerated oxidation condition compare to the real operation temperature of the SOFC, the decrease of the conductivity of the metallic substrate cannot explain the observed degradation.

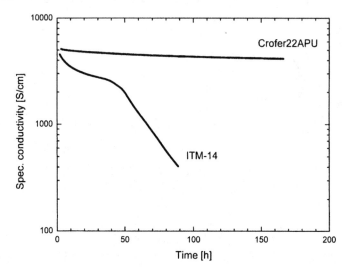

Figure 6: Electrical conductivity of the used metallic substrates at 910°C under a constant current density of 0.63 A/cm² in humidified Ar/4% H₂ atmosphere.

(iii) The materials used for the electrolyte and the cathode layer are YSZ and LSFC, respectively. It is known that during sintering processes the formation of zirconates like $SrZrO_3$ and $La_2Zr_2O_7$ takes place at temperatures above 1000°C. These species have low oxygen ion conductivity[17] and therefore can be responsible for a degradation of the SOFC. For the application of the used cathode layer no additional sintering step is necessary. Therefore, the maximum temperature where such an undesired reaction between the electrolyte and the cathode material can take place will be the operation temperature of the SOFC.

Particular experiments on heat treated YSZ/LSFC powder mixes have shown that the formation of $SrZrO_3$ occurs even at 800°C[18]. To separate the influence of the strontium zirconate formation on the cell performance the LSFC cathode was applied on standard SOFCs which normally only show a negligible degradation rates of the electrochemical performance[19, 20]. Figure 7 shows the cell voltages versus time of the anode supported Ni/YSZ-type single cells with an 8YSZ electrolyte and "non-sintered" LSFC cathode. To avoid direct contact between the YSZ and the LSFC and therefore, avoiding the $SrZrO_3$ formation, one of the tested cells was prepared with a $Ce_{0.8}Gd_{0.2}O_{1.9}$ (CGO) interlayer between the electrolyte and the cathode layer.

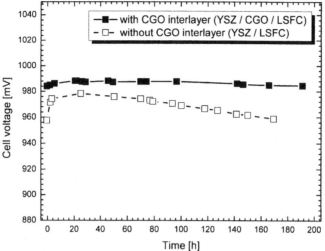

Figure 7: Cell voltage output at 800°C under constant electrical load (0.3 A/cm²) for 16-cm² single cells with a NiO/YSZ substrate. Solid symbols: with a CGO layer between YSZ electrolyte and LSFC cathode; open symbols: without a CGO layer between YSZ electrolyte and LSFC cathode. Fuel gas: H_2 (3% H_2O) = 1 l/min, oxidant: air = 1 l/min).

A clear difference in the degradation rates between SOFCs with and without a CGO interlayer can be seen. Whereas the cell with an interlayer shows only a vanishing aging rate of about 1.7%/1000 h, the cell voltage of the cell without a CGO layer drops with about 16%/1000 h. This experiment clearly illustrates the necessity of an interlayer between the YSZ electrolyte and the LSFC cathode to avoid the formation of $SrZrO_3$ and therefore a loss of cell performance of time. Since the degradation of the APS SOFCs are about one order of magnitude higher compare to the "sintered" cells, the interaction between the electrolyte and the cathode can be excluded as the main mechanism responsible for the observed aging rate on the "APS" cells.

CONCLUSION
SOFC layer systems with a metallic support, an APS anode as well as electrolyte and a screen printed, not-sintered LSFC cathode show high degradation rates of cell performance within the first 200 h of operation in single cell tests. The formation of strontium zirconate between the electrolyte and the cathode layer is possible. But this low oxygen ion conducting interaction zone has to be excluded as the main degradation mechanism of the studied SOFCs. Furthermore, a loss of the electronic conductivity of the metallic substrate due to oxidation could not be identified as the primary mechanism for the rapid aging of the cells.

The strongest influence with respect to the degradation could be seen by varying the used alloy of the metallic substrate or changing the thickness of the deposited anode layer. Therefore, it can be concluded that the high degradation has its origin in an interaction between the metallic substrate and the Ni/YSZ anode. Diffusion of Ni into the substrate and Fe as well as Cr into the anode layer was observed by EDX analysis. Further experiments are necessary to understand how these diffusion processes are influencing the performance of the SOFC.

To improve the lifetime of the SOFC layer system two additional layers are suggested: (i) an CGO layer between the electrolyte and the cathode layer, to avoid the formation of $SrZrO_3$ and (ii) a suitable diffusion barrier coating between the metallic substrate and the anode layer[21]. Since the metallic support is limited with respect to moderate sintering temperatures, the atmospheric plasma-spraying should be considered as the adequate process to apply the additional suggested layers.

ACKNOWLEDGEMENTS

Part of the work has been performed within the CexiCell project which is funded by the European Union under the "Energy Environment and Sustainable Development (1998-2002)" contract number ENK5-CT-2002-00642. The funding by the EU is gratefully acknowledged. The authors also gratefully acknowledge the work of Mr. K.H. Rauwald and Mr. R. Laufs for the manufacturing of the plasma-sprayed coatings, Mrs. A. Hilgers, Mrs. H. Moitroux, Mr. R. Dahl, Mr. P. Lersch, Mr. W. Herzhof, Mr. V. Bader and Mr. M. Kappertz for assistance in sample preparation and characterization, and Mrs. C. Tropartz, Mrs. R. Röwekamp and Mr. H. Wesemeyer for performing the electrochemical measurements.

REFERENCES

[1]H. Itoh, M. Mori, N. Mori, and T. Abe, "Production cost estimation of solid oxide fuel cells", *J. Power Sources*, **49**, 315-332 (1994).

[2]S. Takenoiri, N. Kawokawa, and K. Koseki, "Development of Metallic Substrate Supported Planar Solid Oxide Fuel Cells Fabricated by Atmospheric Plasma Spraying", *J. Thermal Spray Tech.*, **9**, 360-363 (2000).

[3]G. Schiller, T. Franco, M. Lang, P. Metzger, and A.O. Stömer, "Recent results of the SOFC APU development at DLR", *Proc. solid oxide fuel cells IX*, S.C. Singhal and J. Mizusaki, Eds., May 15-20, 2005 (Quebec City/ Canada), The Electrochemical Society, Inc., 2005, p. 66-75.

[4]P. Szabo, M. Lang, T. Franco, and G. Schiller, "Fabrication and Development of Optimized Plasma-Sprayed SOFC Layers for Use in the DLR Spray Concept", *Proc. of 6th European SOFC Forum*, M. Morgensen, Ed., June 28 - July 2, 2004 (Lucerne/Switzerland), European Fuel Cell Forum, 2004, p 259-267.

[5]R. Vaßen, D. Hathiramani, R.J. Damani, and D. Stöver, "Gas-tight zirconia electrolyte layers for SOFCs by atmospheric plasma-spraying", *Proc. solid oxide fuel cells IX*, S.C. Singhal and J. Mizusaki, Eds., May 15-20, 2005 (Quebec City/ Canada), The Electrochemical Society, Inc., 2005, p 1016-1024.

[6]R. Zheng, X.M. Zhou, S.R. Wang, T.-L. Wen, and C.X. Ding, "A study Ni+8YSZ / 8YSZ / $La_{0.6}Sr_{0.4}CoO_{3-d}$ ITSOFC fabricated by atmospheric plasma spraying", *J. Power Sources*, **140**, 217-225 (2005).

[7]D. Hathiramani, A. Mobeen, W. Fischer, P. Lersch, D. Sebold, R. Vaßen, D. Stöver, and R.J. Damani, "Simultaneous deposition of LSM and YSZ for SOFC cathode functional layers by an APS process", *Thermal Spray 2005: Advances in Technology and Application*, E. Lugscheider, Ed., May 2-4, 2005 (Basel/Switzerland), DVS, 2005, p 585-589.

[8]S.C. Singhal, in S.C. Singhal, M. Dokiya (Eds.), Proc. 6th Int. Symp. Solid Oxide Fuel Cells (SOFC-VI), The Electrochemical Society, Pennington, NJ, 1999, p. 39.

[9]D. Simwonis, A. Naoumidis, F.J. Dias, J. Linke, and A. Moropoulou, "Material characterization in support of the development of an anode substrate for solid oxide fuel cells", *J. Mater. Res.*, **12**, 1508-1518 (1997).

[10]T. Franko, M. Lang, G. Schiller, P. Szabo, W. Glatz, and G. Kunschert, "Powder metallurgical high performance materials for substrate-supported IT-SOFCs", *Proc. of 6th European SOFC Forum*, M. Morgensen, Ed., June 28 - July 2, 2004 (Lucerne/Switzerland), European Fuel Cell Forum, 2004, p. 209-217.

[11]P. Lamp, J. Tachtler, O. Finkenwirth, S. Mukerjee, and St. Shaffer, "Development of an Auxiliary Power Unit with Solid Oxide Fuel Cells for Automotive Applications", *Fuel Cells* 2003, No.3, Wiley-VCH Verlag, Weinheim, Germany, p. 1-7.

[12]Y.B. Matus, L.C. De Jonghe, C.P. Jacobson, and S.J. Visco, "Metal-supported solid oxide fuel cell membranes for rapid thermal cycling", *Solid State Ionics,* **176**, 443-449 (2005).

[13]Crofer 22 APU, Material Data Sheet No. 8005, *ThyssenKrupp VDM GmbH.*

[14]W.J. Quadakkers, J. Piron-Abellan, V. Shemet, and L. Singheiser, "Metallic interconnectors for solid oxide fuel cells – a review", *Materials at high temperatures* **20**, 115-127 (2003).

[15]W. Glatz, G. Kunschert, and M. Janousek, "Powder-metallurgical processing and properties of high performance metallic SOFC interconnect materials", *Proc. of 6th European SOFC Forum*, M. Morgensen, Ed., June 28 - July 2, 2004 (Lucerne/Switzerland), European Fuel Cell Forum, 2004, p.1612-1627.

[16]D. Hathiramani, R. Vaßen, D. Stöver, and R.J. Damani, "Comparison of atmospheric plasma sprayed anode layers for SOFCs using different feedstock", *special issue of the J. Thermal Spray Tech.*, International Thermal Spray Conference And Exposition (ITSC), May 15-18, 2006, Seattle/USA, accepted for publication.

[17]K.V.G. Kutty, C.K. Mathews, T.N. Rao, and U.V. Varadaraju, "Oxide ion conductivity in some substituted rare earth pyrozirconates", *Solid State Ionics*, **80**, 99-110 (1995).

[18]A. Mai, M. Becker, W. Assenmacher, D. Hathiramani, F. Tietz, E. Ivers-Tiffée, D. Stöver, and W. Mader, "Time-dependent performance of mixed-conducting SOFC cathodes", *submitted to Solid State Ionics*, special issue to the conference SSI-15, Baden-Baden/Germany (2005).

[19]A. Mai, V.A.C. Haanappel, S. Uhlenbruck, F. Tietz, and D. Stöver, "Ferrite-based perovskites as cathode materials for anode-supported solid oxide fuel cells: Part I. Variation of composition", *Solid State Ionics,* **176**, 1341-1350 (2005).

[20]M. Becker, A. Mai, E. Ivers-Tiffée, and F. Tietz, Long-term measurements of anode-supported solid oxide fuel cells with LSCF cathode under various operating conditions", *Proc. solid oxide fuel cells IX*, S.C. Singhal and J. Mizusaki, Eds., May 15-20, 2005 (Quebec City/Canada), The Electrochemical Society, Inc., 2005, p 514-523.

[21]M. Brandner, M. Bram, D. Sebold, S. Uhlenbruck, S.T. Ertl, T. Höfler, F.-J. Wetzel, H.P. Buchkremer, and D. Stöver, "Inhibition of diffusion between metallic substrates and Ni-YSZ anodes during sintering", *Proc. solid oxide fuel cells IX*, S.C. Singhal and J. Mizusaki, Eds., May 15-20, 2005 (Quebec City/ Canada), The Electrochemical Society, Inc., 2005, p.1235-1243.

EFFECT OF TRANSITION METAL IONS ON THE CONDUCTIVITY AND STABILITY OF STABILISED ZIRCONIA

D. Lybye and M. Mogensen,
Materials Research Department, Risø National Laboratory,
P.O. BOX 49, DK-4000 Roskilde, Denmark

ABSTRACT

Zirconia compounds stabilised with rare-earth metal oxides like yttria, ytterbia and scandia are known to be good oxide ion conductors suitable as electrolyte material in solid oxide fuel cells. However, stabilised zirconia with high oxide ion conductivity is often only metastable at fuel cell operation temperatures and changes in temperature or oxygen partial pressure together with long-term operation are seen to induce partial destabilisation and even phase changes. In order to avoid these effects co-doping has proven helpful. Based on experimental data available in literature, we discus the effect of co-doping with smaller transition metal ions such as Ti-, Fe- and Mn-ions. Many of the ionic radii of the transition metal ions are too small compared to the host lattice ionic radius of zirconium. Here we explore the effect of a) the small ionic radii compared to the large ionic radii of the host lattice and b) the preferred six coordination compared to the desired eight-fold coordination of the fluorite structure. Particular interest is paid to the solubility of the transition metal ions and to the conductivity of the resulting material.

Indium is not a transition metal but due to the size of the ionic radius of the metal, the effect of doping with In is also explored.

INTRODUCTION

Fluorite-structured oxides are of the type MO_2, where M is a relatively large four-valent cation like Zr^{4+} and Ce^{4+}. These may exhibit high oxide ion conductivity when suitably doped with lower valent metal ions, and they are, therefore, used as electrolytes and components in the composite electrodes in solid oxide fuel cells (SOFC), solid oxide electrolyser cells (SOEC) and oxygen sensors (e.g. lambda sensors in car engines). The most common SOFC electrolyte is Y_2O_3-stabilized ZrO_2 (YSZ), but in the recent years much R&D work on SOFCs with doped CeO_2 or zirconia stabilised with Sc^{3+} as electrolyte has taken place.

The understanding of the conductivity of oxide ions in this type of oxides is of major importance, and the influence of various parameters is the subject of numerous papers, see e.g.[1,2,3,4,5,6], in which many significant parameters are discussed. We have recently re-analysed this topic and based on experimental data available in the literature, we argued that lattice distortion (lattice stress and deviation from cubic symmetry) due to ion radii mismatch determines the ionic conductivity to a very large extent, and that lattice distortion is of much greater importance than many other proposed parameters.[7]

The main purpose of this short review is, based on experimental data available in literature, to discus the effect of co-doping with smaller transition metal ions such as Ti-, Fe- and Mn-ions. Many of the ionic radii of the transition metal ions are too small compared to the host lattice ionic radii of zirconium. Here we explore the effect of a) the small ionic radii compared to the large ionic radii of the host lattice and b) the preferred six coordination compared to the desired eight-fold coordination of the fluorite structure. Particular interest is paid to the solubility of the transition metal ions and to the conductivity of the resulting material. The solubility of transition metals varies a lot. The solubility of NiO is less than 0.5 mol % at 1000°C in 8YSZ (8 mol % Y_2O_3 stabilised zirconia)[8] while the solubility of $CrO_{1.5}$ and of FeO_x is calculated to be approximately 0.43 % and 3 %, respectively at 1000 °C in 8YSZ. The solubility of Mn-ions is calculated to 8 % of the metal atoms in 8YSZ at 1000°C[9]. Solubility, phase stability and conductivity are discussed in terms of preferred co-ordination number, ionic radius and temperature.

The ionic radii quoted through out this paper are the ionic radii determined and compiled by Shannon[10] for oxides and partly for fluorites. The radii are based on empirical data with the approximations and averagering necessary for to obtain one value. The first and most important assumption Shannon states is that the ionic radii of both anion and cation are additive to reproduce interatomic distances if one considers coordination number, electronic spin, covalence, repulsive forces, and polyhedral distorsion[10]. It should also be kept in mind that the ionic radii are determined from room temperature data. When discussing conductivity the temperature range is usually from 600 °C to 1000 °C. The thermal vibrations will cause the effective ionic radii to be larger at high temperature.

PREVIOUS ANALYSIS
Matching radius

The solubility of transition metals and indium in yttria stabilised zirconia has been investigated by several groups both experimentally and estimated by calculations. Generally, solubility depends on temperature, but for transition metals it also depends on oxygen partial pressure due to their ability to change valance state with changing oxygen partial pressure. Furthermore, in the fluorite system, solubility is closely dependent on ionic radius and on the valance of the dopant metal ion compared to the host fluorite lattice. For simple solid solutions of one metal oxide into another, the lattice parameter usually follows Vegard's rule to a good approximation; i.e. a linear relationship exists between lattice parameter and the concentration of the solute. The slope of this straight line is termed Vegard's slope. Kim[11] published a set of empirical relationships between concentrations, ionic radius of the metal ion - of oxides (dopants) dissolved in fluorite-structured oxides - and the lattice parameter. In the case of ZrO_2 he finds:

$$a = 5.120 \text{ Å} + \Sigma m_k (0.0212 \, \Delta r_k + 0.0023 \Delta z_k) \text{ Å} \qquad [1]$$

where a (in Å) is the lattice constant of the zirconia solid solution at room temperature, Δr_k (in Å) is the difference in ionic radius ($r_k - r_{Zr4+}$) of the kth dopant and the Zr^{4+}-radius, which in eight-fold coordination is 0.84 Å[10], Δz_k is the valence difference, ($z_k - 4$), and m_k is the mole percent of the kth dopant in the form of MO_x. From eq. [1] the following expressions for Vegard's slopes can be derived:

$$S_V = 0.0212 \, (mol\%)^{-1} (r_k - 0.84\text{Å}) - 2.3 \, 10^{-3}\text{Å/mol\%} \qquad [2]$$

for trivalent dopants, and

$$S_V = 0.0212 \, (mol\%)^{-1} (r_k - 0.84\text{Å}) - 4.6 \, 10^{-3}\text{Å/mol\%} \qquad [3]$$

for divalent dopants. This implies that there exists a "matching" radius, r_m, at which Vegard's slope is zero. According to Kim[11], the value of r_m is 1.057 Å for divalent ions and 0.948 Å for trivalent ions in zirconia.

Kim argues that the solubility limit of a solute depends on the elastic energy, W, which is introduced in the lattice due to differences in ionic radius. The larger the elastic energy per substituted

Table 1. Ionic radii as determined and compiled by Shannon[10]. r_{mx+} are the matching radii determined by Kim[11]. All radii are in Å.

Zr^{4+} (8) = 0.84		
r_{m4+} = 0.84	r_{m3+} = 0.948	r_{m2+} = 1.057
Mn^{4+} (6) = 0.53	Mn^{3+} (6) = 0.58	Mg^{2+} (8) = 0.89
Ti^{4+} (8) = 0.74	Co^{3+} (6) = 0.61	Zn^{2+} (8) = 0.9
	Ti^{3+} (6) = 0.67	Co^{2+} (8) = 0.9
Hf^{4+} (8) = 0.83	Fe^{3+} (8) = 0.78	Fe^{2+} (8) = 0.92
Ce^{4+} (8) = 0.97	Sc^{3+} (8) = 0.87	Mn^{2+} (8) = 0.96
	In^{3+} (8) = 0.92	Ca^{2+} (8) = 1.12
	Yb^{3+} (8) = 0.985	
	Er^{3+} (8) = 1.004	
	Y^{3+} (8) = 1.019	
	Dy^{3+} (8) = 1.027	
	Gd^{3+} (8) = 1.053	

ion, the lower is the solubility. The relation between W and the change in the lattice parameter, Δa, due to formation of a substitution solid solution is given by:

$$W = 6 \, Ga_0 \, (\Delta a)^2 \qquad\qquad [4]$$

where G is the shear modulus. Δa is governed by Vegard's slope for the given solute. This implies that the highest solubilities should be achieved for a Vegard's slope of $S_v = 0$. However, the solubility also depends on temperature and for transition metal ions that have more than one valence state (equals more than one value for the ionic radius) also on oxygen partial pressure.

Table 1 gives the ionic radii determined and compiled by Shannon[10] together with the matching radii calculated by Kim[11]. Examining the matching radius calculated by Kim[11] a little more in detail, there is at least one significant exception from the rule; the Sc_2O_3-Y_2O_3-ZrO_2-system. It should be noted that Sc^{3+} in zirconia (and hafnia) seems to constitute a significant exception from the rules using Kim's value for the zirconia matching radius. The conductivity of scandia-doped zirconia seems to be far higher (by 2 - 3 times)[12,13] and has a significant lower activation enthalpy[3] than any other doped zirconia in this temperature regime even though the radius of Sc^{3+} in 8-fold coordination is only 0.87 Å. Therefore, we have included indium in this review. In^{3+} (0.92 Å) in eight coordination is smaller than the calculated matching radius but in between Sc^{3+} and Yb^{3+} in size, see Table 1. However, as summarised below, the conductivity is in the same range as for yttria stabilised zirconia.

If we compare the ionic radii of the transition metals with the calculated matching radius, the transition metal ions are in most cases significantly smaller. For Cr, Ni, V and Ti, the ionic radius is smaller in all valence states[10]. Ti^{4+}, however, is found in eight coordination[10], so the solubility and influence on conductivity are investigated in the following. For Co, Fe and Mn, the ionic radii in valence state 2+ and with eightfold coordination show a larger value than the radius of Zr^{4+}, but significant lower than the matching radius determined by Kim[11] for valence 2+, see Table 1.

Solubility and structure

Sasaki et al.[14] have determined the phase diagram for the ZrO_2-$InO_{1.5}$ system. The solubility of $InO_{1.5}$ is less than 1 mol % in ZrO_2 at temperatures below 1000 °C. Only above 1315 °C is the cubic phase formed with 25 mol % $InO_{1.5}$. At 1700 °C, an extended solid solubility from 12.4 ± 0.8 to 56.5 ± 3 mol % $InO_{1.5}$ is found in cubic ZrO_2. Below 12.4 mol %, $InO_{1.5}$ is still soluble but the phase is tetragonal. By cooling at rates faster than 10 K/min, the cubic phase can be maintained at lower temperatures.

Naito et al.[15] has determined the solubility limit in the 10 mol% Y_2O_3 stabilised zirconia (10YSZ) – india system. The solubility limit is 35 mol% $InO_{1.5}$ at 1700 °C.

For MnO_x in TZ8Y, Kawada et al.[16] have found the solubility in air to be 5 mol % at 1273 K, 8 mol % at 1473 K, 12 mol % at 1673 K and 15 mol % at 1773 K. This agrees well with calculations by Chen et al.[17], while calculations by Yokokawa et al.[8] point to solubility of 8 mol % at 1273 K with a mixture of Mn^{3+} and Mn^{2+}. Yokokawa et al.[8] also predict an increasing solubility with decreasing oxygen partial pressure due to the reduction of Mn^{3+} to Mn^{2+} which has a better matching radius compared to Zr^{4+} see Table 1. There are, however, some disagreements about the amount of Mn in different valance states when dissolved in ZrO_2 in air. The valance state of Mn-ions in both ZrO_2 and YSZ has been studied by several groups[18,19,20,21]. Occhiuzzi et al.[21] used Electron Paramagnetic Resonance, EPR, to investigate the valance of manganese in monoclinic, tetragonal and cubic zirconia. They suggest that the valance state of manganese ions in zirconia is determined both by crystal structure and by oxygen partial pressure. Occhiuzzi et al.[21] state that Mn^{2+} (high spin $3d^5$) can occupy cationic sites with four, six or eight coordination while Mn^{4+} has a strong preference for 6-coordination. Due to the high Crystal Field Stabilisation Energy (CFSE) in cubic (and tetragonal) zirconia, Mn^{4+} is destabilised in this structure and Mn enters with lower valence state. They have no comments on Mn^{3+}. Apple et al.[20] have determined the valence state by Electron Energy Loss Spectroscopy, EELS. They find it to be a mixture of Mn^{+2} and Mn^{+3}. Due to strong interaction with the oxygen vacancies increasing with increasing Mn-concentration, it is not possible to make any quantitative determination of the Mn^{2+} and Mn^{3+} concentrations.

MnO-stabilised ZrO_2 is confirmed to have the cubic fluorite structure and Mn is present as Mn^{2+} due to treatment in Ar^{22}. Apple et al.[23] has shown that addition of 5 mol % Mn stabilises TZ8Y in the cubic phase. By electron diffraction in the transmission electron microscope (TEM) they find that by addition of 5 mol % MnO_x the forbidden diffraction spots, which normally are present due to incomplete stabilisation, disappear. 2 mol % MnO_2 was not sufficient to obtain this effect.

As expected, Ti^{4+} is too small to stabilise cubic zirconia at temperatures relevant for fuel cell applications. Up to 22 atom % Ti is soluble in tetragonal ZrO_2 at 1500 °C. On cooling the samples to room temperature 100% conversion from tetragonal to monoclinic occurred. The temperature of the monoclinic to tetragonal transition decreases from 1170 °C for ZrO_2 to below

$1000\ ^{\circ}C$ for $Zr_{0.85}Ti_{0.15}O_2$.[24]

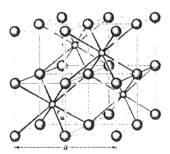

Figur 1 Cubic fluorite structure with the small Zr^{4+}-ion in eight coordination in every second cube formed by oxide ions.

The solubility of TiO_2 in yttria stabilised zirconia depends on the amount of yttria added, the more yttria the more titania can be dissolved. This might be related to the fact that the large yttrium ion can compensate for the missing volume of the smaller titanium. No detailed structural studies were found in literature to confirm this. Tao and Irvine[25] has found a maximum solubility of 18 mol % TiO_2 in the ternary systems Y_2O_3-ZrO_2-TiO_2 and Sc_2O_3-ZrO_2-TiO_2 at $1500\ ^{\circ}C$. In the quaternary system Sc_2O_3-Y_2O_3-ZrO_2-TiO_2, they find a slightly higher solubility of 20 mol % TiO_2. In the Y_2O_3-ZrO_2-TiO_2 cubic system the yttrium content is between 14 and 20 atom %.[24]

In 8YSZ the solubility is found to be approximately 10 mol % TiO_2 depending on the preparation conditions. By a closer examination, more groups find Ti-enriched grain boundaries[26]. Even though XRD-examination shows cubic single phase, TEM-investigations[26] have shown that the cubic symmetry of 8YSZ is lost by addition of TiO_2. Tetragonal precipitates are found in cubic grains after heat treatments in air at temperatures up to $1090\ ^{\circ}C$. Diffraction spots forbidden in fcc structure and with odd, odd, even indices are clearly seen. The same type of tetragonal precipitates has also been observed in Ni-doped 8YSZ after exposure to reducing conditions and was followed by a significant reduction in conductivity[27]. Phase stability over time has been investigated by Kobayashi et al.[28]. They find transition to tetragonal after annealing at temperatures between $800\ ^{\circ}C$ and $1000\ ^{\circ}C$ for 486 hours.

The ability of iron alone to stabilise zirconia in its cubic form has also been investigated[29,30,31]. The solubility of Fe_2O_3 in the ZrO_2 is ~ 2 mol % after calcination at $1100\ ^{\circ}C$[30,31] and the phase is not cubic[29]. Fe addition alone is not enough to stabilise zirconia.

Several authors[32,33,34] have claimed that small amounts of Fe_2O_3 are incorporated as interstitials rather than in solid solution. Figure 1 shows the cubic fluorite structure and it would be possible for small transition metal ion to sit as interstitials in the oxygen cubes where no Zr is present. The theory is supported by the fact that the cell constant increases even though, the ionic radius of Fe^{3+} is smaller than the radius of Zr^{4+}[10]. The limit of interstitial solid solution is around 0.6 mol % Fe_2O_3. Higher amounts are incorporated by substitution[33]. By addition of 1, 3 and 5 cation % Fe to Tosoh 8YSZ, Slitaty and Marques[35] find that the cell constant decreases, though, not in a simple way[35]. SEM/EDS analysis showed a significant amount of Fe-rich secondary phase in the 5 cation % doped Tosoh 8YSZ[35] sintered at $1450\ ^{\circ}C$. Wilhelm and Howarth[32] find a solubility limit of (4.35 mol % Fe_2O_3) 7 cation % Fe in ultra fine 8YSZ sintered at $1550\ ^{\circ}C$. However, this is only possible, if excess Fe_2O_3 is added, meaning that FeO_x is found as segregations. From their measurements of the lattice parameter, it seems that the solubility without having FeO_x segregations is about 2 mol % Fe_2O_3 corresponding to 4 cation % Fe. This is in accordance with the results of Slitaty and Marques[35].

Yokokawa[9] has calculated the solubility of Fe in YSZ as a function of oxygen partial pressure at 1273 K. The calculated solubility of Fe^{3+} in air is ca. 3 cation %. The solubility increases as Fe^{3+} is reduced to Fe^{2+}. The maximum solubility is 5.6 cation % at an oxygen partial pressure between $2.5 \ 10^{-14}$ to 10^{-15} bar. The solubility decreases rapidly after this maximum and at an oxygen partial pressure of $< 10^{-18}$ bar, Fe is reduced to metallic iron and the solubility is zero.

Ionic conductivity.

Both the oxide ion conductivity and the activation energy are strong functions of dopant type as well as concentration (see e.g. Minh and Takahashi[36]). Some data, which are useful for the present discussion, are collected in Table 2.

The ionic conductivities in In-doped zirconia are comparable to those of Y-doped zirconia[37]. The optimum conductivity in the ZrO_2-$InO_{1.5}$ system is found close to the eutectoid composition of 23.5 ± 1 mol % $InO_{1.5}$[14]. Gauckler et al.[38] found that the cubic phase of 25 mol % $InO_{1.5}$ doped ZrO_2 has the highest conductivity in the ZrO_2-$InO_{1.5}$ system at temperatures above 650 °C. The conductivity is almost the same as that in 8 mol% stabilised ZrO_2 at 1000 °C. Below 650 °C, the tetragonal phase with 15 mol % $InO_{1.5}$ has the highest conductivity.

The conductivity of 10YSZ doped with In decreases with increasing In-content, and the activation energy increases with increasing $InO_{1.5}$ concentration[15]. This phenomenon corresponds to the activation energy change with the Y_2O_3 concentration in YSZ[39]. Due to the higher activation energy, the conductivity of 20 mol % $InO_{1.5}$-doped 10YSZ is almost comparable to that of 10YSZ at 1400 °C. The conductivity of 30 mol % $InO_{1.5}$ -doped 10YSZ found by Naito et al.[15] is the same as the conductivity found for 35 mol% $InO_{1.5}$-ZrO_2 [38] from 600 to 725 °C. In other words, In^{3+} has an effect very similar to Y^{3+} on the oxide ion conductivity of zirconia

The conductivity of In-doped 10YSZ at 1000 °C and 1200 °C as a function of oxygen partial pressure has been investigated by Naito et al.[15]. They find that for $InO_{1.5}$ concentrations above 20 mol %, conductivity increases with decreasing oxygen partial pressure resulting from n-type electronic conductivity. The increase appears at oxygen partial pressures below 10^{-10} atm depending on the In-concentration; this means that the n-type conduction is due to reduction of In^{3+} to In^{1+} since YSZ is stable down to 10^{-25} atm[12]. The total conductivity of 20 mol% $InO_{1.5}$ -doped 10YSZ is comparable to the conductivity of 10 mol% TiO_2-doped 10YSZ.[15, 40]

When doping yttria stabilised zirconia with manganese, the ionic conductivity decreases even though the cubic structure is stabilised. Kawada et al.[16] find that 4 mol % Mn-doped 8YSZ exhibits p-type conduction at high oxygen partial pressures and n-type conductivity at oxygen partial pressures above 10^{-10} Pa. The electronic contributions to the conductivity are low and Mn, Y-codoped zirconia is still a good electrolyte. Appel et al.[23] have investigated the long term stability in air at 850°C and at 1000°C, see Figure 2. They find that at 1000°C, 1 cation % and 2 caiont % Mn is not enough to stabilise the cubic structure. The conductivity curves have the same shape as for undoped 7.9YSZ (7.9 mol % Y_2O_3 stabilised zirconia), decreasing with time. The conductivity of 5 cation % Mn doped 7.9YSZ starts at a lower value, but is stable or even increasing slightly with time. At 800°C, the 4 curves are almost parallel, but the decrease in conductivity with time is lower for the Mn-doped samples and lowest for the addition of 5 cation % Mn. In Sc and Y co-doped ZrO_2 addition of 2 cation % Mn has been found to reduce the conductivity with 20 %[41].

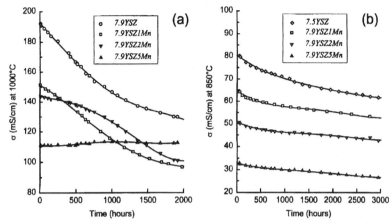

Figur 2 Conductivity as a function of time for Mn-doped 8YSZ determined by Appel et al.[23] at a) 1000 °C and b) 850 °C.

Addition of TiO_2 to stabilised zirconia introduces electronic n-type conduction at low oxygen partial pressures (< 10^{-12} atm), in fact TiO_2-doped zirconia has been considered as redox stable anode for solid oxide fuel cells [25,26,42,43]. The introduction of n-type electronic conduction under reducing conditions is at the expense of the ionic conductivity, which is lowered considerable (1000 °C; σ_{8YSZ} = 0.16 S/cm; $\sigma_{8YSZ+10 \, mol\% \, TiO2}$ = 0.023 S/cm)[26]. Good correlations are observed between activation energy and concentration of either Ti or Y if the concentration of the other element is kept constant. Activation energy increases with Y content but decreases with Ti content. The composition dependence of the activation energy seems very closely related to the dependence of unit cell parameter (also related to composition)[44].

Kharton et al.[45] find that small additions of Fe do not significantly affect the conductivity. The impedance measurements obtained at 400°C by Slitaty and Marques[35] contradicts this. They find that addition of only 1 cation % Fe increases both the bulk and grain boundary resistance. However, by further additions the largest increase in resistivity appears in the bulk. The conductivity of the 5Fe8YSZ (sintered at 1450 °C) composition is approximately 50% of the conductivity of pure 8YSZ. Wilhelm and Howarth[32] find that at high temperature, the conductivity only decreases slightly with increasing addition of Fe_2O_3 to 8YSZ. At low temperatures, electronic conduction is introduced and conductivity increases with iron content at temperatures blow 500°C. The contradiction between the results obtained by Slitaty and Marques[35] and Wilhelm and Howarth[32] could be a result of different preparation methods and the fact that Wilhelm and Howarth[32] probably have segregated iron in some form in their samples.

DISCUSSION

Table 2 gives some data on conductivity and solubility mentioned in the previous sections. The samples are prepared by different routes and investigated under different conditions.

Seen from the electrolyte point of view, the addition of a small 4-valent ion like Ti^{4+} to zirconia should not be beneficial at all. The vacancy concentration is not increased and the small Ti^{4+} ion is

predicted to induce lattice stresses and increase trapping of vacancies and actually, the ionic
Table 2. Compilation of preparation temperature, dopant concentration and corresponding conductivity for the materials described in the text.

ZrO_2	Ion added	Dopant conc.	Temperature °C (hold time h)	Conductivity at 1000 °C	Oxygen partial pressure	ref
	In	25 mol% $InO_{1.5}$	1600	0.15	air	Gauckler et al.[38]
+10Y	In	35 mol% $InO_{1.5}$	1700 (5)		Air	Naito et al.[15]
+10Y	In	30 mol% $InO_{1.5}$	1700 (5)	0.011 0.03	Air 10^{-16}	Naito et al.[15]
+ 8Y	Mn	5 mol % MnO_x	1350 (8)	0.111 0.113	Air 2000 hours	Appel et al.[23]
+ 8Y	Mn	2 mol % MnO_x	1350 (8)	0.144 0.114	Air 2000 hours	Appel et al.[23]
+ 8Y	Fe	4.4 mol % Fe_2O_3	1550	0.10 – 0.07	air	Wilhelm and Howarth[32]
+ 8Y	Fe	5 mol % Fe_2O_3	1450	0.07	air	Slilaty and Marques[35]
+8Y	Ti	10 mol % TiO_2	1600 (48)	0.023 0.06	Air 10^{-19} atm	Lindegaard et al.[26]
+ 20 Sc	Ti	20 mol % TiO_2	1350 (8)	0.144 0.114	Air	Tao and Irvine[25]
+10Y + 10Sc	Ti	20 mol % TiO_2	1550	0.1 – 0.07	10^{-18} atm	Tao and Irvine[25]
	Y	18 mol % $YO_{1.5}$	1550 (1)	0.11	Air	Männer et al.[46]
	Sc	20 mol% $ScO_{1.5}$	1500	0.30	Air	Ishihara et al.[47]

conductivity is decreased when Ti^{4+} is added. The solubility of Ti^{4+} increases with increasing Y-content. This points to the assumption that solubility is a matter of average lattice radius. Meaning that adding a larger and a smaller ion to a given lattice may give an average radius matching the lattice.

Data on the dopant concentration, c^*, of three valent ions giving the maximum conductivity and the resulting conductivity are given in Figure 3. Maximum c^* is found for In^{3+}, which has the ionic radius closest to the matching radius found by Kim[11], see Table 1. It is also seen that the further from the matching radius the lower is c^*. The conductivity follows almost the same pattern. Here the highest conductivity is found for the dopant radius closest to the matching radius, but larger than the matching radius. There is, however, the significant exception, Sc^{3+}. The conductivity obtained by adding Sc^{3+} is double the conductivity of the next best material. What makes Sc^{3+} so significantly better than al other three valent ions? Our postulate is that it is a combination of two factors, an ionic radius close to the radius of Zr^{4+} and a low atomic weight. The atomic weight of Sc is 44.96, while it is 91.22 for Zr and 88.91 for Y. Yb that gives the next highest conductivity has an atomic weight of 173.04. As mentioned previously, the atomic radii stated here are for room temperature data. When heating to 1000 °C, this will induce thermal vibrations of the ions and the vibration can be interpretated as a higher effective ionic radii. A light ion as Sc might vibrate relatively more than heavy ions, bringing the effective radius closer to the matching radius. Strong vibration will also

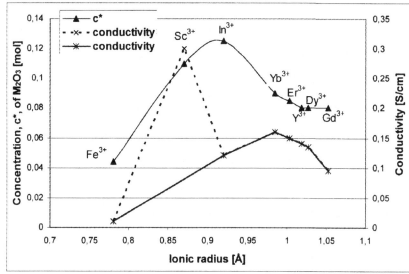

Figur 3 Concentration, c*, of three valent ions at which maximum conductivity for the system is obtained and the corresponding maximum conductivity as function of ionic radii. Notice how the conductivity of Sc^{3+} deviates from the rest of the data.

reduce the tendency to trap vacancies due to less interaction. No literature data on thermal factors from structural analysis at high temperature has been found.

Examining Figure 3 a little closer, it is seen that $Mn^{3}+$ is not included. Only Fe^{3+} and In^{3+} are included. According to the data by Shannon[10], Mn^{3+} is not found in eight coordination. The ionic radius is only given for Mn^{3+} in six coordination. This does not mean that it is impossible to find it in eight coordination (this is reported previously), but Mn^{3+} is small and prefers six coordination. Mn is, therefore, to a large extent found as Mn^{2+} in zirconia and another matching radius is valid. The behaviour of Mn is therefore, not directly comparable to the behaviour of Fe.

CONCLUSION

In summary, we postulate that:

a) The dopant concentration, c*, at which maximum conductivity occurs is dependent on how close the radius of the dopant is to the matching radius. The closer, the higher is c*.

b) Conductivity depends on lattice strains and the resulting interaction with vacancies.

c) The high conductivity of Sc-doped zirconia at elevated temperatures is due to the low atomic weight of Sc compared to other ions large enough to stabilise zirconia.

ACKNOWLEDGEMENTS
This work was carried out under the EU-contract: PIP-SOFC NNE5-2001-00791.

REFERENCES

[1] W. van Gool, "Relationship between Structure and Anomalously Fast Ion Diffusion" in *Fast ion transport in solids*, W. van Gool, Editor, p. 201-16, North-Holland, Amsterdam-London, (1972).
[2] M. Pouchard and P. Hagenmuller, "Solid Electrolytes as a Materials Problem", in *Solid Electrolytes*, P. Hagenmuller and W. van Gool, Editors, p.191-99, Academic Press, N.Y., (1978).
[3] J.A. Kilner and R.J. Brook, "A study of oxygen ion conductivity in doped non-stoichimetric oxides" *Solid State Ionics* **6**, 237-52 (1982).
[4] J.A. Kilner and J.D. Faktor, in *Progress in Solid electrolytes*, T.A. Wheat, A.K. Kuriakose, Editors, Publication ERP/MSL 83-94 (TR), p. 347, Energy, Mines and Resources, Ottawa, Canada (1983).
[5] J. Kilner, "Fast anion transport in solids", *Solid State Ionics*, **8**, 201 (1983).
[6] P.J. Shlichta, "A crystaalgrphic search program for oxygen-conducting electrolytes", *Solid State Ionics* **28-30**, 480 (1988).
[7] M. Mogensen, D. Lybye, N. Bonanos, P.V. Hendriksen and F.W. Poulsen, "Factors controlling the oxide ion conductivity of fluorite and perovskite structured oxides", *Solid State Ionics*, **174**, 279-86 (2004).
[8] H. Yokokawa, N. Sakai, T. Kawada and M. Dokyia, "Thermodynamic analysis on Solubilities and Valence State of Transition Metal Ions in Yttria Stabilized Zirconia", *ISSI Letters* **2** [2] 7-8 (1991).
[9] H. Yokokawa, "Phase Diagrams and Thermodynamic Properties of Zirconia Based Ceramics", *Key Eng. Mat.* **153-154**, 37-74 (1998).
[10] R.D. Shannon, "Revised Effective Ionic Radii and Systematic Studies of Interatomic distances in Halides and Chalcogenides", *Acta Cryst.*, **A32**, 751 (1976).
[11] D.-J. Kim, "Lattice Parameters, Ionic Conductivities, and Solubility Limits in Fluorite-Structure MO_2 Oxide (M = Hf^{4+}, Zr^{4+}, Ce^{4+}, Th^{4+}, U^{4+}) Solid Solutions", *J. Am. Ceram. Soc.*, **72**, 1415-21 (1989).
[12] T.H. Etsell and S.N. Flengas, "The electrical properties of solid oxide electrolytes", *Chem. Rev.*, **70** [3] 339-76 (1970).
[13] M.F. Trubelja and V.S. Stubican,"Electrical Conductivity in the ZrO_2-Rich Region of Several M_2O_3-ZrO_2 Systems", *Solid State Ionics* **49**, 89 (1991).
[14] K. Sasaki, P. Bohac and L.J. Gauckler, "Phase Equilibria in the System ZrO_2-$InO_{1.5}$", *J. Am. Ceram. Soc.* **76** [3] 689-98 (1993)
[15] H. Naito, H. Yugami and H. Arashi, "Electrical properties of ZrO_2-In_2O_3-Y_2O_3 and its application to a membrane for gas separation" *Solid State Ionics* **90**, 173-76 (1996).
[16] T. Kawada, N. Sakai, H. Yokokawa, and M. Dokyia, "Electrical properties of Transition-metal-doped YSZ", *Solid State Ionics* **53-56**, 418-25 (1992).
[17] M. Chen, B. Hallstedt and L. Gauckler, "Thermodynamic modeling af phase equilibria in the Mn-Y-Zr-O system", *Solid State Ionics* **176**, 1457-64 (2005).
[18] I. Voigt and A. Feltz, "Reactivity of manganese oxide-doped Y_2O_3-stabilized ZrO_2", *Solid State Ionics* **63-65**, 31-36 (1993).

[19] K. Sasaki, P. Murugaraj, M. Haseidl, J. Maier, in *U. Stimming (ED.), Fifth International Symposium on solid Oxide Fuel Cell (SOFC-V)*, The Electrochemical Society, Pennington, NJ, 1190 (1999).

[20] C. Appel, G.A. Botton, A. Horsewell, W.M. Stobbs, "Chemical and Structural Changes in Manganese-Doped Yttria-Stabilized Zirconia Studied by Electron Energy Loss Spectroscopy Combined with Electron diffraction", *J. Am. Ceram. Soc.*, **82**, 429-35 (1999).

[21] M. Occhiuzzi, D. Cordisschi, R. Dragone,"Manganese ions in the monoclinic, tetragonal and cubic phases of zirconia: an XRD and EPR study", *Phys Chem, Chem. Phys.* **5**, 4938-45 (2003).

[22] R.L. Shultz, and A. Muan,"Phase equilibria in system MnO-FeO-ZrO_2-SiO_2", *J. Am. Ceram. Soc.*, **54**, 504 (1971).

[23] C.C. Appel, N. Bonanos, A. Horsewell and S. Linderoth, "Ageing behaviour of zirconia stabilised by yttria and maganese oxide", *J. Mat. Sci.* **36**, 4493-4501 (2001).

[24] A.J. Feighery, J.T.S. Irvine D.P. Fagg and A. Kaiser, "Phase Relations at 1500°C in the Ternary System ZrO_2- Y_2O_3-TiO_2", *J. Solid State Chem.*, **143**, 273-76, (1999).

[25] S. Tao and J.T.S. Irvine, "Optimization of Mixed Conducting Properties of Y_2O_3-ZrO_2-TiO_2 and Sc_2O_3- Y_2O_3-ZrO_2-TiO_2 Solid Solutions as Potential SOFC Anode Materials", *J. Solid State Chem.*, **165**, 12-18 (2002).

[26] T. Lindegaard, C. Clausen and M. Mogensen, "Electrical and eletrochemical properties of $Zr_{0.77}Y_{0.13}Ti_{0.1}O_{1.93}$", in *High Temperature Electrochemical Behaviour of Fast Ion and Mixed Conductors* Ed: F.W. Poulsen et al., Risø National Laboratory, Roskilde, 311-18 (1993).

[27] S. Linderoth, N. Bonanos, K.V. Hansen and J.B. Bilde-Sørensen, "effect of NiO-to-Ni Transformation on Conductivity and Structure of Yttria-Stabilized ZrO_2", *J. Am. Ceram. Soc.*, **84** [11] 2652-56 (2001).

[28] K. Kobayashi, Y. Kai, S. Yamaguchi, T. Kawashima and Y. Igushi, "Partial Conductivity of YSZ Doped with 10 mol % TiO_2", *Kor. J. Ceram.*, **4** [2] 114-21 (1998).

[29] S. Davison, R. Kershaw, K. Dwight and A. Wold, "Preparation and Characterization of Cubic ZrO_2 Stabilized by Fe(III) and Fe(II)", *J. Solid State Chem.*, **73**, 47-51 (1988).

[30] G. Stefanic, B. Grzeta and S. Muxic, "Influence of oxygen on the thermal behavior of the ZrO_2-Fe_2O_2 system", *Mat. Chem. Phys.*, **65**, 216-21 (2000).

[31] G. Stefanic, B. Grzeta, K. Nomura, R. Trojko and S. Muxic, "The influence of thermal treatment on phase development in ZrO_2-Fe_2O_2 and HfO_2-Fe_2O_2 systems", *J. Alloys Comp.* **327**, 151-60 (2001).

[32] R.V Wilhelm and D.S. Howarth, "Iron oxide-doped ytria-stabilized zirconia ceramic – iron solubility and electrical conductivity", *Am. Ceram. Soc. Bull.*, **58** [2], 228-32 (1979).

[33] M.M.R Boutz., A.J.A Winnubst., F.Hartgers and A.J Burggraaf., "Effect of additives on densification and deformation of tetragonal zirconia", *Mater. Sci.* **29** 6374-82(1994)

[34] M. Hartmanova, F.W. Poulsen, F. Hanic, K. Putyera, D. Tunega, A.A. Urusovskaya, and T.V. Oreshnikova, "Influence of Copper- and Iron-doping on cubic yttria-stabilized zirconia." *J. Mater.Sci.* **29** 2152-58 (1994).

[35] R.M. Slilaty and F.M.B. Marques, Bol. Soc. Esp. Cerám. Vidrio. 35 [2], 109 (1996).

[36] N.Q. Minh and T. Takahashi, *Science and Technology of Ceramic Fuel Cells*, p. 78, Elsevier, Amsterdam, (1995).

[37] D.K. Hohke, "Ionic conductivity of $Zr_{1-x}In_{2x}O_{2-x}$," *J. Phys. Chem. Solids*, **41**, 777-84 (1980).

[38] L.J. Gauckler, K Sasaki, H. Heinrich, P. Bhoac and A. Oriukas, "Ionic Conductivity of Tetragonal- and Cubic-ZrO_2 Doped with In_2O_3", *Science and Technology of Zirconia V* 555-66 (1993).

[39] D.W. Strickler and W.G. Carlson, "Electrical conductivity in the ZrO2-rich Region of Several M2O3-ZrO2-Systems", *J. Am. Ceram. Soc.*, 47, 122 (1964).

[40] H. Naito and H. Arashi, "Electrical properties of ZrO_2-TiO_2-Y_2O_3 system," *Solid State Ionics* **53-56**, 436-41 (1992).

[41] D. Lybye, Y.-L. Liu, M. Mogensen and S. Linderoth, "Effect of impurities on the conductivity of Sc and Y co-doped ZrO_2", In: *Solid Oxide Fuel Cells IX.* Ed: S.C. Singhal and J. Mizusaki, 954-63 (2005).

[42] S.S. Liou and W.L. Worrell, "Electrical Properties of Novel Mixed-Conducting Oxides", Appl. Phys., **A40**, 25-31 (1989).

[43] S.S. Liou and W.L. Worrell, "Mixed-Conducting Oxides, Electrodes for Solid Oxide Fuel Cells". In: *Proceedings of First International Symposium on Solid Oxide Fuel Cells.* Ed: S.C. Singhal, 81-89 (1989).

[44] J.T.S. Irvine, A.J. Feighery, D.P. Fagg and S. Garcia-Martin, "Structural studies on the optimisation of fast oxide ion transport", *Solid State Ionics* **136-137**, 879-85 (2000).

[45] V.V. Kharton, E.N. Naumovich and A.A. Vecher, "Research on the electrochemistry of oxygen ion conductors in the former Soviet Union. I. ZrO2-based ceramic materials." *J. Solid State Electrochem.* **3** (2), 61-81 (1999).

[46] R. Männer, E. Ivers-tiffée and W. Wersing, "Characterization of YSZ electrolyte materials with various yttria contents", in F. Grosz et al, Proc. Second Int. Symp. on Solid Oxide Fuel Cells, Athens, Greece, 715-725, (1991).

[47] T. Ishihara, N.M. Sammes and O. Yamamoto, "Electrolytes" in *Solid Oxide Fuel Cells*, Ed: S.C. Singhal and K. Kendall, Elsevier, 83-112, 2003.

THERMOPHYSICAL PROPERTIES OF YSZ AND Ni-YSZ AS A FUNCTION OF TEMPERATURE AND POROSITY

M. Radovic, E. Lara-Curzio, R. M. Trejo, H. Wang and W. D. Porter
Metals and Ceramics, Oak Ridge National Laboratory
Oak Ridge, TN 37831-6069

ABSTRACT

The thermal diffusivity, heat capacity, thermal conductivity, coefficient of thermal expansion and elastic properties of Ni-YSZ and YSZ are reported in the temperature interval between 20 °C and 1000 °C. Specific heat capacity (C_p) and thermal diffusivity (α) were determined by differential scanning calorimetry (DSC) and by the laser flash method, respectively, while thermal conductivity was calculated from its relationship with c_p, α and the density of the material. The coefficients of thermal expansion (CTE) were determined using a thermomechanical analyzer (TMA), while elastic properties (Young's and shear modulus), were determined by resonant ultrasound spectroscopy (RUS). The effect of temperature and porosity on the thermophysical properties of Ni-YSZ and YSZ is discussed.

INTRODUCTION

The reliability of Solid Oxide Fuel Cells (SOFCs) is determined by the distribution of strength of their components and the state of stress to which these are subjected. Therefore, to ensure reliable operation, it is necessary to utilize materials that possess the required mechanical integrity while they perform their functional requirements (e.g.- gas transport, ionic conductivity[1]). Stresses arise from various sources, which can be categorized as: residual stresses, which result from the mismatch in the thermoelastic properties of the constituents; fabrication stresses arising from constraints and external forces needed to package the fuel cell and; operational stresses, which arise from temperature gradients that result from the gas flow profiles and the electrochemical activity of the cell. Recent advances in simulation tools have made possible the implementation of multi-physics models to predict the thermal, electrochemical and mechanical performance of SOFCs [2-4]. However, the implementation of such models requires knowledge of the physical and mechanical properties of the constituents.

In this paper we report on the thermophysical properties of Yttria Stabilized Zirconia (YSZ) and Nickel-YSZ cermet (Ni-YSZ) that are widely used as electrolyte and anode materials for SOFC. Thermophysical properties were studied as a function of porosity in the 20-1000 °C temperature range. Properties of interest include specific heat capacity, thermal diffusivity, and thermal conductivity that are important for simulation of temperature gradients in SOFC stacks. The thermophysical properties such as coefficient of thermal expansion and elastic moduli (Young's and shear moduli), that are necessary for modeling of thermally induced stresses in the SOFC stack were determined as well for YSZ electrolyte and Ni-YSZ anode materials.

MATERIALS

The Ni-YSZ materials used in this investigation were prepared from a powder mixture of 75 mol% NiO and ZrO_2 stabilized with 8 mol% Y_2O_3 (NiO-YSZ). To investigate the effect of porosity on the thermomechanical properties of Ni-YSZ materials, different amounts of organic pore former (0- 30 vol% of rice starch) were added to the powder mixture to obtain test specimens with different levels of porosity. Fully dense YSZ test specimens were prepared from ZrO_2 powder stabilized with 8 mol% Y_2O_3. Green test specimens were prepared by tape casting NiO-YSZ and YSZ slurries into 250 μm thick layers, which were subsequently laminated. Discs with nominal diameter of 25.4 mm were hot-knifed from the laminated green tapes and sintered

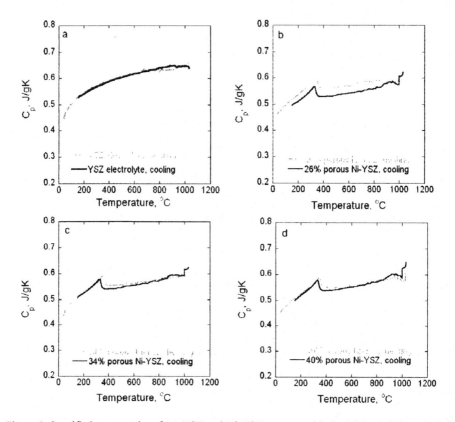

Figure 1. Specific heat capacity of (a) YSZ and Ni-YSZ cermets with (b) 26%, (c) 34% and (d) 40% porosity.

at 1400 °C in air for 2 hours. NiO-YSZ test specimens were subsequently reduced in a gas mixture of 4%H_2 and 96%Ar at 1000 °C for 30 minutes to obtain Ni-YSZ cermet test specimens. The weight of the test specimens was determined before and after reduction to confirm that NiO had been completely reduced to metallic Ni. Their relative porosities were determined by alcohol immersion .

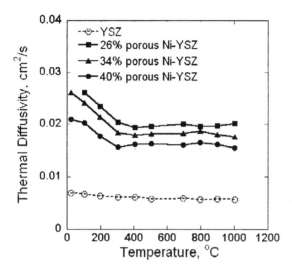

Figure 2. Thermal diffusivity of YSZ and Ni-YSZ as a function of temperature.

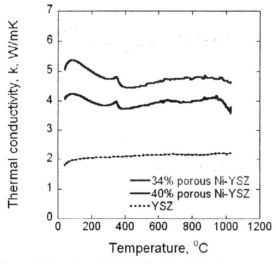

Figure 3. Thermal conductivity of YSZ and Ni-YSZ as a function of temperature.

PROCEDURES, RESULTS AND DISCUSSION

Heat Capacity

The heat capacity of the SOFC materials was determined by DSC (Netzsch DSC 404C*) between 20 and 1000 °C according to ASTM standard E 1269-01[5]. The test specimens used for these measurements were disks with diameter of 6 mm and thickness of ≈1 mm that had been core-drilled from larger Ni-YSZ and YSZ discs. The measurements were carried out in argon to avoid oxidation of Ni.

Figure 1 shows a plot of specific heat capacity, C_p as a function of temperature for YSZ and Ni-YSZ test specimens of different porosities. The peak in the curve near 350 °C, which was observed both on heating and cooling, is known as the "lambda" transition, because of its resemblance to the Greek letter "lambda". This transition occurs at the Curie temperature of Ni, at which its behavior changes from ferromagnetic to paramagnetic[6]. It was found that for the range of porosities examined, porosity had no effect on the magnitude of the specific heat capacity of Ni-YSZ.

Thermal Diffusivity

Thermal diffusivity was measured by the Laser Flash method using an Anter Co. unit (Mod. Flashline 5000*). According to this technique, the front face of a small disk-shaped test specimen is subjected to a very short burst of radiant energy emitted by a laser, with the irradiation time being of the order of one millisecond or less. The resulting temperature rise of the rear surface of the test specimen is recorded and measured and thermal diffusivity values are computed from the temperature rise versus time data by using the expression:

$$\alpha = 1.38 l^2 / (\pi^2 t_{1/2})$$

where l is the thickness of the specimen, $t_{1/2}$ is a specific time at which the rear surface temperature reaches half its maximum value[7]. The test specimens used for these measurements were 12-mm diameter and ≈1 mm thickness disks. The tests were carried out in an argon atmosphere from room temperature to 1000°C.

Figure 2 presents plots of thermal diffusivity as a function of temperature for YSZ and Ni-YSZ with various levels of porosity. It was found that the thermal diffusivity of YSZ decreases slightly with temperature. However, it was found that the thermal diffusivity of Ni-YSZ decreases with increasing porosity and temperature up to 300°C above which it remains almost constant up to 1000°C. Similar behavior was reported previously for pure Ni where a deflection at 350°C was related to the Curie temperature[8].

Thermal Conductivity

The thermal conductivity of Ni-YSZ and YSZ materials was determined according to the following relationship:

$$k = \rho \cdot C_p \cdot \alpha$$

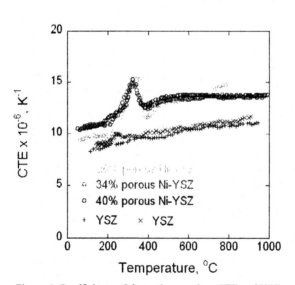

where ρ is the density of the test specimen. Figure 3 shows plots of thermal conductivity versus temperature for YSZ and Ni-YSZ with various levels of porosity. It was found that the thermal conductivity of Ni-YSZ decreases with increasing porosity. The small peak in the thermal conductivity curves at the Curie temperature of ≈350°C can be observed.

Thermal Expansion

The thermal expansion of YSZ in air and Ni-YSZ cermets in 4%H_2/96%Ar gas mixture were determined using a thermomechanical analyzer (TA Instruments Mod. TMA

Figure 4. Coefficients of thermal expansion, CTE, of YSZ and Ni-YSZ as a function of temperature.

Q400*). Test specimens of Ni-YSZ with various levels of porosity were evaluated. 3 x 3 x ≈1 mm test specimen geometries were used to determine their thermal expansion behavior in the 50-1000°C temperature range. Coefficients of thermal expansion were calculated using the following relationship:

$$CTE = \frac{d\varepsilon}{dT} = \frac{d}{dT}\left[\frac{(l_T - l_0)}{l_0}\right]$$

where ε is the thermal expansion, and l_0 and l_T are heights of the test specimen at room temperature and temperature T, respectively.

The coefficient of thermal expansion of YSZ and Ni-YSZ cermets of different porosities are plotted in Figure 4 as a function of temperature. It was found that the CTE of YSZ increases slightly with temperature from ≈8.5x10⁻⁶ K⁻¹ at 100°C to ≈10.5x10⁻⁶ K⁻¹ at 950°C. The coefficient of thermal expansion of Ni-YSZ cermets is higher than that of YSZ and changes between ≈10x10⁻⁶ K⁻¹ at 100°C to ≈14 x10⁻⁶ K⁻¹ at 950°C independently of porosity. It was found that the coefficient of thermal expansion exhibited a peak at the Curie temperature of ≈300 °C[9].

Elastic Moduli

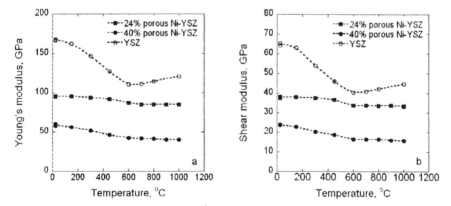

Figure 5. Elastic moduli, (a) Young's modulus and (b) shear modulus, of YSZ and Ni-YSZ cermets as a function of temperature.

Elastic properties, namely Young's, E, and shear, G, moduli of YSZ and Ni-YSZ cermets were determined by resonant ultrasound spectroscopy, RUS. RUS is based on measuring the resonance peaks - that are dependent on density, elastic moduli and shape - of a freely suspended solid[10]. The specimens are supported by three piezoelectric transducers, one is used to generate an elastic wave of constant amplitude and varying frequency, and the other two to detect the resulting signals. To obtain the spectra at elevated temperatures the test specimen was supported by semi-spherical tips of sapphire rods placed inside the furnace. The other ends of the sapphire rods were attached to the transducers, which were located outside the furnace and maintained at room temperature. RUS of YSZ and Ni-YSZ disc-shaped test specimens were carried out in air and 4%H$_2$/96%Ar gas mixture, respectively. A multidimensional algorithm (Quasar International*, Albuquerque, NM) that minimizes the root-mean-square (RMS) error between the measured and calculated resonant peaks was used for the determination of the elastic constants from a single frequency scan.

Young's and shear moduli of YSZ and Ni-YSZ cermets of different porosity are plotted in Figure 5 as a function of temperature. The magnitude of the elastic constants at room temperature is in good agreement with previously published results[11-13] for porous Ni-YSZ cermets and YSZ. It was found that the elastic properties of both YSZ and Ni-YSZ change with temperature in a non-linear manner. Both E and G initially decrease with temperature between 25 and 600 ^{oC}. Above 600 ^{oC}, E and G increase slightly or remain constant with temperature up to 1000 ^{oC}. However, non-linear changes of elastic moduli with temperature are more pronounced in the case of YSZ. Such a behavior can be attributed to the oxygen vacancies order-disorder transition in YSZ. This transition was reported to occur in the 400-600 °C temperature range[14] depending on the testing frequency. Rearrangement of oxygen vacancies at the transition temperatures and their hopping around Y cations causes significant increase in mechanical loss (damping) and consequently significant drop in elastic moduli in that temperature range.

SUMMARY
The thermophysical properties such as heat capacity, thermal diffusivity, thermal conductivity, thermal expansion and elastic moduli of YSZ and Ni-YSZ cermets of different porosity were fully characterized in the 20-1000°C temperature range. It was found that all these properties change non-linearly with temperature. The peaks in heat capacity, thermal conductivity and CTE as well as the deflection point in thermal diffusivity that occur at ≈350°C for Ni-YSZ cermets were attributed to the ferromagnetic to paramagnetic transition of Ni, i.e. Curie temperature. However, the significant drop in elastic moduli in 400-600°C temperature range that was observed in both, YSZ and Ni-YSZ, is attributed to the oxygen vacancies order-disorder transition in YSZ.

ACKNOWLEDGMENTS
This research work was sponsored by the US Department of Energy, Office of Fossil Energy, SECA Core Technology Program at ORNL under Contract DE-AC05-00OR22725 with UT-Battelle, LLC. The authors are grateful for the support of NETL program managers Wayne Surdoval and Travis Shultz. The authors are indebted to Beth Armstrong and Claudia Walls of ORNL for help with specimen's preparation.

REFERENCES
[1]Minh N. Q. and Takahashi T., "Science and Technology of Ceramic Fuel Cells", Elsevier, Amsterdam (1995)
[2]Ziegler C., Schmitz A., Tranitz M, Fontes E. and Schumachera J. O., "Modeling Planar and Self-Breathing Fuel Cells for Use in Electronic Devices", J. of The Electrochemical Soc. 151 (2004) A2028-A2041
[3]E. Lara-Curzio, M. Radovic, R. Trejo, "Reliability and Durability of Materials and Components for Solid Oxide Fuel Cells", FY 2005 Annual Report, Office of Fossil Energy Fuel Cell Program, (2005), http://204.154.137.14/technologies/coal_and_power systems/ distributed_generation/seca/refshelf.html

[4]Pradhan S.K., Mazumder S.K., Radovic M., Lara-Curzio E., Luttrell C. and Hartvigsen J. "Analyses if Electrically-induced Thermal Effect of Lad Transient on Planar Solid Oxide Fuel Cell", submitted to J. of Power Sources, (2006)

[5]ASTM standard: E 1269-01, Standard Test Method for Determining Specific Heat Capacity by Differential Scanning Calorimetry

[6]Touloukin Y.S. and Byuco E.H., "Specific Heat – Metallic Elements and Alloys", Plenum Publishing Corporation, New York, 1970

[7]Plummer W.A., Differential Dilatometry- A Powerful Tool, AIP Conference Proceedings, Series Editor: Hugh C. Wolfe, Number 17, Editors: R. E. Taylor and G. L. Denman, American Institute of Physics, New York, 1974.

[8]Sanchez-Lavega A. and Salazar A., "Thermal Diffusivity Measurements in Opaque Solids by the Mirage Techniques in the Temperature Range from 300 to 1000 K", J. Appl. Phys. 76 (1994) 1462-1468

[9]Mori M., Yamamoto T., Itoh H., Inaba H. and Tagawa H., "Thermal Expansion of Nickel-Zirconia Anodes in Solid Oxide Fuell Cells during Fabrication and Operation", J. Electrochem. Soc. Vol 145 (1998) 1374-1381

[10]Radovic M., Lara-Curzio E. and Rieser L., "Comparison of Different Experimental Techniques for Determination of Elastic Properties of Solids", *Mater. Sci. Eng.*, 368 (2004) 56-70

[11]Radovic M. and Lara-Curzio E., "Mechanical Properties of Tape Cast Nickel-based Anode Materials for Solid Oxide Fuel Cells Before and After Reduction in Hydrogen", Acta. Mater. 52 (2004) 5747

[12]Radovic M. and Lara-Curzio E., "Changes of Elastic Properties of SOFC Anode during the Reduction of NiO to Ni in Hydrogen", J. Am. Ceram. Soc. 87 (2004) 2242

[13]Radovic M., Lara-Curzio E., Trejo R., Armstrong B. and Walls C., "Elastic Properties, Equibiaxial Strength and Fracture Toughness of 8mol%YSZ Electrolyte for SOFC" Ceramic Engineering and Science Proceedings, Vol. 25, No. Issue 3, pp. 287-292 (2005)

[14]Weller M., Damson B. and Lakki A., "Mechanical Loss of Cubic Zirconia", J. of Alloys and Compounds 310 (2000) 47-53

[*] Certain commercial equipment, instruments, or materials are identified in this paper in order to specify the experimental procedure adequately. Such identification is not intended to imply recommendation or endorsement by Oak Ridge National Laboratory, nor is it intended to imply that the materials or equipment identified are necessarily the best available for the purpose.

PHYSICAL PROPERTIES IN THE Bi_2O_3-Fe_2O_3 SYSTEM CONTAINING Y_2O_3 AND CaO DOPANTS

Hsin-Chai Huang, Yu-Chen Chang, and Tzer-Shin Sheu
Department of Materials Science and Engineering
I-Shou University
Kaohsiung, Taiwan

ABSTRACT

Dopants Y_2O_3 and CaO were added into the Bi_2O_3-Fe_2O_3 system to observe the formation of perovskite structure, $BiFeO_3$, so-called 113 phase. The content of Y_2O_3 or CaO dopant was up to 10~13 mol%. The perovskite sturature was likely to form in the Fe_2O_3-rich composition region, but not in the Bi_2O_3-rich composition region. At high sintering temperatures, T=850°C, γ-Bi_2O_3 appeared in the Bi_2O_3-Fe_2O_3-Y_2O_3 system, but not in the Bi_2O_3-Fe_2O_3-CaO system.

From an electrical conductivity measurement, a sintered sample with the composition of $Bi_{0.4675}Fe_{0.45}Ca_{0.0825}O_{1.5}$ had a higher ionic conductivity up to 1×10^{-1} S/cm at 700°C. Most solid electrolytes had an activation energy of 0.5eV for a charge carrier at 30~400°C, but a much lower activation energy at 400~700°C. The electrical conduction mechanisms were further studied.

INTRODUCTION

Bismuth-oxide-containing ceramics have been well known to have high oxygen ionic conductivity, because of the presence of face-centered cubic bismuth oxide (δ-Bi_2O_3) at high temperature. For obtaining δ-Bi_2O_3 in a wide temperature range, several alloy-design techniques are used to stabilize it at lower temperature.[1-6] Except for the stabilized δ-Bi_2O_3 phase, the perovskite structure is also considered as an excellent electrical conductor. In the Bi_2O_3-MnO_2 system, the perovskite structure $BiMnO_3$ can be increased its conductivity by using the Sr substitution.[7] The increasing electrical conductivity is due to the formation of point defects like hole.[8]

A two-phase composite $BiFeO_3$ plus $Bi_2Fe_4O_9$ or one of each phase was alloy-designed in the Bi_2O_3-Fe_2O_3-M_xO_y (M= Ca, or Y) system, with the composition of $Bi_{1-x-y}Fe_xM_yO_{3/2-z}$ (x=0.45-0.55, and y=0-0.12) in this study. By slightly varying the composition of Fe^{3+} (or Fe^{2+}), Ca^{2+}, and Y^{3+}, it is hoped that the electrical properties of $BiFeO_3$ perovskite and its related composites can be systematically studied as a function of sintering temperature, oxygen vacancy, and cation dopants. Furthermore, sintering temperature, high-temperature phase transformation, and microstructures of different perovskite-containing composites are also observed and discussed in this study.

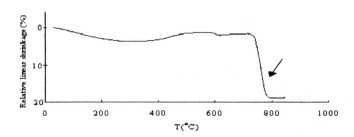

Fig. 1.   A typical sintering curve for the green powder compact with the composition of $Bi_{0.425}Fe_{0.5}Ca_{0.075}$.

EXPERIMENTAL PROCEDURES

Starting powders or chemicals were Bi_2O_3 (99.9% pure, Nihon Shinyaku, Japan), Fe_2O_3 (99.9% pure, Cerac, USA), Y_2O_3 (99.99% pure, Johnson Matthey), and CaO (99% pure, Aldrich). According to each specific composition of $Bi_{1-x-z}Fe_x(Y, Ca)_zO_{1.5}$ (x= 0.45-0.55, and z= 0-0.12), these starting powders were well mixed with ethanol alcohol. After mixing, these powders were dried at T=100°C to remove the solvent. A batch of dried powders was cold-pressed to form a cylindrical powder pellet under a uniaxial pressure of 50MPa. Cold-pressed green powder compacts were sintered at 650-850°C for 0.5-2 h under different mixed-gas atmospheres, such as N_2/O_2 or air. After sintering, some of sintered specimens were further heat-treated at 700°C for 20-50h to observe any phase transformation or phase decomposition at high temperature.

For determining the optimal sintering temperature, a dilatometric unit was used to determine the sintering curve of each green powder compact at T=25-850°C and with a heating rate of 10°C/min. The temperature with a maximum shrinkage rate in each sintering curve was defined as the optimal sintering temperature. Differential thermal analysis (DTA, Perkin Elmer DTA7) was used to determine any chemical reactions or phase transformations at high temperatures. Phase existence of the sintered or heat-treated samples was determined by the X-ray diffraction method at room temperature. Microstructure was observed by using a scanning electron microscope (Hitachi S-2700). An LCR meter (HP model 4284B) was used to measure the electrical conductivity of the sintered or heat-treated specimens at 25-750°C in air.

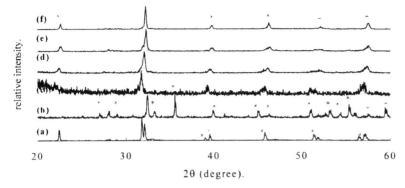

Fig.2.   X-ray diffraction patterns for samples (a) Y2a, (b) Y2d, (c) C1a, (d) C2a, (e) C2b, and (f) C2x.   Symbol ' + ' is for 113 phase, symbol '×' for 249 phase, and symbol '*' for γ phase.

RESULTS AND DISCUSSION

(A) Sintering behavior of green samples

A typical sintering curve of green powder compact in the Bi₂O₃-Fe₂O₃-CaO system is shown in Fig. 1.   For this particular green powder compact, an early shrinkage started at T=725°C, and the maximum shrinkage rate was approximately located at T=780°C, as marked by an arrow symbol in Fig. 1.   Therefore, the optimal sintering temperature of this powder compact was defined at T=780°C.   Therefore, the powder compacts in the Bi₂O₃-Fe₂O₃-CaO and Bi₂O₃-Fe₂O₃-Y₂O₃ systems were sintered at 750-850°C.

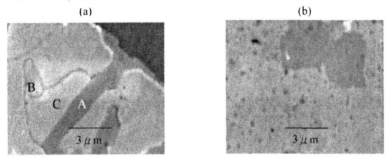

Fig. 3.   SEM micrographs for samples (a) Y3d, and (b) C2a.   These samples were thermally etched to reveal their microstructures.

Phase existence of sintered or heat-treated specimens is listed in Table I. X-ray diffraction patterns of these sintered or heat-treated specimens are selectively shown in Fig. 2. Most sintered or heat-treated specimens contained γ-Bi_2O_3 ,$BiFeO_3$, and $Bi_2Fe_4O_9$ phases. However, a single perovskite phase, $BiFeO_3$, was not being observed in the sintered specimens, even the sintered specimen with an exact composition of $BiFeO_3$. With the long heat treatment at 700°C for 20-50h, a single $BiFeO_3$ phase was not being found.

Table I.   Phase existence of sintered specimens with or without any heat treatments.

Sample index	Nominal Composition (mol%)	Sintering temperature (°C)	Heat treatments	Phases
Y1a	$Bi_{0.36}Fe_{0.55}Y_{0.09}$	750	NO	$BiFeO_3$+$Bi_2Fe_4O_9$
Y1b	$Bi_{0.36}Fe_{0.55}Y_{0.09}$	750	700°C, 20h	$BiFeO_3$+$Bi_2Fe_4O_9$
Y1c	$Bi_{0.36}Fe_{0.55}Y_{0.09}$	750	700°C, 50h	$BiFeO_3$+$Bi_2Fe_4O_9$
Y1d	$Bi_{0.36}Fe_{0.55}Y_{0.09}$	850	NO	γ-Bi_2O_3+$BiFeO_3$+$Bi_2Fe_4O_9$
Y2a	$Bi_{0.4}Fe_{0.5}Y_{0.1}$	750	NO	$BiFeO_3$+$Bi_2Fe_4O_9$
Y2b	$Bi_{0.4}Fe_{0.5}Y_{0.1}$	750	700°C, 20h	$BiFeO_3$+$Bi_2Fe_4O_9$
Y2c	$Bi_{0.4}Fe_{0.5}Y_{0.1}$	750	700°C, 50h	$BiFeO_3$+$Bi_2Fe_4O_9$
Y2d	$Bi_{0.4}Fe_{0.5}Y_{0.1}$	850	NO	γ-Bi_2O_3+$BiFeO_3$+$Bi_2Fe_4O_9$
Y3a	$Bi_{0.44}Fe_{0.45}Y_{0.11}$	750	NO	$BiFeO_3$+$Bi_2Fe_4O_9$
Y3b	$Bi_{0.44}Fe_{0.45}Y_{0.11}$	750	700°C, 20h	$BiFeO_3$+$Bi_2Fe_4O_9$
Y3c	$Bi_{0.44}Fe_{0.45}Y_{0.11}$	750	700°C, 50h	$BiFeO_3$+$Bi_2Fe_4O_9$
Y3d	$Bi_{0.44}Fe_{0.45}Y_{0.11}$	850	NO	γ-Bi_2O_3+$BiFeO_3$+$Bi_2Fe_4O_9$
C1a	$Bi_{0.3825}Fe_{0.55}Ca_{0.0675}$	750	NO	$BiFeO_3$+$Bi_2Fe_4O_9$
C1b	$Bi_{0.3825}Fe_{0.55}Ca_{0.0675}$	750	700°C, 20h	$BiFeO_3$+$Bi_2Fe_4O_9$
C1c	$Bi_{0.3825}Fe_{0.55}Ca_{0.0675}$	750	700°C, 50h	$BiFeO_3$+$Bi_2Fe_4O_9$
C1d	$Bi_{0.3825}Fe_{0.55}Ca_{0.0675}$	850	NO	$BiFeO_3$+$Bi_2Fe_4O_9$
C2a	$Bi_{0.425}Fe_{0.5}Ca_{0.075}$	750	NO	$BiFeO_3$+$Bi_2Fe_4O_9$
C2b	$Bi_{0.425}Fe_{0.5}Ca_{0.075}$	750	700°C, 20h	$BiFeO_3$+$Bi_2Fe_4O_9$
C2c	$Bi_{0.425}Fe_{0.5}Ca_{0.075}$	750	700°C, 50h	$BiFeO_3$+$Bi_2Fe_4O_9$
C2d	$Bi_{0.425}Fe_{0.5}Ca_{0.075}$	850	NO	$BiFeO_3$+$Bi_2Fe_4O_9$
C3a	$Bi_{0.4675}Fe_{0.45}Ca_{0.0825}$	750	NO	$BiFeO_3$+$Bi_2Fe_4O_9$
C3b	$Bi_{0.4675}Fe_{0.45}Ca_{0.0825}$	750	700°C, 20h	$BiFeO_3$+$Bi_2Fe_4O_9$
C3c	$Bi_{0.4675}Fe_{0.45}Ca_{0.0825}$	750	700°C, 50h	$BiFeO_3$+$Bi_2Fe_4O_9$
C3d	$Bi_{0.4675}Fe_{0.45}Ca_{0.0825}$	850	NO	$BiFeO_3$+$Bi_2Fe_4O_9$

(B) Microstructure and Electrical Conductivity

SEM micrograph for sample Y3d is shown in Fig. 3(a). This sample was sintered at 850°C for 1h, and it contained γ-Bi$_2$O$_3$, BiFeO$_3$, and Bi$_2$Fe$_4$O$_9$ three phases, as listed in Table I. From electron dispersive X-ray spectrum (EDXS) composition analysis, the dark area, marked by the letter "A", was identified to be the Bi$_2$Fe$_4$O$_9$ phase; the white area, marked by the letter "B", was the γ-Bi$_2$O$_3$ phase; and the gray area, marked by the letter "C", was the BiFeO$_3$ phase. SEM micrograph for sample C2a is shown in Fig. 3(b). It contains BiFeO$_3$ and Bi$_2$Fe$_4$O$_9$ two phases. From EDXS composition analysis, the gray area is the BiFeO$_3$ phase, and the dark area is the Bi$_2$Fe$_4$O$_9$ phase.

(a)   **(b)**

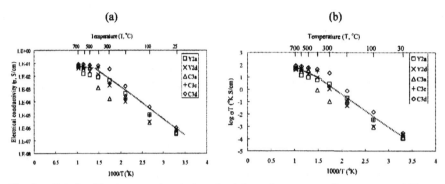

Fig. 4   Relationships between electrical conductivity and temperature for (a) σ versus 1/T, and (b) log(σT) versus 1/T.

The relationships between electrical conductivity and temperature for the sintered or heat-treated specimens are shown in Fig. 4(a). Electrical conductivity is divided into two temperature-dependent regimes, 25-400°C and 400-700°C. At high temperatures, at T=400-700°C, samples C3c and C3d have a higher electrical conductivity than the other specimens. Why these two specimens have a higher electrical conductivity is probably due to the creation of oxygen vacancies in the BiFeO$_3$ and Bi$_2$Fe$_4$O$_9$ two phases, through the introduction of CaO dopant. The electrical conductivity for sample C3d with the composition of Bi$_{0.4675}$Fe$_{0.45}$Ca$_{0.0825}$O$_{1.5}$ is up to 1*10^{-1} S/cm at 700°C.

Relationship between log(σT) and 1/T is shown in Fig. 4(b), where σ is electric conductivity and T is temperature (°K). From the relationship between electrical conductivity and temperature, the activation energy of a charge carrier at 25-400°C is 0.42-0.61eV. A possible charge carrier for these solid electrolytes is due to oxygen vacancy. Calculated activation energy for the charge carrier at different temperatures is listed in Table II. It indicates that the activation energy at higher temperature is much lower than that at lower temperature. Apparently, the electrical conduction mechanism was changed at 400°C.

Table II. Calculated activation energy for the charge carrier at different temperatures.

Sample	30-300(400)°C	400(500)-700°C
Y2a	0.56 eV	0.39 eV
Y2d	0.52 eV	0.26 eV
C3a	0.42 eV	0.14 eV
C3c	0.49 eV	0.17 eV
C3d	0.61 eV	0.13 eV

CONCLUSIONS

With CaO and Y_2O_3 oxide dopants, $BiFeO_3$-containing composites with a composition of $Bi_{1-x-z}Fe_x(Y, Ca)_zO_{1.5}$ (x= 0.45-0.55, and z= 0-0.12) were sintered at 700-850°C in air or N_2/O_2 mixed gases. Most sintered or heat-treated specimens contained $BiFeO_3$ and $Bi_2Fe_4O_9$ phases. Except for these two phases, γ-Bi_2O_3 phase was sometimes being observed in the samples with the Y_2O_3 dopant after sintered or heat-treated at 850°C. However, a single $BiFeO_3$ phase was not found in these sintered and heat treated specimens. Electrical conductivity of the $BiFeO_3$-containing solid electrolytes was divided into two temperature-dependent regimes, 25-400°C and 400-700°C. A solid electrolyte with a composition of $Bi_{0.4675}Fe_{0.45}Ca_{0.0825}O_{1.5}$ had a higher ionic conductivity of $1*10^{-1}$ S/cm at 700°C. From the calculated activation energy, a major charge carrier was probably due to the oxygen vacancy for these $BiFeO_3$-containing solid electrolytes at low temperatures. However, at high temperatures, it had a much lower activation energy, and the type of major charge carrier was further investigated.

REFERENCES

[1]B.C.H. Steele, "Appraisal of $Ce_{1-y}Gd_yO_{2-2/y}$ electrolytes for IT-SOFC Operation at 500°C," *Solid State Ionics*, **129**, 95-110 (2000).

[2]T. Takahashi, H. Iwahara, and T. Arao, "High Oxide Ion Conduction in the Sintered Oxides of the system Bi_2O_3-Y_2O_3," *J. App. Electrochem.*, **5**, 187-95 (1975).

[3]M.J. Verkerk, K. Keizer, and A.J. Burgraff, "High Oxygen Ion Conduction in Sintered Oxides of the Bi_2O_3-Er_2O_3 System," *J. Appl. Electrochem.*, **10**, 81-90 (1980).

[4]V. Joshi, S. Kulkarni, J. Nachlas, J. Diamond, and N. Weber, "Phase Stability and Oxygen Transport Characteristics of Yttria- and Niobia- Stabilized Bismuth Oxide," *J. Mater. Sci.*, **25**,

1237-45 (1990).

[5]P. Su and A.V. Virkar, "Ionic Conductivity and Phase Transformation in Gd_2O_3-Stabilized Bi_2O_3," *J. Electrochem. Soc.*, **139** [6],1671-77 (1992).

[6]D. Liu, Y. Liu, S.Q. Huang, and X. Yao, "Phase Structure and Dielectric Properties of Bi_2O_3-ZnO-Nb_2O_5-Based Dielectric Ceramics," *J. Am. Ceram. Soc.*, **76** [8], 2129-32 (1995).

[7]H. Chiba, T. Atou, and Y. Syono, "Magnetic and Electrical Properties of $Bi_{1-x}Sr_xMnO_3$: Hole Effect on Ferromagnetic Perovskite $BiMnO_3$," *J. Am. Ceram. Soc.*, **132**, 139-143(1997).

[8]Jae-Hyun Park, Patrick M. Woodward, "Synthesis, structure and optical properties of two new Perovskites: $Ba_2Bi_{2/3}TeO_6$ and $Ba_3Bi_2TeO_9$," *International Journal of Inorganic Materials*, **2**, 153-66 (2000).

ELECTRICAL PROPERTIES OF $Ce_{0.8}Gd_{0.2}O_{1.9}$ CERAMICS PREPARED BY AN AQUEOUS PROCESS

Toshiaki Yamaguchi*
National Institute of Advanced Industrial Science and Technology (AIST)
Shimo-shidami, Moriyama-ku
Nagoya, 463-8560, Japan

Yasufumi Suzuki, Wataru Sakamoto, and Shin-ichi Hirano
Department of Applied Chemistry, Graduate School of Engineering, Nagoya University
Furo-cho, Chikusa-ku
Nagoya, 464-8603, Japan

ABSTRACT

Gd-doped ceria (GDC) ceramics were synthesized using the ammonia coprecipitation method. By selecting the Ce(IV) compound as a cerium source, the nano-sized GDC precursor could be prepared without any compositional deviations. The GDC powders calcined above 800°C can be sintered to 97% of the theoretical result at 1600°C. The ionic conductivity of GDC ceramics from the Ce(IV) source was found to be 14.6mS/cm at 700°C, which was superior to that of 6.9mS/cm at 700°C for GDC ceramics from Ce(III) source, due to its lower formation of defect associations.

INTRODUCTION

Solid oxide fuel cells (SOFC) have received much attention recently because it is environmentally harmless and has good energy efficiency. For various applications, the ceria-based solid solutions have been extensively studied in order to reduce the SOFC operation temperature, because of their superior ionic conductivity.[1] Reducing the SOFC operation temperature makes it possible to have more selections in materials for SOFC, reduce cost, and solve degradation problems. Gd- or Sm-doped ceria is considered a promising electrolyte for SOFC operated at intermediate temperatures.[2-5] Various wet chemical processes, such as hydrothermal synthesis and homogeneous precipitation, have been used to prepare fine doped-ceria precursors with good sinterability.[6-9] Many researchers have selected the Ce(III) compounds as a starting cerium source. Complicated chemical reactions using oxidants are required for the oxidation from Ce(III) to Ce(IV) during the hydration and polymerization processes of cerium ions. In contrast, the Ce(IV) compounds were only used for preparation of the nano-scale CeO_2 crystalline, and fabrication of the doped-ceria electrolytes using Ce(IV) compounds has not been reported on much so far.[10-12] In this study, we studied the effectiveness of the Ce(IV) compound as a cerium source for preparation of the doped-ceria precursor, and compared the results with those of the Ce(III) compound.

EXPERIMENTAL PROCEDURE

GDC powders were fabricated via an aqueous process using $Ce(SO_4)_2 \cdot 4H_2O$ (Kishida Chemical Co.), or $Ce(NO_3)_3 \cdot 6H_2O$ (Kishida Chemical Co.) and $Gd(NO_3)_3 \cdot 5.2H_2O$ (Kojundo Chemical Co.) as starting materials. $Ce(SO_4)_2$ and $Gd(NO_3)_3$, which correspond to $Ce_{0.8}Gd_{0.2}O_{1.9}$ stoichiometry, were dissolved in distilled water (Ce(IV) solution), and aqueous ammonia was added to this solution as a precipitant. The pH of the solution was adjusted to 7.5. The mixed solution was then aged at room temperature for 24h. The prepared GDC precipitate was washed using aqueous ammonia, and then collected by filtration. The GDC precursor was also prepared from a solution made from $Ce(NO_3)_3$ and $Gd(NO_3)_3$ (Ce(III) solution) using the same method as mentioned above. The GDC precipitates were calcined at various temperatures for 2h. After grinding, the calcined powders were cold isostatically pressed at 300MPa into pellets. These pellets were then sintered at 1600°C for 6h. The prepared ceramics and powders were characterized by X-ray diffraction (XRD) analysis using a monochrometer (Rigaku, RAD-2B) with CuKα radiation. Thermogravimetry and differential thermal analysis (TG-DTA, Rigaku, TAS-300) were performed in air. The Raman spectra were recorded using an NR1100 (JASCO). The powder compositions were investigated by ICP-AES (Jarrell Ash, Plasma AtomComp. MK II). The densities of the GDC ceramics were measured by the Archimedes method using water. The microstructures of the prepared ceramics and powders were observed using a scanning electron microscope (SEM; JEOL, JSM-6100). Electrical conductivities were measured using a complex impedance method as functions of temperature and oxygen partial pressure. Oxygen partial pressure was adjusted by changing the mixing ratio of O_2, Ar, H_2 and water vapor.

RESULTS AND DISCUSSIONS

Aqueous process for $Ce_{0.8}Gd_{0.2}O_{1.9}$ precipitates

Figure 1 displays the XRD profiles of GDC precipitates using the two different cerium sources, Ce(III) and Ce(IV). The precipitate from the Ce(IV) solution was in an X-ray amorphous state, as shown in Fig. 1(b), while the diffraction pattern of the GDC precipitate from the Ce(III) solution exhibited a fluorite structure with a crystallite size of about 8nm (Fig. 1(a)).

ICP-AES revealed that the Ce/Gd ratio of the precipitate from the Ce(IV) solution was in good agreement with that of the initial composition. In contrast, the precipitate from the Ce(III) solution had a lower Ce/Gd ratio than 7/3, which is different from the initial ratio of Ce/Gd= 8/2. In addition, the GDC yield from the Ce(III) solution was less than 80%, and the GDC yield from the Ce(IV) solution was 98%. In the case of Ce(IV), reactions from Ce^{4+} to CeO_2 do not require any changes of cerium's valence.[10-12] On the other hand, as previously reported, when the Ce(III) compounds were selected as a cerium source, oxidants should be added into the precursor solution beforehand.[6,8,9] It can be presumed that the valence change from Ce^{3+} to Ce^{4+} accompanying CeO_2 formation affects the deviation of Ce/Gd ratio and low GDC yield from Ce(III) solution, currently the detailed reactions are under investigation.

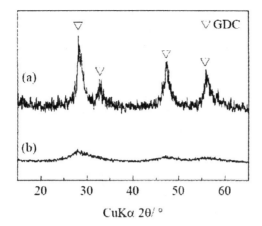

Figure 1
XRD profiles of GDC precipitates prepared from
(a) Ce(III) and (b) Ce(VI) solutions

Figure 2 shows the TG and DTA curves of GDC precipitates from Ce(III) and Ce(IV). For the precipitate from Ce(IV), the DTA curve has a strong endothermic peak at about 100°C and a very weak exothermic peak at about 550°C, and the TG curve exhibits a heavy weight loss below 150°C and then a gradual weight loss until 800°C. The endothermic peak with about 28% weight loss is mainly due to the dehydration of the GDC precipitates, and an exothermic peak at about 550°C corresponds to the crystallizing process of $Ce_{0.8}Gd_{0.2}O_{1.9}$. The gradual weight loss up to 800°C was attributed to the elimination of residual sulfate ions. In contrast, the precipitate from Ce(III) shows a strong exothermic peak with about 10% weight loss, which indicates the removal of the residual water, and a relatively strong exothermic peak with about 10% weight loss due to the elimination of the residual nitrate ions. Similarly to the case of Ce(IV), the gradual weight loss could also be observed up to 800°C.

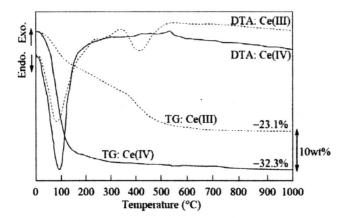

Figure 2
TG-DTA curves of GDC precipitates prepared from
Ce(III) and Ce(IV) solutions

Figure 3 illustrates the XRD profiles of precipitate from the Ce(IV) solution calcined at various temperatures for 2h. The fluorite single phase was obtained by the calcination of GDC precipitate from Ce(IV) above 600°C, as shown by curves (a)~(c) in Fig. 3. This result is consistent with the crystallization peak observed at about 550°C in the DTA curve (Fig. 2). The as-prepared precipitate at room temperature had coagulated in almost round shapes with sizes below 50nm. After a heat-treatment at 800°C for 2h, sizes of the GDC powders increased to 0.1μm~0.5μm. On the other hand, the as-prepared precipitate from the Ce(III) solution had already crystallized into the fluorite single phase, and with increasing the calcination temperature the crystallinity of the powders increased. The grain sizes of GDC powders (III) prepared at room temperature and calcined at 800°C were about 20nm and 0.5μm, respectively.

Figure 4 compares the Raman spectra of the GDC precipitates (III) and (IV) heated under different conditions. All samples had an F_{2g} mode around 465cm[-1], which is attributed to a symmetric stretching mode of the oxygen atoms around each cation.[13,14] There is also a broad band ~570cm[-1] in all samples, which is attributed to the presence of oxygen vacancies. For Raman spectrum on the as-prepared precipitate from the Ce(IV) solution (Fig. 4(f)), a small shift to lower frequency, broadening, and an increase in its asymmetry were all observed due to its insufficient grain size and crystallinity, while that from the Ce(III) solution with high crystallinity had a relatively sharp F_{2g} peak around 465cm[-1] (Fig. 4(c)).[15]

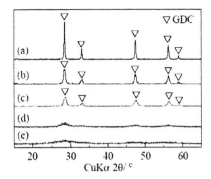

Figure 3
XRD profiles of GDC powders calcined at various
temperatures for 2 h: (a) 1000°C. (b) 800°C. (c) 600°C.
(d) 400°C. and (e) as-prepared precipitate

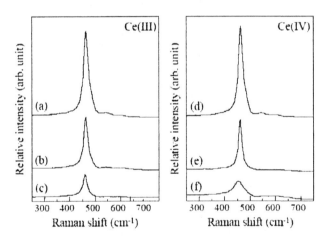

Figure 4
Raman spectra of GDC powders: (a) (d) heated at 1600°C
for 6 h after calcination at 800°C for 2 h. (b) (e) calcined
at 800°C for 2 h. and (c) (f) as-prepared precipitate

Electrical properties of $Ce_{0.8}Gd_{0.2}O_{1.9}$ ceramics

The GDC powders calcined at higher temperatures ($\geq 800°C$) were densified to 98.5% of the theoretical result by sintering at 1600°C for 6h. On the other hand, when calcined below 600°C, the GDC ceramics were not fully densified, since the residual sulfate ions were eliminated during the sintering process, as shown in TG-DTA data (Fig. 2). In the case of the Ce(III) source, GDC powders calcined above 800°C reached a maximum density of 97.4% after the sintering process. Figure 5 shows SEM pictures of GDC ceramics (III) and (IV) sintered at 1600°C for 6h after calcination at 800°C for 2h. GDC ceramics (IV) had a larger grain size from 2μm to 10μm than that of GDC ceramics (III), which were from 1μm to 8μm. This result agrees with a previous report that with increasing Gd content the grain growth and densification of ceria were depressed.[16]

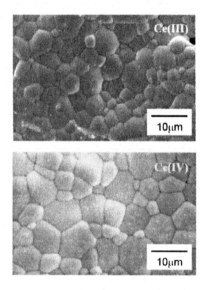

Figure 5
Microstructures of GDC ceramics (III) and (IV) sintered
at 1600°C for 6 h after calcination at 800°C for 2 h

The ionic conductivities of GDC ceramics (III) and (IV) sintered at 1600°C for 6h after calcination at 800°C for 2h were investigated. The bulk conductivities derived from the impedance measurements are shown in Fig. 6. GDC ceramics (IV) had a superior ionic conductivity to GDC ceramics (III) over the whole temperature range of 300°C-700°C, and the conductivity at 700°C for GDC ceramics (IV) was found to be 14.6mS/cm, which was higher than that of 6.9mS/cm for GDC ceramics (III). This is because that the oxide ion's mobility of GDC ceramics (III) decreases due to the formation of defect associations.[7, 16-19] The activation energies of the ionic transportation of GDC ceramics (III) and (IV) were estimated to be 0.95eV and 0.97eV, respectively. In addition, it is well known that at high $P(O_2)$ the conductivity is predominantly ionic, and at lower $P(O_2)$

GDC ceramics become a mixed conductor of ions and electrons due to the reduction of Ce^{4+} to Ce^{3+}.[20] As shown in Fig. 7, GDC ceramics (IV) showed an excellent reduction-resistive property, which is comparable to those previously reported, unlike the GDC ceramics (III).[5,21] Currently, the analysis of the differences in the reduction mechanism of GDC ceramics (III) and (IV) is under investigation.

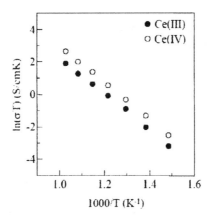

Figure 6
Ionic conductivities of the dense GDC ceramics (III) and (IV) sintered at 1600°C for 6 h after calcination at 800°C for 2 h

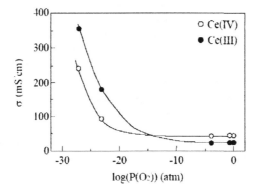

Figure 7
P(O$_2$) dependence of ionic conductivities at 700°C of GDC ceramics (III) and (IV) sintered at 1600°C for 6 h after calcination at 800°C for 2 h

CONCLUSION

Fully densified GDC ceramics were successfully synthesized via the simple aqueous process using a Ce(IV) source. The nano-sized GDC precipitate from the Ce(IV) solution was obtained in 98% yield and was in an X-ray amorphous state. The ionic conductivity of GDC ceramics (IV) was found to be 14.6mS/cm at 700°C, which was higher than that of 6.9mS/cm for GDC ceramics (III).

ACKNOWLEDGEMENT

The authors are very grateful to Dr. T. Shimura of Nagoya University for their assistance with the characterization of the electrical conductivities.

*Former Address: Department of Applied Chemistry, Graduate School of Engineering, Nagoya University

REFERENCES

[1] H. Inaba, and H. Tagawa, "Ceria-based Solid Electrolytes- Review," *Solid State Ionics*, **83**, 1-16 (1996).

[2] G. M. Christie, and F. P. F. van Berkel, "Microstructure- Ionic Conductivity Relationships in Ceria- Gadolinia Electrolytes," *Solid State Ionics*, **83**, 17-27 (1996).

[3] K. Zheng, B. C. H. Steele, M. Sahibzada, and I. S. Metcalfe, "Solid Oxide Fuel Cells on Ce(Gd)O_{2-x} Electrolytes," *Solid State Ionics*, **86-88**, 1241-44 (1996).

[4] C. Xia, and M. Liu, "Microstructures, Conductivities, and Electrochemical Properties of $Ce_{0.9}Gd_{0.1}O_2$ and GDC-Ni Anodes for Low-temperature SOFCs," *Solid State Ionics*, **152-153**, 423-30 (2002).

[5] T. Matsui, M. Inaba, A. Mineshige, and Z. Ogumi, "Electrochemical Properties of Ceria-based Oxides for Use in Intermediate-temperature SOFCs," *Solid State Ionics*, **176**, 647-54 (2005).

[6] R. S. Torrens, N. M. Sammes, and G. A. Tompsett, "Characterization of $(CeO2)_{0.8}(GdO_{1.5})_{0.2}$ Synthesized Using Various Techniques," *Solid State Ionics*, **111**, 9-15 (1998).

[7] E. Suda, B. Pacaud, Y. Montardi, M. Mori, M. Ozawa, and Y. Takeda, "Low-temperature Sinterable $Ce_{0.9}Gd_{0.1}O_{1.95}$ Powder Synthesized through Newly-devised Heat-treatment in the Coprecipitation Process," *Electrochemistry*, **10**, 866-72 (2003).

[8] J. G. Cheng, S. W. Zha, J. Huang, X. Q. Liu, and G. Y. Meng, "Sintering Behavior and Electrical Conductivity of $Ce_{0.9}Gd_{0.1}O_{1.95}$ Powder Prepared by the Gel-casting Process," *Materials Chemistry and Physics*, **78**, 791-95 (2003).

[9] T. S. Zhang, J. Ma, L. B. Kong, P. Hing, Y. J. Leng, S. H. Chan, and J. A. Kilner, "Preparation and Electrical Properties of Dense Submicron-grained $Ce_{0.8}Gd_{0.2}O_{2-\delta}$ Ceramics," *J. Mater. Sci. Lett.*, **22**, 1809-11 (2003).

[10] W. P. Hsu, L. Ronnquist, and E. Matijevic, "Preparation and Properties of Monodispersed Colloidal Particles of Lanthanide Compounds. 2. Cerium (IV)," *Langmuir*, **4**, 31-37 (1988).

[11] N. Audebrand, J. P. Auffrédic, and D. Louër, "An W-ray Powder Diffraction Study of the Microstructure and Growth Kinetics of Nanoscale Crystallites Obtained from Hydrated Cerium Oxides," *Chem. Mater.*, **12**, 1791-99 (2000).

[12] M. Hirano, and M. Inagaki, "Preparation of Monodispersed Cerium (IV) Oxide Particles by Thermal Hydrolysis: Influence of the Presence of Urea and Gd Doping on Their

Morphology and Growth," *J. Mater. Chem.*, **10**, 473-77 (2000).

[13]J. R. McBride, K. C. Hass, B. D. Poindexter, and W. H. Weber, "Raman and X-ray Studies of $Ce_{1-x}RE_xO_{2-y}$, where RE= La, Pr, Nd, Eu, Gd, and Tb," *J. Appl. Phys.*, **76**, 2435-41 (1994).

[14]C. Peng, Y. Wang, K. Jiang, B. Q. Bin, H. W. Liang, J. Feng, and J. Meng, "Study on the Structure Change and Oxygen Vacation Shift for $Ce_{1-x}Sm_xO_{2-y}$ Solid Solution," *J. Alloy. Compd.*, **349**, 273-78 (2002).

[15]S. Wang, W. Wang, J. Zuo, and Y. Qian, "Study of the Raman Spectrum of CeO_2 Nanometer Thin Films," *Mater. Chem. Phys.*, **68**, 246-48 (2001).

[16]Z. Tianshu, P. Hing, H. Huang, and J. Kilner, "Ionic Conductivity in the CeO_2-Gd_2O_3 System ($0.05 \leq$ Gd/Ce$\leq$ 0.4) Prepared by Oxalate Coprecipitation," *Solid State Ionics*, **148**, 567-73 (2002).

[17]J. Cheng, S. Zha, X. Fang, X. Liu, and G. Meng, "On the Green Density, Sintering Behavior and Electrical Property of Tape Cast $Ce_{0.9}Gd_{0.1}O_{1.95}$ Electrolyte Films," *Mater. Res. Bull.*, **37**, 2437-46 (2002).

[18]S. Zha, C. Xia, and G. Meng, "Effect of Gd (Sm) Doping on Properties of Ceria Electrolyte for Solid Oxide Fuel Cells," *J. Power Sources*, **115**, 44-48 (2003).

[19]A. Sin, Y. Dubitsky, A. Zaopo, A. S. Aricò, L. Gullo, D. L. Rosa, S. Siracusano, V. Antonucci, C. Oliva, and O. Ballabio, "Preparation and Sintering of $Ce_{1-x}Gd_xO_{2-x/2}$ Nanopowders and Their electrochemical and EPR Characterization," *Solid State Ionics*, **175**, 361-66 (2004).

[20]S. Wang, H. Inaba, H. Tagawa, M. Dokiya, and T. Hashimoto, "Nonstoichiometry of $Ce_{0.9}Gd_{0.1}O_{1.95-x}$," *Solid State Ionics*, **107**, 73-79 (1998).

[21]S. P. S. Badwal, F. T. Ciacchi, and J. Drennan, "Investigation of the Stability of Ceria-Gadolinia Electrolytes in Solid Oxide Fuel Cell Environments," *Solid State Ionics*, **121**, 253-62 (1999).

STRUCTURAL STUDY AND CONDUCTIVITY OF BaZr$_{0.90}$Ga$_{0.10}$O$_{2.95}$

Istaq Ahmed and Elisabet Ahlberg
Department of Chemistry, Gothenburg University,
Gothenburg, SE-412 96, Sweden.

Sten Eriksson
Department of Environmental Inorganic Chemistry, Chalmers University of Technology,
Gothenburg, SE-412 96, Sweden.

Christopher Knee, Maths Karlsson, Aleksandar Matic and Lars Börjesson
Department of Applied Physics, Chalmers University of Technology,
Gothenburg, SE-412 96, Sweden.

ABSTRACT

Traditional solid state sintering has been used to prepare the perovskite BaZr$_{0.9}$Ga$_{0.1}$O$_{2.95}$. Analysis of X-ray powder diffraction data shows that a decrease of the unit cell parameter a was observed compared to the undoped BaZrO$_3$, which confirms successful substitution of Ga^{3+} for Zr^{4+} at the B site. Rietveld analysis of room temperature neutron powder diffraction data confirmed cubic symmetry (space group Pm-3m) for both as- prepared and deuterated BaZr$_{0.90}$Ga$_{0.10}$O$_{2.95}$ samples. The strong O-H stretch band (2500-3500 cm^{-1}) in the infrared absorbance spectrum clearly manifests the presence of protons in the pre-hydrated material. The bulk and total conductivities of pre-hydrated BaZr$_{0.9}$Ga$_{0.1}$O$_{2.95}$ are 1.17×10^{-5} and 3.55×10^{-6} Scm^{-1}, respectively at 400°C, which are more than one order of magnitude higher than for dried a sample at the same temperature. In contrast, the total conductivity of pre-hydrated and dried samples is similar at higher temperature, e.g. T > 800 °C. The higher activation energy (e.g. 0.7 eV) for pre-hydrated sample compared to typical value (0.4-0.5 eV) of proton conduction may suggest that the protons are trapped in the material.

INTRODUCTION

Proton conducting solid electrolytes has a wide range of technological applications in fuel cells, batteries, gas sensors, hydrogenation/dehydrogenation of hydrocarbons, electrolysers, etc[1-3]. The investigation of proton conductivity in perovskite-ceramics started more than two decades ago, and Yb doped SrCeO$_3$[4], Nd doped BaCeO$_3$[5] and Y doped BaZrO$_3$[6-8] are examples of systems that exhibit good proton conducting properties under humid or hydrogen containing atmosphere at elevated temperatures.

Literature data compiled by Norby[2] show the presence of a "gap", within a certain intermediate temperature (approximately 500 to 800 K) range, in which no materials show high proton conductivity. Narrowing this gap would be very beneficial from a technological point of view. The development and improvement of material conduction properties necessitates an understanding of the proton transport mechanism.

By infrared (IR) spectroscopy, Omata et al.[9] investigated the O-H stretch region in several hydrated perovskites and found quite interesting different spectral features of Ga^{3+} doped BaZrO$_3$ compared to the other trivalent dopants like Y^{3+} and In^{3+}. The short O-O separation distance in the

Ga^{3+} doped perovskites (ionic radius 0.62Å) compared to the larger O-O distance in Y^{3+} (ionic radius 0.90 Å), and In^{3+} doped (ionic radius 0.80 Å) analogues may explain the strong hydrogen bonding in these materials[10]. The relationship between the degree of hydrogen bonding and the proton conductivity is not clear. Sites of strong hydrogen bonding may act as traps and hinder proton transfer or enhance the probability for proton transfer and thereby improve the proton conductivity[11]. This motivated us to investigate the structural features and the conductivity of pre-hydrated and dried $BaZr_{0.90}Ga_{0.10}O_{2.95}$ samples in inert (Ar) atmosphere.

The combined techniques of X-ray powder diffraction (XRPD), Neutron powder diffraction (NPD), Infrared Spectroscopy and Electrochemical measurements (Impedance analysis) have been performed in this study. Hydrated samples have been investigated by IR and Impedance analysis, while deuterium substituted samples have been used in the NPD experiments due to the small coherent neutron cross-section of protons.

EXPERIMENTAL

Samples were prepared by solid state sintering. Appropriate amounts of $BaCO_3$, Ga_2O_3 and ZrO_2 have been mixed in order to obtain the desired compositions. The oxides were heated to 800 °C over night to remove moisture prior to weighing. To ensure thorough mixing, ethanol (99.5%) was added during the milling, which was performed manually using an agate mortar and pestle. The finely ground material was fired at 1000 °C for 8 h and subsequently ground and pelletized using a 13 mm diameter die under a pressure of 8 tons. The pellets were sintered at 1200 °C in air for 72 h and finally reground and compacted under similar conditions and re-fired at 1500 °C for 48 h. After the final sintering the pellets showed a density corresponding to ≈ 91% of the theoretical value. In order to remove the protons from the as-prepared material, the samples were dried by annealing at 800°C in 5×10^{-6} mbar pressure overnight. Before starting deuteration we performed dynamic thermogravimetric analysis (TGA) under humid condition (the data are not shown in this paper). The results from TGA show that maximum mass gain started at 300 °C. Based on this result deuteration and protonation reactions were performed by annealing the powder (for NPD) and the pellet (for impedance) samples at 300°C under an Ar gas flow (12mL/min) saturated with D_2O and H_2O vapour, at 76.2 °C for 168h, respectively.

The X-ray diffraction measurements were carried out at ambient temperature using a Guinier-Hägg camera ($CuK\alpha_1$ = 1.5406 Å) with silicon (NBS 640b) as internal standard. The programs $TREOR^{12}$ and $Checkshell^{13}$ were used for indexing and refinement of the lattice parameters. Neutron powder diffraction (NPD) data were collected on the NPD diffractometer at the R2 Research Reactor at Studsvik, Sweden. A double monochromator system consisting of two parallel copper crystals in (220) orientation were aligned to give a wavelength of 1.470(1) Å and the step scan covered a 2θ range 4° -139.92° with a step size of 0.08°. A vanadium can was used as the sample holder. The obtained data sets were refined by the Rietveld method[14] using the Fullprof software[15] and diffraction peak shapes described by a pseudo-Voigt function. Background intensities were described by a Chebyshev polynomial with six coefficients.

The impedance was measured from 4.5 MHz to 1 Hz using a Solatron 1260 frequency response analyser in the stand alone mode. The sine wave amplitude was 1 V rms. The conductivity cell used was a ProboStat™ from Norwegian Electro Ceramics AS (NorECs)[16]. Approximately 0.3 cm^2 of the oxide electrode surface was covered with conducting platinum paste and platinum grid to assure good ohmic contacts. Impedance measurements were performed from 150 to 1100 °C (heating cycle) and down from 1100 to 150°C (cooling cycle) in 50 °C steps on the pre-hydrated sample under inert (Ar) atmosphere. For the dried sample the measurement was performed only during the heating cycle under but under the same inert atmosphere. After reaching the desired temperature, 10 minutes elapsed before the impedance spectra was recorded. Two time constants were generally observed corresponding to the bulk and grain boundary conduction, respectively. Typical impedance spectra are shown in Fig. 1 together with the equivalent circuit used in the data analysis. For the second time constant, which represent grain boundary conduction, a constant phase element (CPE) was used. The capacitance values were 2.6×10^{-11} Fcm^{-2} for bulk conductivity, 8×10^{-10} F cm^{-2} for grain boundary conduction.

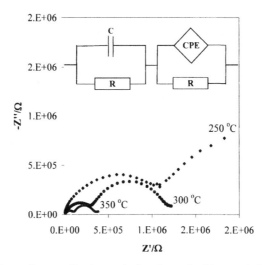

Fig. 1. Complex impedance diagrams for the pre-hydrated sample. Two semi-circles are clearly observed and are related to bulk conduction (high frequencies) and grain boundary conduction (low frequencies).

The IR experiments were performed in an inert atmosphere at room temperature in diffuse reflectance mode using a Bruker VECTOR 22 FTIR spectrometer, equipped with a KBr beamsplitter, a deuterated triglycerine sulfate detector, and a diffuse reflectance device (Graseby Specac, "Selector"). A wrinkled aluminium foil was used as reference. The vibrational spectra were derived by taking the logarithm of the ratio between the reference spectrum and the sample spectrum.

RESULTS

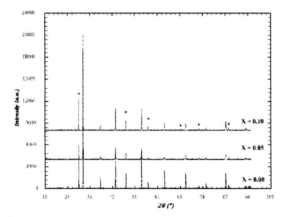

Fig. 2 X-ray diffraction patterns of BaZr$_{1-x}$Ga$_x$O$_{3-0.5x}$ (x = 0.00, 0.05 and 0.10). Peaks marked with an asterisk (*) correspond to the silicon standard used.

Fig. 2 shows the room temperature XRPD pattern of as-prepared BaZr$_{1-x}$Ga$_x$O$_{3-0.5x}$ (x = 0.00, 0.05 and 0.10). Indexing of these patterns clearly indicates that all samples possess cubic symmetry of space group Pm-3m. The indexed unit-cell parameters (a) have been determined to 4.1924(7) Å, 4.1853(3) Å and 4.1842(2) Å for x = 0.00, 0.05 and 0.10, respectively.

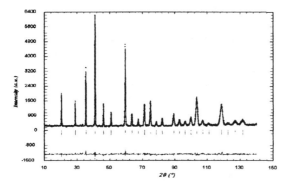

Fig. 3 Observed room temperature neutron powder diffraction profile of as-prepared BaZr$_{0.90}$Ga$_{0.10}$O$_{2.95}$ (points), calculated pattern (solid line) and difference pattern (at bottom). The tick marks show the positions of the Bragg peaks predicted by the structural model.

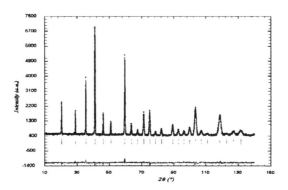

Fig. 4 Observed room temperature neutron powder diffraction profile of deuterated BaZr$_{0.90}$Ga$_{0.10}$O$_{2.95}$ (points), calculated pattern (solid line) and difference pattern (line below). The tick marks show the position of Bragg peaks predicted by the structural model.

Fig. 3 and Fig. 4 show Rietveld fits of NPD data collected at room temperature for as-prepared and deuterated $BaZr_{0.90}Ga_{0.10}O_{2.95}$, respectively. A summary of the Rietveld structural analysis is shown in Table I. Rietveld refinements of as-prepared and deuterated samples confirm their cubic symmetry and space group (Pm-3m). The unit cell parameters of as-prepared and deuterated samples are 4.1879(2) and 4.1880(2), respectively (see Table I). These values are the identical within the experimental uncertainty, which is unlike the expansion often seen for deuterated phases[17-19].

Table I. Summary of results obtained from Rietveld analysis of neutron powder diffraction data for as-prepared and deuterated $BaZr_{0.90}Ga_{0.10}O_{2.95}$, collected at 295 K on the NPD diffractometer at Studsvik.

	(as-prepared)	(deuterated)
Space group	Pm-3m	Pm-3m
Lattice parameter (Å)	4.1879(2)	4.1880 (2)
Isotropic thermal parameter (B_{iso})		
Ba on $1(b)$ ($\frac{1}{2},\frac{1}{2},\frac{1}{2}$)	0.37(4)	0.38 (5)
Zr on $1(a)$ (0,0,0)	0.35(4)	0.40(4)
Ga on $1(a)$ (0,0,0)	0.35(4)	0.40(4)
O on $3(d)$ $\frac{1}{2}$,0,0	0.74(3)	0.78(4)
Site Occuapncies (O)	2.98(2)	3.01(2)
Distance (Å)		
Ba-O	2.96131(1)	2.96135(1)
Zr/Ga-O	2.09396(1)	2.09399(1)
O-O	2.96131(1)	2.96135(1)
χ^2	1.57	1.47
R_{wp}	6.2	4.90
R_{Bragg}	3.99	5.18
No of fitted parameters	19	19

The isotropic temperature factors (Table I) for both samples were comparable to each other, although a slight increase was noticed for the deuterated sample. The refined oxygen site occupancy for the as-prepared sample was $O_{2.98(2)}$, which was in fair agreement with the expected value of $O_{2.95}$ (based on charge neutrality). The oxygen site occupancy achieved for the deuterated sample was $O_{3.01(2)}$, i.e. the oxygen vacancies were as expected filled by O-H groups[20]. This is in excellent agreement with the expected oxygen content of 3.00.

According to our recent investigation[21] on a closely related material deuterium ions are found close to the oxygen site on the Wyckoff 12h position (0.5, 0.2, 0.0) or possibly the 24k position (0.54, 0.21, 0) within space group Pm-3m. We used similar starting values for deuterated

$BaZr_{0.90}Ga_{0.10}O_{2.95}$ in our attempts to locate deuterium ions within the structure. However, the refined positions and occupancies did not converge well enough to draw any definite conclusions.

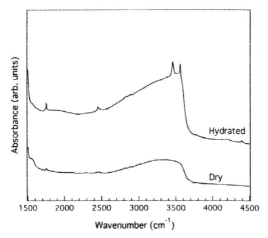

Fig. 5 Infrared absorbance spectra of dry and hydrated samples of $BaZr_{0.90}Ga_{0.10}O_{2.95}$.

A clear manifestation of the presence of protons in the hydrated material is the strong O-H stretch band, 2500-3500 cm^{-1}, in the infrared absorbance spectrum (see Fig. 5). Also shown in this figure is the spectrum of the dry sample where the O-H stretch band is clearly weaker. The signature of an O-H stretch band also in the dry sample is a result of the difficulty to remove all protons during annealing at 800 °C in vacuum. Note that the dissociation of water molecules during hydration of the perovskite material is confirmed by the absence of a water bend band in the spectrum, which should occur at around 1640 cm^{-1}.

Fig. 6 and Fig. 7 show the Arrhenius plot of total and bulk conductivity for dried and pre-hydrated $BaZr_{0.90}Ga_{0.10}O_{2.95}$ respectively. Table II shows the summary of the conductivity results at different temperatures. Our experimental results are quite comparable to the work of Iwahara et al.[6] where a value of 10^{-4}-10^{-5} Scm^{-1} were observed for $BaZr_{0.95}Ga_{0.05}O_{2.975}$ in reducing atmosphere (H_2) specially at elevated temperature i.e. at T > 600 °C. There is a limitation in the instrumental setup of about 5×10^8 ohm, corresponding to a conductivity of 10^{-9} Scm^{-1} for the samples used in this work. That is why it was difficult for us to extract the true conductivity values at low temperatures. At temperatures higher than 600 °C it was difficult to separate the bulk and total conductivity.

From Table II it can be seen that the bulk conductivity for the pre-protonated sample is more than one order of magnitude higher than for the dried sample at 400 °C, and for the same pre-protonated sample in the cooling cycle. For the total conductivity at 400 °C the difference is more than two orders of magnitude. At temperature higher than 800 °C the total conductivity is similar for both samples.

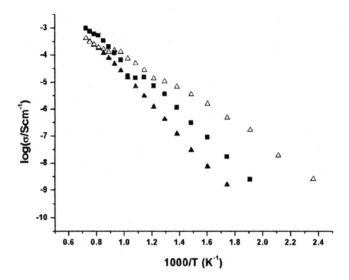

Fig. 6 Total conductivity of BaZr$_{0.90}$Ga$_{0.10}$O$_{2.95}$. $\triangle$: pre-protonated heating cycle, $\blacktriangle$: pre-protonated cooling cycle and $\blacksquare$: dried sample heating cycle.

Table II Summary of conductivity results for BaZr$_{0.90}$Ga$_{0.10}$O$_{2.95}$.

Sample type	σ at 400°C(Scm^{-1})/ E$_a$(eV)	σ at 800°C(Scm^{-1})/ E$_a$(eV)	σ at 1100°C(Scm^{-1})/ E$_a$(eV)
Protonated (Heating cycle)	3.55×10^{-6} (total) /0.73 1.17×10^{-5} (bulk) /0.72	1.48×10^{-4}(total) /0.75	4.08×10^{-4} (total) /1.0
Protonated (Cooling cycle)	2.98×10^{-8} (total) / 1.1 2.53×10^{-7} (bulk) /1.2	4.65×10^{-5} (total) /1.1	4.16×10^{-4} (total) / 1.1
Dried (Heating cycle)	3.10×10^{-7} (total) / 1.0 2.66×10^{-6} (bulk) / 0.86	1.15×10^{-1} (total) /1.1	9.47×10^{-1} (total) / 1.1

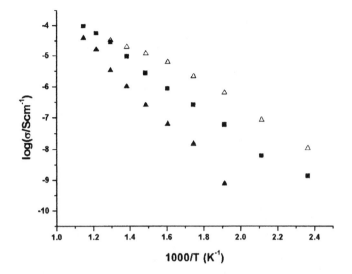

Fig. 7 Bulk conductivity of BaZr$_{0.90}$Ga$_{0.10}$O$_{2.95}$. $\triangle$: pre-protonated heating cycle, $\blacktriangle$: pre-protonated cooling cycle and $\blacksquare$: dried sample heating cycle.

DISCUSSIONS

Incorporation of trivalent dopant ions into BaZrO$_3$ at B sites (Zr^{4+}) with the corresponding creation of oxygen vacancies is very important for the proton incorporation. The unit cell parameter of Ga^{3+} doped BaZrO$_3$ was found to be lower compared to the un-doped BaZrO$_3$ (see page 4). The lower value is due to the partial substitution of the smaller Ga^{3+} compared to Zr^{4+}. In contrast, the unit cell parameters obtained by Omata et al.[9] were 4.1937(1) Å for un-doped BaZrO$_3$ and 4.1946(3) Å for BaZr$_{0.95}$Ga$_{0.05}$O$_{3-a}$. However, if the smaller Ga^{3+} is replacing some Zr^{4+}, the volume of the unit cell must decrease in BaZr$_{0.95}$Ga$_{0.05}$O$_{3-a}$ compared to the un-doped BaZrO$_3$, hence a lower unit cell parameter, a, is expected. The unit cell parameter obtained in the present study indeed confirms that Ga^{3+} successfully substituted Zr^{4+} at B site positions. Consequently, the charge compensating oxygen vacancies (V$_O^{\cdot\cdot}$) were created due to substitution of trivalent Ga for tetravalent Zr ions. However, electronic holes can be formed in an oxidizing atmosphere[22-23]. Often an expansion of the unit cell parameter after deuteration of the dried sample is observed[17-19]. This expansion is a result of filling of oxygen vacancies (V$_O^{\cdot\cdot}$) by hydroxyl groups (OH$_O^{\cdot}$) in the presence of water vapour according to the following reaction,

$$H_2O\,(g) + O_O{}^x + V_O{}^{..} \leftrightarrow 2OH_O{}^{.} \text{ ----------- (1)}$$

However, in the present study the refined unit cell parameters of as-prepared and deuterated samples are the same within experimental uncertainty (see Table 1). For 10 mol% Ga^{3+} substituted $BaZrO_3$, a maximum 0.05 moles of oxygen vacancies ($V_O{}^{..}$) can be formed. Our refined oxygen occupancies for the as-prepared sample showed that only 0.02 moles of $V_O{}^{..}$ were in fact obtained. This discrepancy may be a result of filling of some oxygen vacancies while the sample was cooling in the furnace under ambient atmosphere. The actual filling of oxygen vacancies during deuteration was smaller than expected, which makes it less probable to observe any expansion of the unit cell parameter.

The IR results (Fig. 5) clearly show that the content of protons increased after exposing the sample to water vapour. These results also show that it is very difficult to remove all the protons from dry samples. As already mentioned above, the dry samples were prepared by annealing the as-prepared samples at 800 °C under vacuum. This might be a result of "trapped" protons. This also indicates that some of the oxygen vacancies are already filled by the above mentioned hydroxyl groups ($OH_O{}^{.}$) prior to the hydration. The partial filling of oxygen vacancies by hydroxyl groups ($OH_O{}^{.}$) is often observed during cooling[11, 17, 24]. However, after treating the dry sample at 300 °C in D_2O vapour for 168 h, a value of 3.01(2) for the oxygen occupancy was obtained from Rietveld refinement (see Table I). This means that all the oxygen vacancies were filled according to reaction 1. The deuterium positions were not located successfully within the structure. The concentration and stability of deuterium ions are very important to locate their positions. At 10 mol% Ga^{3+} substituted $BaZrO_3$ full occupation of the oxygen sites and corresponding incorporation of twice as many deuterium(D) ions would lead to a maximum of only 0.1 mol D ($BaZr_{0.90}Ga_{0.10}O_{2.9}(OD)_{0.1}$) per mol of $BaZr_{0.90}Ga_{0.10}O_{2.95}$. At such low deuterium concentration it is quite difficult to locate their position reliably. Moreover, at ambient temperatures the deuterium ions rotate around the oxygen ions and are not fixed at one position. As suggested by Knight[25] at higher doping which corresponds to the higher concentration of deuterium into the structure is required to obtain accurate refinement of deuterium positions, thermal parameters, occupancies etc. However, high-resolution NPD data collected at low temperature can be useful for further studies. Reducing the thermal movement may reduce the number of possible sites for D to occupy and it may thus be possible to locate the species. This is an approach, which has already been successfully used for deuterated $BaZr_{0.50}In_{0.50}O_{2.75}$ at 5 K by Ahmed et al.[21] and for protonated $BaCe_{0.9}Y_{0.1}O_{2.95}$ at 4.2K by Knight[25].

The higher value of conductivity obtained in the intermediate temperature range (250-550 °C) for the protonated $BaZr_{0.90}Ga_{0.10}O_{2.95}$ compared to the dried sample is due to the presence of a large number of protons within the sample, which have been incorporated into the material according to equation 1. These protons act as charge carriers. Also the dried sample contains some protons as revealed by the IR results (Fig. 5), yielding a higher conduction compared to the conduction of the pre-protonated sample in the cooling cycle. With increasing temperature the protons leave the samples and at higher temperatures, T > 800 °C, the conductivities are very similar for both samples (Fig. 6). Similar results have been observed in the literature for several systems [18, 24, 26-32]. Also the TGA on the pre-protonated powder sample under inert atmosphere (the data are not shown in this paper) showed a gradual mass loss, which started at ~ 300°C and continued up to 750 °C during the heating cycle, which may be due to proton loss from the material, while no mass gain was observed during the cooling cycle. All these observations clearly show that proton conduction is not dominating at temperatures higher than 800°C.

In order to confirm this behaviour, impedance measurements were performed during the cooling cycle from 1100 to 150 °C under identical inert atmosphere for the pre-protonated sample. The conductivity values of pre-protonated sample during the cooling cycle was more than two orders of magnitude lower than for the heating cycle at a temperature below 450 °C (Fig. 6-7). When the pre-protonated sample reached the maximum temperature (1100 °C) used, the protons already had left the sample and the conductivity is similar to that obtained for the dried sample. Since the experiment was performed in inert atmosphere no proton up-take from the atmosphere was possible. Thus, the conductivity of a proton free sample is measured in the cooling cycle, which is also reflected in the higher value of the activation energy.

Typically, the hole and oxygen-ion conductivities have higher activation energies and tend to dominate at elevated temperatures[33]. The higher activation energies (0.75-1.1 eV) found for temperatures higher than 800 °C, for the pre-protonated sample (heating and cooling cycles) and the dried sample also indicate that at higher temperatures the conductivity is not due to protons, but rather to holes or oxide-ion conduction. The activation energy for proton conduction (for temperatures less than 450 °C) as measured for the pre-protonated sample during heating cycle, was 0.72 eV. This value is higher than the typical values for perovskite proton conductors (0.4-0.5 eV) [6-7, 34]. This may be a result of trapping of protons within the material. Generally, in perovskite-type oxides, protonic defects are known to diffuse by the Grotthuss diffusion mechanism[35], which involves both a proton-transfer step (hopping from one oxygen to another oxygen) and a reorientation step. Short O-O separation distances (Table I) due to partial replacement of large Zr^{4+}(ionic radius 0.72 Å) by small Ga^{3+} (ionic radius 0.62Å) resulted in strong hydrogen bonding[10]. This strong hydrogen bonding may be beneficial for the proton transfer step while it traps the proton during the reorientation process[11]. The higher activation energy may suggest that the proton reorientation step is the rate-limiting step rather than the proton transfer step, in agreement with the work on similar materials, for example the In^{3+} doped $BaZrO_3$ [11]. In contrast, in acceptor-doped $BaCeO_3$ the proton-transfer step was found to be rate-limiting [36]. Further work is necessary to sort out the detailed mechanism and quasi-elastic neutron scattering may be useful in this respect.

CONCLUSIONS

By using a standard solid-state synthesis route it was possible to obtain single-phase $BaZr_{0.90}Ga_{0.10}O_{2.95}$ samples. Rietveld analysis of room temperature NPD data indicates that the structure of the as-prepared samples and samples treated with D_2O vapour show cubic symmetry (space group Pm-3m). IR spectra clearly show the presence of a large number of protons in the hydrated sample compare to the dried sample. Higher activation energy and lower conductivity compared to other proton conducting perovskites such as the Y^{3+} doped $BaZrO_3$ or $BaCeO_3$ were obtained. These results indicate that strong hydrogen bonding is present due to the small ionic radius of Ga^{3+}, resulting in trapping (immobilising) of the protons within the structure

ACKNOWLEDGEMENTS
This work was supported by the Swedish Research Council, and the National Graduate School in Material Science at Chalmers University of Technology, Gothenburg, Sweden

REFERENCES

[1] H. Iwahara, "Technological Challenges in the Application of Proton Conducting Ceramics" *Solid State Ionics* **77**, 289-298 (1995).

[2] T. Norby, "Solid State Protonic Conductors: Principles, Properties, Progress and Prospects" *Solid State Ionics* **125**, 1-11 (1999).

[3] Ph. Colomban (Ed), "Proton Conductors: Solids, Membranes and Gels-Materials and Devices", *Cambridge University Press, Cambridge 1992.*

[4] H. Iwahara, H. Uchinda, S. Tanaka, "High Temperature Type Proton Conductor Based on $SrCeO_3$ and its Application to Solid Electrolyte Fuel Cells" *Solid State Ionics* **9-10**,1021-1025 (1983).

[5] H. Iwahara, H. Uchinda H. K. Ono, K. Ogaki, "Proton Conduction in Sintered Oxides Based on $BaCeO_3$" *J. Electrochem. Soc.* **135**, 529-533 (1988).

[6] H. Iwahara, T. Yajima, T. Hibino, K. Ozaki, H. Susuki, "Protonic Conduction in Calcium, Strontium and Barium Zirconates" *Solid State Ionics* **61**, 65-69 (1993).

[7] K.D. Kreuer, "Aspect of the Formation and Mobility of Protonic Charge Carriers and the Stability of Perovskite-type Oxides" *Solid State Ionics* **125**, 285-302 (1999).

[8] K.D. Kreuer, St. Adams, W.Munch, A.Fuchs, U.Klock, J.Maier, "Proton Conducting Alkaline Earth Zirconates and Titanates for High Drain Electrochemical Applications" *Solid State Ionics* **145**, 295-306 (2001).

[9] T. Omata, M.Takagi, S. Otsuka-Yao-Matsuo "O-H Stretching Vibrations of Proton Conducting Alkaline-Earth Zirconates " *Solid State Ionics* **168**, 99-109 (2004).

[10] A. Novak "Hydrogen bonding in Solids" *Struct.Bond.* **18**, 177-216 (1974)

[11] M. Karlsson, A. Matic, L.Börjesson, D. Engberg, M. Björketun, P. Sundell, G. Wahnström, I. Ahmed, S-G Eriksson, P. Berastegui, "Vibrational Properties of Protons in Hydrated $BaIn_xZr_{1-x}O_{3-x/2}$ " *Physical Review B* **72**, 094303-1-7(2005).

[12] P.E. Werner, L. Eriksson, M.Westdahl, " TREOR, a semi-exhaustive trial-and-error powder indexing program for all symmetries" *J. Appl. Crystallogr* **18**, 367-370(1985).

[13] J.Lougier, B. Bochu, "Checkshell: Graphical Powder diffraction Indexing celland Space group Assignment Software", http://www.inpg.fr/LMGP.

[14] H.M.Rietveld, "Profile refinement method for nuclear and magnetic structures" *J. Appl. Crystallogr.* **2**, 65-71 (1969).

[15] J.Rodriuez-Carvajal, "Recent advances in magnetic structure determination by neutron powder diffraction" *Physica B* **192**, 55-69 (1993).

[16] www.norecs.com.

[17] I. Ahmed, S-G Eriksson, P. Berastegui, L-G. Johansson, M. Karlsson, A. Matic, L. Börjesson, D. Engberg, "Synthesis and Structural Characterization of Perovskite type Proton Conducting $BaZr_{1-x}In_xO_{3-0.5x}$ $(0.0 \le x \le 0.75)$" *will be submitted to Solid State Ionics in revised form.*

[18] D.J.D. Corcoran, J.T.S. Irvine, "Investigation into $Sr_3CaZr_{0.5}Ta_{1.5}O_{8.75}$, a Novel Proton Conducting Perovskite Oxide" *Solid State Ionics* **145**, 307-313 (2001).

[19]I.Sosnowska, R. Przenioslo, W. Schäfer, W. Kockelmann, R. Hempelmann, K. Wysocki, "Possible Deuterium Positions in the High-Temperature Deuterated Proton Conductor Ba$_3$Ca$_{1+y}$Nb$_{2-y}$O$_{9-\delta}$ Studied by Neutron and X-ray Powder Diffraction" *J. Alloys and Comp.* **328**, 226-230 (2001).

[20]W-K. Lee, A.S. Nowick, "Protonic Conduction in Acceptor-Doped KTaO$_3$ Crystal" *Solid State Ionics* **18**, 989-993 (1986).

[21]I. Ahmed, C.S. Knee, S-G Eriksson, P. Berastegui, L-G. Johanssson, P.F. Henry, M. Karlsson, A. Matic, L. Börjesson, D. Engberg, "Location of Deuterium Sites in Proton Conducting Perovskite BaZr$_{0.5}$In$_{0.5}$O$_{2.5}$(OD)$_{0.5}$ at 5 K" *will be submitted to Solid State Ionics.*

[22]N. Bonanos, B.Ellis, M.N. Mahmood, "Oxide Ion Conduction in Ytterbium-Doped Strontium Cerate" *Solid State Ionics* **28-30**, 579-584 (1988).

[23]I. Kosacki, H.L. Tuller, "Mixed Conductivity in SrCe$_{0.95}$Yb$_{0.05}$O$_3$ Protonic Conductors" *Solid State Ionics* **80**, 223-232 (1995).

[24]J.F. Liu, A.S. Nowick, "The Incorporation and Migration of Protons in Nd-doped BaCeO$_3$" *Solid State Ionics* **50**, 131-138 (1992).

[25]K. S. Knight, "Powder Neutron Diffraction Studies of BaCe0.9Y0.1O$_{2.95}$ and BaCeO$_3$ at 4.2 K: a Possible Structural Site for the Proton" *Solid State Ionics* **127**, 43-48 (2000).

[26]R. Mukundan, Eric L. Brosha, S. A. Birdsell, A.L. Costello, F.H. Garzon, R.S. Willms "Tritium Conductivity and Isotope Effect in Proton-Conducting Perovskites" *Journal of the Electrochemical Society,* **146 (6)** , 2184-2187,(1999).

[27]R. Haugsrud, T. Norby "Proton Conduction in Rare-earth Ortho-niobates and Ortho-tantalates" *Nature Materials,* **5(3)**, 193-196, (2006).

[28] T. Norby, M.Wideroe, R. Gloeckner, Y. Larring, "Hydrogen in Oxides" *Dalton Transactions,*19, 3012-3018 (2004).

[29] T. Norby, "The Promise of Protonics" *Nature, 410*, 877-878 (2001).

[30]D. Lybye, N. Bonanos, "Proton and Oxide ion Conductivity of Doped LaScO$_3$" *Solid State Ionics* **125**, 339-344 (1999).

[31]S. Kim, K.H. Lee, H. L. Lee "Proton Conduction in La$_{0.6}$Ba$_{0.4}$ScO$_{2.8}$" *Solid State Ionics* **144**, 109-115 (2001).

[32]F.M.M. Snijkers, A. Buekenhoudt, J.J. Luyten, J. Cooymans, M. Mertens, "Proton Conductivity in Perovksite Type Yttrium Doped Barium Hafnate" *Scripta Materialia* **51**, 1129-1134 (2004).

[33]T. Yajima and H. Iwahara, "Studies on Proton Behaviour in Doped Perovskite-Type Oxides: (II) Dependence of Equilibrium Hydrogen Concentration and Mobility on Dopant Content in Yb-Doped SrCeO$_3$" *Solid State Ionics* **53-54**, 983-988 (1992).

[34]K.D. Kreuer, "Proton-Conducting Oxides" *Ann. Rev.Mater.Res..* **33**, 333-359, (2003).

[35]K.D. Kreuer, "Proton Conductivities: Materials and Applications" *Chem.Mater.* **8**, 610-641, (1996).

[36]K.D. Kreuer, T. Dippel, Y.M. Baikov, J. Maier, "Water Solubility, Proton and Oxygen Diffusion in Acceptor Doped BaCeO$_3$: A Single Crystal Analysis " *Solid State Ionics* **86-88**, 613-620 (1996).

HYDROGEN FLUX IN TERBIUM DOPED STRONTIUM CERATE MEMBRANE

Mohamed M. Elbaccouch[*] and Ali T-Raissi

Florida Solar Energy Center
University of Central Florida
1679 Clearlake Road
Cocoa, FL 32922-5703
*Tel: (321) 638-1504
Fax: (321) 504-3438
Email: melbaccouch@fsec.ucf.edu

ABSTRACT

Hydrogen production membranes of $SeCe_{0.95}Tb_{0.05}O_{3-\delta}$ are synthesized using the liquid-phase method. The membranes are of perovskite-type structure with mixed ionic-electronic conductivities. The membranes are dense providing 100% hydrogen selectivity. Hydrogen permeability is studied as a function of temperature, hydrogen partial pressure, hydrogen dry conditions, and water vapor pressure. Also, the influence of nickel deposition on hydrogen flux is evaluated.

INTRODUCTION

Hydrogen production and separation using mixed ionic-electronic conductors of perovskite-type structure ($A^{2+}B^{4+}O_3$) are of great interest to the coal, natural gas, and petrochemical industries. The doped perovskite structure for hydrogen production has several attractive characteristics including, 100% pure hydrogen, external electrical power is not required, high thermal and energy efficiencies, cost-effective compared to palladium membranes, simple and flexible compared to pressure swing adsorption, and on-site hydrogen production.

Above 550 °C, the perovskite ceramic oxides exhibit ionic and electronic conductivities when the B-sites are doped with rare earth trivalent cations.[1] Ion transport in perovskite structures include three fundamental steps, a gas-solid interfacial reaction where hydrogen dissociate into protons and electrons, ion migration through the solid lattice, and solid-gas interfacial reaction where protons and electrons form hydrogen atoms.[2]

The characteristics of ionic conductivity in perovskite-type ceramic oxides were first discovered in 1981 by Iwahara et al.[1] Since then, various doped perovskite ceramics have been investigated including the respective cerates ($SrCeO_3$[2-4] and $BaCeO_3$[5-7]) and strontium zirconate ($SrZrO_3$[4,8]). Evidence of proton conductivity in perovskite materials have been the focus of many research programs.[9-12] $SrCeO_3$-based material is chosen for this investigation because it displays one of the highest proton conductivity.

In this work we report hydrogen permeability data of terbium doped strontium cerate oxide ($SrCe_{0.95}Tb_{0.05}O_{3-\delta}$) as a function of temperature (600-900 °C), hydrogen partial pressure (P_{H2}), and water vapor partial pressure (P_{H2O}). Also, the effect of metallic nickel impregnation on hydrogen flux is evaluated.

EXPERIMENTAL

Chemicals

Strontium nitrate ($Sr(NO_3)_2$) (99.995%) was purchased from Aldrich Chemical Co. 1.5 N cerium nitrate ($Ce(NO_3)_4.31H_2O$) solution, terbium nitrate ($Tb(NO_3)_3.6H_2O$) (99.9%), and citric acid (99%) were purchased from Alfa Aesar. Nitric acid (HNO_3) was obtained from Fisher Scientific. Distilled water was supplied by the Florida Solar Energy Center (FSEC). All chemicals were used as-received without further purifications.

Membrane Fabrication

Stoichiometric molar ratios of the metal nitrate precursors are mixed in distilled water to form 0.2 M solution. Nitric acid and excess citric acid (50% of total moles of metal ions) are added to the solution followed by stirring for 5 h in a reflux mode at 100 °C to initiate the polymerization process.[13] Water is evaporated on a hot plate and the solution is condensed to a gel-like material. The gel material is dried at 110 °C for 24 h to form a sponge-like brittle material. A self-ignition step is followed for about 40 min at 400 °C. The resulting powder is sieved twice to 90 μm and calcined at 600 and 800 °C for 24 hr to eliminate all organic residues. The powder is ground again to 90 μm and pressed under 130 Mpa hydraulic pressure into disks. Final sintering is carried out at 1200 °C for 24h.

Hydrogen Permeation Apparatus

A schematic diagram of the permeation unit is represented in Figure 1.[14] The set up consists of three interconnected alumina tubes (Coors Tek) used as the feed gas chamber (22"L x 1" D), sweep gas inlet chamber (15" L x 0.5" D), and sweep gas outlet chamber (17" L x 0.25" D). A ceramic sealant is used to obtain a gas-tight seal between the membrane disk and the sweep gas inlet alumina tube. The inlet feed gas and the argon sweep gas are introduced from opposite sides of the membrane and their flow rates are monitored using volumetric flow meters (Gilmont GF-1060). The upstream and downstream flow rates are maintained at atmospheric pressure and vacuum respectively. A gas chromatograph (SRI Instruments 8610) is used to analyze the product stream.

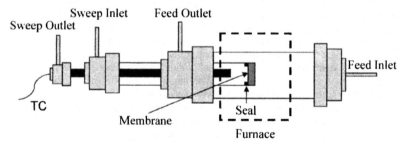

Figure 1. Hydrogen membrane permeation apparatus.

RESULTS AND DISCUSSION

Leakage in the membrane disk due to pin-holes or incomplete membrane seal is checked by introducing helium in the feed stream as a tracer and measuring helium content in the product stream. The total hydrogen permeation rate (mol/s) is calculated from the hydrogen content in the argon sweep gas and argon flow rate. Hydrogen flux is calculated by dividing the permeation rates by the effective surface area (mol/cm^2 s).

Figure 2 represents hydrogen flux as a function of temperature (600-900 °C) for SrCe$_{0.95}$Tb$_{0.05}$O$_{3-\delta}$. The figure indicates that hydrogen flux increases linearly with temperature. The ceramic sealant and the gas-tight requirement of the membrane disk limit the upper temperature of the measurements to 900 °C.

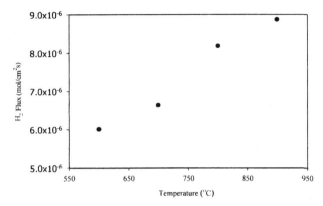

Figure 2. Temperature dependence of hydrogen flux for SrCe$_{0.95}$Tb$_{0.05}$O$_{3-\delta}$.

Figure 3 shows the influence of P_{H2} at different temperatures under dry conditions on hydrogen flux for SrCe$_{0.95}$Tb$_{0.05}$O$_{3-\delta}$. The figure shows that hydrogen flux increases as P_{H2} increases.

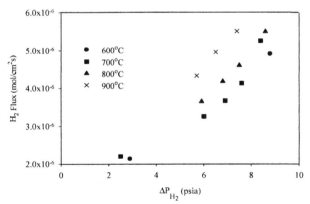

Figure 3. Influence of ΔP_{H2} at different temperatures under dry conditions on hydrogen flux for SrCe$_{0.95}$Tb$_{0.05}$O$_{3-\delta}$.

The influence of P_{H2O} on hydrogen flux at 800 °C as a function of ΔP_{H2} of $SrCe_{0.95}Tb_{0.05}O_{3-\delta}$ is given in Figure 4. Hydrogen permeability decreases with increasing P_{H2O} due to increase in P_{O2}. It is expected that the flux for 100% dry hydrogen conditions to be higher than that of wet hydrogen conditions.

The hydrogen flux at 800 °C and 100% hydrogen dry conditions for $SrCe_{0.95}Tb_{0.05}O_{3-\delta}$ impregnated with nickel (2.3 mm) is 1.12×10^{-5} mol/(cm^2 s), while the hydrogen flux for a similar membrane disk (1.5 mm thick) without nickel is 8.18×10^{-6} mol/(cm^2 s). Although the nickel membrane sample has a larger thickness, it has a higher permeation rate. That is because nickel impregnation enhances hydrogen permeability by enhancing the catalytic activity at the solid-gas interfaces.

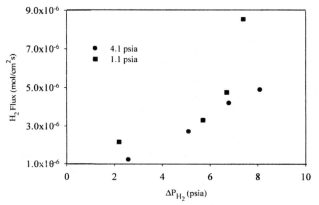

Figure 4. Influence of ΔP_{H2O} on hydrogen flux at 800°C for $SrCe_{0.95}Tb_{0.05}O_{3-\delta}$.

CONCLUSIONS

We developed high temperature dense ceramic oxide membranes of $SrCe_{0.95}Tb_{0.05}O_{3-\delta}$ with mixed ionic electronic conductors for hydrogen production and separation. Hydrogen permeability for $SrCe_{0.95}Tb_{0.05}O_{3-\delta}$ is presented as a function of temperature (600-900 °C), P_{H2}, and P_{H2O}. For the infinite hydrogen selectivity, the membrane needs to be dense, free of pin-holes (i.e. N_2 like molecules do not diffuse through), and free of cracks. Results indicate that hydrogen flux increases linearly with temperature. Also, hydrogen flux increases as P_{H2} increases. 100% dry hydrogen condition has higher H_2 permeability than wet H_2 conditions. Nickel impregnation enhances hydrogen permeability by enhancing the catalytic activity at the solid-gas interface. Experiments are in progress to generate H_2 flux data for $SrCeO_3$-based materials doped with yttrium, neodymium, and gadolinium trivalent cations.

ACKNOWLEDGEMENTS

This work was supported by a grant provided by the National Aeronautics and Space Administration (NASA) - Glenn Research Center under contract No. NAG32751.

REFERENCES

[1]H. Iwahara, T. Esaka, H. Uchida, and N. Maeda, "Proton Conduction in Sintered Oxides and its Application to Steam Electrolysis for Hydrogen Production," *Solid State Ionics* **3/4**, 359-363 (1981).

[2]S. –J Song, E. D. Wachsman, J. Rhodes, S. E. Dorris, and U. Balachandran. Hydrogen, "Permeability of $SrCe_{1-x}M_xO_{3-\delta}$ (x = 0.05, M = Eu, Sm)," *Solid State Ionics* **167**, 99-105 (2004).

[3]X. Qi, and Y.S. Lin, "Electrical Conducting Properties of Proton-Conducting Terbium-Doped Strontium Cerate Membrane," *Solid State Ionics* **120**, 85-93 (1999).

[4]S. Hamakawa, L. Li, A. Li, and E. Iglesia, "Synthesis and Hydrogen Permeation Properties of Membranes Based on Dense $SrCe_{0.95}Yb_{0.05}O_{3-\delta}$ Thin Films," *Solid State Ionics* **48**, 71-81 (2002).

[5]R.C.T. Slade, and N. Singh, "Generation of Charge Carriers and H/D Isotope Effect in Proton-Conducting Doped Barium Cerate Ceramics," *J. Mater. Chem.* **1(3)**, 441-445 (1991).

[6]K. Takeuchi, C.-K Loong, J.W. Richardson Jr., J. Guan, S.E. Dorris, and U. Balachandran, "The Crystal Structure and Phase Transition in Y-Doped BaCeO3: Their Dependence on Y Concentration and Hydrogen Doping," *Solid State Ionics* **138**, 63-77 (2000).

[7]V. Agarwal, and M. Liu, "Preparation of Barium Cerate-Based Thin Films Using a Modified Pechini Process," *J. mat. Sci.* **32**, 619-625 (1997).

[8]J. Muller, K.D. Kreuer, J. Maier, S. Matsuo, and M. Ishigame, "A Conductivity and Thermal Gravimetric Analysis of a Y-Doped SrZrO3 Single Crystal," *Solid State Ionics* **97**, 421-427 (1997).

[9]J. Guan, "Ceramic Membranes of Mixed Ionic-Electronic Conductors for Hydrogen Separation," Ph.D. Thesis, Georgia Institute of Technology, June, 1986.

[10]K.D. Kreuer, Proton-Conducting Oxides, Annual Review Mater. Res. 33 (2003) 333-359.

[11]H. Iwahara, "High Temperature Proton Conductors Based on Perovskite-Type Oxides," in: Proton Conductors - solids, Membranes and Gels-Materials and Devices, P. Colomban, Cambridge University Press, Cambridge, 1992, Chapter 8, P 122-137.

[12]S. M. Haile, G. Stanefe, and K.H. Ryu, "Non-Stoichiometry, Grain Boundary Transport and Chemical Stability of Proton Conducting Perovskites," *J. Membr. Sci.* **36**, 1149-1160 (2001).

[13]D. Dionysiou, X. Qi, Y. S. Lin, G. Meng, and D. Peng, "Preparation and Characterization of Proton Conducting Terbium Doped Strontium Cerate Membranes," *J. membr. Sci.* **154**, 143-153 (1999).

[14]F. Lau, and S. Doong, "Coal to Hydrogen: A Novel Membrane Reactor for Direct Extraction," Gas Technology Institute Presentation at GCEP Energy Workshops at Stanford University, April 26-27 (2004).

A MECHANICAL-ELECTROCHEMICAL THEORY OF DEFECTS IN IONIC SOLIDS

Narasimhan Swaminathan and Jianmin Qu
G.W. Woodruff School of Mechanical Engineering, Georgia Institute of Technology

801 Ferst Drive NW

Atlanta, Georgia, 30332-0405

ABSTRACT

In this paper we present a coupled thermodynamic formulation to predict stresses, in an ionic solid due to diffusion of charged defects in an electrochemical potential gradient. Chemical expansion is considered primarily by treating compositional strains as eigen strains. Two material properties are introduced a) A second order tensor that represents the eigen strains in the solid due to non-stoichiometry and b) A fourth order tensor that represents the variations in elastic properties due to non-stoichiometry. A general theory is first developed for a steady state, isothermal condition, while considering typical electrochemical reactions at the interface and the bulk of a typical oxide ion conductor by considering all the major defects that are known to operate. Two geometries typical of solid electrolytes (planar and tubular) are considered, involving diffusion of vacancies and electrons. The governing equations are solved for the resulting stresses due to chemically induced strains resulting from a deviation in the stoichiometric composition of the solid. The results show the influence of considering the coupled problem on the distribution of defects, electrostatic potential, and the current voltage characteristics.

INTRODUCTION

Problem Statement

Ionic solids are used as electrolytes in solid oxide fuel cells. Ions are transported due to a gradient in their electrochemical potential. The formulation of a diffusion problem in electrolytes involves the chemical potential as a function of temperature and concentration only. However due to difference in formation volume of the defects and the host atoms, the chemical potential is also a function of self-stress generated due to the volumetric change. The stress generated may be a consequence of non-uniform composition, due to boundary constraints or both. A previous analysis[1] considers this effect while neglecting the electrostatic potential. This does not allow us to examine the electrical behavior of the solid under self stressed diffusion. It is the aim of this work to examine the effect of self stress on diffusion of charged ionic defects.

Experimental investigations[2] show the dependence of elastic properties and fracture toughness on the deviation from stoichiometric composition for a few ionic solids although no numerical results considering these factors are reported in the literature. During functioning, the solid may be subjected to external loads (mechanical and thermal) and it is important to consider the composition dependent properties to predict the mechanical behavior.

Stress Dependent Electrochemical Potential

We write the internal energy density Π of a network solid as state function of the strain $\left(\varepsilon_{ij}\right)$, concentrations of all components $\left(\overline{\rho}_\alpha\right)$, and the entropy density s. All densities are measured per unit volume in the reference state. It is to be noted that the internal energy density is a function of all the structural elements making up the solid. Equilibrium conditions developing from such a formulation will result in the diffusion potential[3] as the driving force for diffusion. We can however consider that the gradient in the concentration of the host ions (Oxygen ions) of a particular defect (Vacancies) in the sub-lattice being negligible (uniform concentration throughout the solid) and consider only the chemical potential of the defects as the driving force.

The total differential of the strain energy density may be written as,

$$d\Pi = \sigma_{ij}d\varepsilon_{ij} + Tds + \frac{1}{V_m}\sum_\alpha \tilde{\mu}_\alpha d\rho_\alpha \tag{1}$$

Where

$$\sigma_{ij} = \left(\frac{\partial\Pi}{\partial\varepsilon_{ij}}\right)_{s,\rho_\alpha}, T = \left(\frac{\partial\Pi}{\partial s}\right)_{\varepsilon_{ij},\rho_\alpha}, \tilde{\mu}_\alpha = \left(\frac{\partial\Pi}{\partial\overline{\rho}_\alpha}\right)_{\varepsilon_{ij},s,\rho_{\beta\neq\alpha}}, \rho = V_m\overline{\rho} \tag{2}$$

σ_{ij}, is the Cauchy's stress tensor, T, the absolute temperature and $\tilde{\mu}_\alpha$, the electrochemical potential. V_m, is the molar volume of the ionic solid.

The Legendre transform of the internal energy $\Pi(\varepsilon_{ij},s,\rho_\alpha)$ yields the grand potential

$$\Theta(\sigma_{ij},s,\rho_\alpha) = \Pi(\varepsilon_{ij},s,\rho_\alpha) - \sigma_{ij}\varepsilon_{ij} \tag{3}$$

So that,

$$d\Theta(\sigma_{ij},s,\rho_\alpha) = -\varepsilon_{ij}d\sigma_{ij} + Tds + \frac{1}{V_m}\sum_\alpha \tilde{\mu}_\alpha d\rho_\alpha \tag{4}$$

By differentiating the above equation twice with respect to σ_{ij} and ρ_α leads to the Maxwell relation,

$$\frac{1}{V_m}\left(\frac{\partial\tilde{\mu}_\alpha}{\partial\sigma_{ij}}\right)_{\rho_{\beta\neq\alpha}} = -\left(\frac{\partial\varepsilon_{ij}}{\partial\rho_\alpha}\right)_{\sigma_{ij}} \tag{5}$$

We now make use of the elastic constitutive law relating the total strain $\left(\varepsilon_{ij}\right)$, compositional strain $\left(\varepsilon_{ij}^c\right)$, elastic strain $\left(\varepsilon_{ij}^E\right)$ and the compliance $\left(S_{ijkl}\right)$,

$$\varepsilon_{ij}^E = \varepsilon_{ij} - \varepsilon_{ij}^c = S_{ijkl}\sigma_{kl} \tag{6}$$

together with the relation,

$$\varepsilon_{ij}^c = \sum_\alpha \varepsilon_{ij}^\alpha = \sum_\alpha \eta_\alpha \Delta\rho_\alpha \delta_{ij} \tag{7}$$

to reformulate (5) into,

$$\frac{1}{V_m}\left(\frac{\partial \tilde{\mu}_\alpha}{\partial \sigma_{ij}}\right)_{\rho_{\beta\neq\alpha}} = -\eta_\alpha\delta_{ij} - \frac{1}{2}\frac{\partial S_{ijkl}}{\partial \rho_\alpha}\sigma_{kl} = -\eta_\alpha\delta_{ij} - \frac{1}{2}s_{ijkl}\sigma_{kl} \tag{8}$$

η_α, is the Vegard's coefficient for defect species α. It is shown experimentally[4] in Ref.4 that strains are linearly related to $(P_{O_2})^{-1/4}$, where P_{O_2} is the partial pressure of oxygen. For low defect concentrations, since non-stoichiometry is proportional to $(P_{O_2})^{-1/4}$ the linear relation in (7) is valid[5].

By integrating (8) with respect to σ_{ij}, one has

$$\tilde{\mu}_\alpha(\rho_\alpha,\sigma) = \tilde{\mu}_\alpha(\rho_\alpha,0) + V_m\tau_\alpha \tag{9}$$

where $\tilde{\mu}_\alpha(\rho_\alpha,0)$ is the electrochemical potential of defect α at zero stress and

$$\tau_\alpha = -\eta_\alpha\sigma_{kk} - \frac{1}{2}s_{ijkl}^\alpha\sigma_{kl}\sigma_{ij} \tag{10}$$

is the effect of stress on the electrochemical potential. s_{ijkl}^α is the change of the elastic compliance with respect to deviation of species α from the stoichiometric composition.

The electrochemical potential of the species α with equivalent charge z_α is given as a combination of the chemical potential μ_α and the electrostatic potential ϕ as,

$$\tilde{\mu}_\alpha = \mu_\alpha + z_\alpha F\phi \tag{11}$$

Using (11) in (9) we have,

$$\tilde{\mu}_\alpha = \mu_\alpha(\rho_\alpha,0) + V_m\tau_\alpha + z_\alpha F\phi \tag{12}$$

The chemical potential at zero stress can be expressed as a function of the defect concentration ρ_α and the activity coefficient γ_α, i.e.,

$$\mu_\alpha(\rho_\alpha,0) = \mu_\alpha^0 + RT\ln(\gamma_\alpha\rho_\alpha) \tag{13}$$

where μ_α^0 is the chemical potential at some standard state. The activity coefficient takes into account the interaction among the defects and is considered a unity for a dilute solution where such interactions are negligible.

Diffusion of Defects

We consider vacancies $(\alpha = v)$, oxygen interstitials $(\alpha = O)$, electrons $(\alpha = e)$ and holes $(\alpha = h)$ as the major defects. The current density of a charged specie can be given as,

$$\mathbf{J}_\alpha = -\frac{\rho_\alpha D_\alpha}{RT}\nabla\tilde{\mu}_\alpha \tag{14}$$

where $D_\alpha = (z_\alpha F R \vartheta_\alpha T)/V_m$, and ∇ is the spatial gradient operator, and ϑ_α is the mobility of the species. Substitution of (12) into (14) leads to

$$\mathbf{J}_\alpha = -D_\alpha\left(1+\frac{\partial \ln\gamma_\alpha}{\partial \ln\rho_\alpha}\right)\nabla\rho_\alpha - \frac{V_m\rho_\alpha D_\alpha}{RT}\nabla\tau_\alpha - \frac{z_\alpha\rho_\alpha D_\alpha F}{RT}\nabla\phi \tag{15}$$

The continuity condition for each type of defect may be given as,

$$\frac{\partial\rho_\alpha}{\partial t} = -\frac{1}{z_\alpha F}\nabla\cdot\mathbf{J}_\alpha + (G_\alpha - R_\alpha) \tag{16}$$

where G_α and R_α are the generation and recombination rates, respectively, for defect α. Combining (16) for the oxygen interstitials and oxygen vacancies, and similarly for the electrons and electron holes results in,

$$\left(\frac{\partial\rho_O}{\partial t} + \frac{\partial\rho_v}{\partial t}\right) = \frac{1}{2F}\nabla\cdot(\mathbf{J}_O + \mathbf{J}_v) + (G_O - G_v) - (R_O - R_v) \tag{17}$$

$$\left(\frac{\partial\rho_e}{\partial t} + \frac{\partial\rho_h}{\partial t}\right) = \frac{1}{F}\nabla\cdot(\mathbf{J}_e + \mathbf{J}_h) + (G_e - G_h) - (R_e - R_h) \tag{18}$$

At this juncture we introduce the two typical defect equilibrium reactions that are known to occur in an ionic conductor with the assumed defect species. We have,
1) The electron-hole equilibrium reaction,

$$e' + h^\bullet = nil \tag{19}$$

2) The Frenkel equilibrium reaction,

$$O_O^X + V_i^X = O_i'' + V_O^{\bullet\bullet} \tag{20}$$

Note: The Kroger Vink notation has been used to represent the charged defects.
Each reaction can be represented by a mass action law that relates the concentrations,

$$[e'][h^\bullet] = K_e \quad , \quad \frac{[O_i''][V_O^{\bullet\bullet}]}{[O_O^X][V_i^X]} = K_F \tag{21}$$

with, K_e and K_F, being the corresponding equilibrium constant for reactions (19) and (20).

Due to the equilibrium of electrons and electron holes (19) and between vacancies and interstitials (20), the recombination and the generation terms of the ionic defects (vacancies and interstitials) and electronic defects (electrons and holes) vanish, i.e.,

$$\left(G_O - G_v\right) - \left(R_O - R_v\right) = 0, \left(G_e - G_h\right) - \left(R_e - R_h\right) = 0 \tag{22}$$

In other words, the sources and sinks for these defects must be located at the surface of the solid[6]. Therefore,(17) and (18) can be simplified under steady state conditions to,

$$0 = \frac{1}{2F}\nabla\bullet\left(\mathbf{J}_O + \mathbf{J}_v\right) \;,\;\; 0 = \frac{1}{F}\nabla\bullet\left(\mathbf{J}_e + \mathbf{J}_h\right) \tag{23}$$

THE BOUNDARY VALUE PROBLEM (BVP)

Governing Differential and algebraic equations

The BVP that needs to be solved involves the following unknown quantities. They are the four defect concentrations, ρ_e, ρ_h, ρ_v and ρ_O, the electrostatic potential ϕ, and the three displacement components u_i. The required equations involve both differential as well as algebraic equations.

For the displacement field we have the three, well known mechanical equilibrium equations written in terms of displacement,

$$\left(C_{jikl}u_{k,l}\right)_{,j} - \left(C_{jikl}\varepsilon_{kl}^c\right)_{,j} = 0 \tag{24}$$

with C_{jikl} being the stiffness tensor.

The diffusion equations (23) together with the mass action laws (21) constitute four more equations. The other equation that completes the description of the problem is of two forms, either of which could be used depending on the approximation that is valid for the concerned solid.

One of them is the differential equation for the electrostatic potential, which is a consequence of gauss law and has the following form,

$$-\kappa_\varepsilon\nabla^2\phi = \frac{F}{\varepsilon_0 V_m}\sum_\alpha z_\alpha\rho_\alpha \tag{25}$$

where $F = 9.648\times10^4 C/mol$ is the Faraday constant, $\varepsilon_0 = 8.854\times10^{-12} farad/m$ is the electric permittivity of free space, κ_ε is the dielectric constant of the solid. The sum should include all charged defects in the solid including the immobile defects.

The other form is the electroneutrality condition and is useful when the concentrations are small. This is given as,

$$\sum_{\alpha} z_{\alpha} \rho_{\alpha} \approx 0 \tag{26}$$

Note that (26) does not mean that $\nabla^2 \phi = 0$ within the solid. This is because F/ε_0 is typically orders of magnitude greater than the total charge so that the right hand side of (25) does not vanish even if (26) is approximately satisfied. Thus either (25) or (26) may be used as the eighth equation.

Boundary conditions

The mechanical boundary conditions involve specification of displacement on the regions $\partial_u \Omega$ and tractions in the regions $\partial_p \Omega$. Where, $\partial_u \Omega + \partial_p \Omega = \partial \Omega$, represents the entire boundary of the solid occupying domain Ω. These are given as,

$$u_i\big|_{\partial_u \Omega} = \bar{u}, \sigma_{ij} n_j\big|_{\partial_p \Omega} = p_j \tag{27}$$

where n_j represents the normal of the boundary $\partial_p \Omega$.

The specification of the concentrations at the boundary require the gas solid equilibrium reaction,

$$O_O^X (\text{solid}) \underset{\text{reduction}}{\overset{\text{oxidation}}{\rightleftarrows}} \{V_O^{\bullet\bullet} + 2e'\} (\text{solid}) + \frac{1}{2} O_2 (\text{gas}) \tag{28}$$

The mass action law of the above reaction,

$$K_{S1} = [V_O^{\bullet\bullet}][e']^2 [O_2]^{1/2} \tag{29}$$

along with the electroneutrality condition (26) can be used to specify the concentrations at the boundaries. It is important to note that the equilibrium constants are generally obtained by equating the electrochemical potentials. Now that we have modified the electrochemical potentials to include stress also, they are functions of the stress states and hence contribute to the non-linearity in the problem as will be specified later.

The specification of the electrostatic potential on the either ends may be given by,

$$\phi\big|_{S_{II}} - \phi\big|_{S_I} = V + \frac{RT}{F} \ln\left(\frac{\rho_e\big|_{S_{II}}}{\rho_e\big|_{S_I}} \right) \tag{30}$$

Where, V is the applied voltage and, S_{END}^c (END= I or II) represents the conducting boundary of the solid.

EXAMPLES

Planar Case

We consider, as the first example, one dimensional through thickness (x_1) diffusion in, a thin planar electrolyte (with elastic modulus E and Poisson's ratio v) of thickness h, conducting vacancies and electrons having a stoichiometric vacancy concentration of ρ_v^0. A low partial pressure of oxygen is maintained at end I, while a higher partial pressure is maintained at end II. For the sake of simplicity we neglect any variation in elastic properties due to composition and the compositional strain due to electrons. The planar electrolyte is assumed to be plane stressed, thus giving the following expression for the in-plane stress $(x_2 - x_3$ plane),

$$\sigma_{22} = \sigma_{33} = -\frac{E}{(1-v)}\sum_\alpha \eta_\alpha \Delta\rho_\alpha(x_1) \tag{31}$$

Using (31) in (12) and then using (23) we obtain the current density for electrons and vacancies as,

$$hJ_v = -D_v\frac{d\rho_v}{dx} - \frac{2\eta_v^2\rho_v D_v \bar{E} V_m}{RT}\frac{d\rho_v}{dx} - \frac{2\rho_v D_v F}{RT}\frac{d\phi}{dx} \tag{32}$$

$$hJ_e = -D_e\frac{d\rho_e}{dx} + \frac{\rho_e D_e F}{RT}\frac{d\phi}{dx} \tag{33}$$

where $x = x_1/h$.

Using the local electroneutrality condition (26) with the immobile defects ρ_A and eliminating $(d\phi/dx)$ from (32) and (33) we get,

$$\left(6\rho_v - \rho_A + 2(2\rho_v - \rho_A)\rho_v\hat{\eta}_v\right)\frac{d\rho_v}{dx} = \rho_A\hat{J}_v - 2\rho_v(\hat{J}_v + \hat{J}_e) \tag{34}$$

Where, $\hat{\eta}_v = \dfrac{\eta_v^2 \bar{E} V_m}{RT}$ and $\hat{J}_\alpha = \dfrac{hJ_\alpha}{D_\alpha}$

The differential equation (34) can be integrated to obtain,

$$\left(\rho_v(x) - \rho_v(0)\right)\left[A_1\left(\rho_v(x) + \rho_v(0)\right) + A_2\right] + \rho_A A_3 \ln\left(\frac{\hat{J}_v\rho_A - 2A\rho_v(x)}{\hat{J}_v\rho_A - 2A\rho_v(0)}\right) = 2A^3 x \tag{35}$$

Where,

$$A = \hat{J}_e + \hat{J}_v, \quad A_1 = -2A^2\hat{\eta}_v, \quad A_2 = 2A(\rho_A\hat{J}_e\hat{\eta}_v - 3A), \quad A_3 = A(\hat{J}_e - 2\hat{J}_v) + \rho_A\hat{J}_e\hat{J}_v\hat{\eta}_v \tag{36}$$

Next, eliminating dx between (32) and (33) and integrating yields,

$$-2A\hat{J}_e\hat{\eta}_v\left(\rho_v(x)-\rho_v(0)\right)-A_3\ln\left(\frac{\rho_A\hat{J}_v-2A\rho_v(x)}{\rho_A\hat{J}_v-2A\rho_v(0)}\right)=2A^2\left(\hat{\phi}(x)-\hat{\phi}(0)\right) \tag{37}$$

where $\hat{\phi}=\dfrac{F\phi}{RT}$ is the non-dimensional electrostatic potential. Eliminating the ln term between (37) and (35) leads to,

$$\rho_A\left(\hat{\phi}(x)-\hat{\phi}(0)\right)=-\left(\rho_v(x)-\rho_v(0)\right)\left[\hat{\eta}_v\left(\rho_v(x)+\rho_v(0)\right)+3\right]-Ax \tag{38}$$

Equation (35) provides the solution (albeit implicit) to the vacancy concentration, while (37) gives the electrostatic potential explicitly in terms of the concentration. Furthermore, one can also easily compute the electron concentration through the local electroneutrality condition, assuming the immobile defect concentration is given. However, these solutions contain the unknown constants $\hat{J}_e$ and $\hat{J}_v$ (constants due to steady state assumption). To determine these constants, we (37)evaluate at $x=1$,

$$-2A\hat{J}_e\hat{\eta}_v\left(\rho_v''-\rho_v'\right)-\left(A(\hat{J}_e-2\hat{J}_v)+\rho_A\hat{J}_e\hat{J}_v\hat{\eta}_v\right)\ln\left(\frac{\hat{J}_v\rho_A-2A\rho_v''}{\hat{J}_v\rho_A-2A\rho_v'}\right)=2A^2\left(\hat{\phi}''-\hat{\phi}'\right) \tag{39}$$

where $\rho_v'=\rho_v(0)$, $\rho_v''=\rho_v(1)$, $\hat{\phi}'=\hat{\phi}(0)$, and $\hat{\phi}''=\hat{\phi}(1)$. It is interesting to note that the sum of the non-dimensional fluxes is completely determined by the boundary conditions. Thus, A may be calculated from the boundary conditions. We then calculate the ionic or electronic current densities from(39). With this, the distribution of vacancies and potential may be easily obtained from (35) and(37).

The specification of boundary concentrations needs special attention. As mentioned previously, since the electrochemical potentials are also functions of stresses, the equilibrium constants will be functions of stresses. This will need to be solved together with the governing differential equations and hence makes the problem non-linear. For the planar and the tubular cases solved in this paper, it can be easily shown that the relation is,

$$\exp\left(-\hat{\eta}_v\left(\rho_v'-\rho_v''\right)\right)=\frac{\rho_v'\left(\rho_e'\right)^2\sqrt{P^I}}{\rho_v''\left(\rho_e''\right)^2\sqrt{P^{II}}} \tag{40}$$

In the solved examples we specify the concentration and partial pressure at one end (*I*) and find the concentration for a given partial pressure at the end (*II*) by satisfying the above relation.

Tubular case

In the second example we consider a thin tubular structure of inner radius a, outer radius b and thickness $h=b-a$. The mechanical properties are identical to the planar electrolyte. The solutions are similar to the planar case if plane strain (zero axial strain) axis-symmetric analysis is performed on the tube. The trace of the stress then reduces to the same form as the planar case and hence the solutions for concentrations and potential are similar. The vacancy concentration is given, by

$$\left(\rho_v(x)-\rho_v^I\right)\left[A_1\left(\rho_v(x)+\rho_v^I\right)+A_2\right]+\rho_A A_3\ln\left(\frac{\hat{J}_v\rho_A-2A\rho_v(x)}{\hat{J}_v\rho_A-2A\rho_v^I}\right)=2A^3\left(x-\hat{a}\right) \qquad (41)$$

where $\rho_v^I=\rho_v(\hat{a})$, $\hat{\phi}^I=\hat{\phi}(\hat{a})$, and $\hat{a}=a/h$.

The radial, hoop and axial stresses for this case are given by the following equations which are derived using standard linear elasticity.

$$\sigma_r=\frac{\bar{E}\eta_v\bar{\rho}_v}{(\hat{b}^2-\hat{a}^2)}\left(1-\frac{\hat{a}^2}{x^2}\right)-\frac{\bar{E}\eta_v}{x^2}\int_{\hat{a}}^x\xi\Delta\rho_v(\xi)d\xi-p,$$

$$\sigma_\theta=\frac{\bar{E}\eta_v\bar{\rho}_v}{(\hat{b}^2-\hat{a}^2)}\left(1+\frac{\hat{a}^2}{x^2}\right)-\bar{E}\eta_v\Delta\rho_v+\frac{\bar{E}\eta_v}{x^2}\int_{\hat{a}}^x\xi\Delta\rho_v(\xi)d\xi-p \qquad (42)$$

$$\sigma_z=\frac{2v\bar{E}\eta_v\bar{\rho}_v}{(\hat{b}^2-\hat{a}^2)}-\bar{E}\eta_v\Delta\rho_v-2vp,\bar{\rho}_v=\int_{\hat{a}}^{\hat{b}}\xi\Delta\rho_v(\xi)d\xi$$

p is the total pressure on inner and outer surfaces of the tube, and is taken to be 2atm.

RESULTS AND DISCUSSION

Planar Case

We conduct a parametric study for different values of η_v $((0\leq\eta_v\leq0.06)$, see Table2[5]).The plot showing the distribution of vacancy concentration and electrostatic potential for various values of η_v is shown in Figure 1. It can be seen that there is a strong interaction between self stress and diffusion, for the considered example. The material values used for the analysis are shown in Table 1. The mechanical properties correspond to 20% doped Gadolinium doped Ceria.

Table I

$D_e,D_v\left[C/(cm\ s)\right]$	ρ_A	ρ_v^0	$E(GPa)$	v	$h(cm)$	ψ^v,ψ^e (S cm^{-1})	$P_{O_2}^{II}/P_{O_2}^{I}$	ρ_v^I	T (C)
$-1.46,0.034$	0.2	0.1	107	0.31	0.1	0.118,0.248	10^{18}	0.22	900

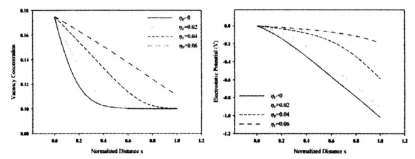

Figure I. Distribution of vacancy and electrostatic potential

For the plots in Fig. 1 applied voltage was zero and it can be seen that the boundary concentration and electrostatic potential at end *II* is affected by the values of η_v due to(40). It can be noticed that the influence of η_v on end *II* is to increase the concentration. This essentially reduces the ionic current since the value of the average gradient for a given voltage is lower.

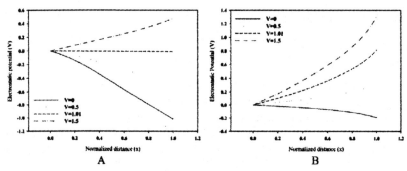

Figure II. Electrostatic Potential Distribution A) $\eta_v = 0$ **B)** $\eta_v = 0.06$

In Fig II, we illustrate the effect of η_v on the distribution of electrostatic potential. In FigIIA, the distribution is zero for an applied voltage of 1.01 volts. The voltage at which the potential vanishes within the electrolyte is reduced for higher values of η_v as can be seen from Fig IIB.

In the subsequent Figure, we show the influence of η_v on the open circuit voltage (OCV). For the values considered, the OCV was 0.87 (Fig. IIIA), for un-coupled condition. This is obtained by plotting the normalized current ($J_a h / \psi^\alpha$, where ψ^α is the conductivity of

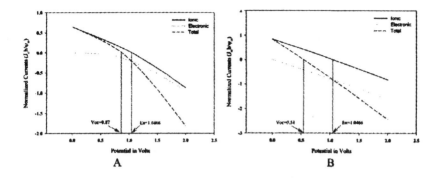

Figure III. Current Voltage characteristics A) $\eta_v = 0$ **B)** $\eta_v = 0.06$

species α) vs. the applied voltage. It can be seen that the voltage at which the total current vanishes (OCV) is reduced to 0.54 for $\eta_v = 0.06$. It can also be noticed that the voltage at which the ionic current vanishes (1.0466V) is independent of η_v. This can be deduced from,(39), (40) and (30) showing that for, $\hat{J}_v = 0$, $V = E_n = (RT/4F)\ln(P''/P') =$ Nernst Voltage .

Figure IV shows the distribution of stresses, within the planar electrolyte, for different values of applied voltages.

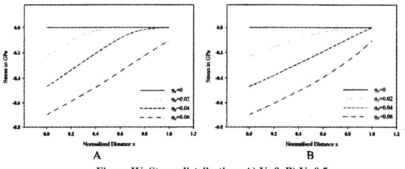

Figure IV. Stress distributions A) V=0, B) V=0.5

The stress values are compressive on the lower partial pressure side I which is concordance with the assumption that an increase in vacancy concentration causes expansion. The stress on end II increases from 0 to more compressive values. This is due to higher deviations from stoichiometry, due to the effect of η_v on concentrations.

Tubular Case

We show the distribution of hoop and axial stress in Figure V. The radial stresses were insignificant and hence is not shown. The plots were obtained for an applied voltage of 0.5 volts.

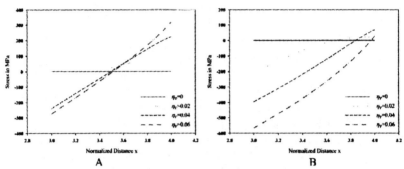

Figure V. A) Hoop Stress B) Axial Stress

CONCLUSION

A linear theory that couples stress and diffusion has been formulated and solved for ionic solids, by considering simple examples. The effect of self stress on the concentration and potential distribution has been demonstrated.

Acknowledgements

This work was partially supported by the US Department of Energy under contract No. DE-AC26-02NT41571).

REFERENCES

1. Krishnamurthy, R.; Sheldon, B. W., Stresses due to oxygen potential gradients in non-stoichiometric oxides. *Acta Materialia* **2004**, 52, (7), 1807.
2. Wang., Y.; Duncan, K. L.; Wachsman, E. D. In *The effect of lattice vacancy concentration on mechanical properties of fluorite related oxides*, 29th International Conference on Advanced Ceramics and Composites, ACerS, Cocoa Beach, FL, 2005; 2005.
3. Larche, F.; Cahn, J. W., Linear theory of thermochemical equilibrium of solids under stress. **1973**, 21, (8), 1051.
4. Yasuda, I.; Hishinuma, M. In *Electrical conductivity, dimensional instability and internal stresses of CeO2-Gd2O3*, Electro Chem Soc, 1998; 1998; pp 178-187.
5. Atkinson, A.; Ramos, T. M. G. M., Chemically-induced stresses in ceramic oxygen ion-conducting membranes. *Solid State Ionics, Diffusion & Reactions* **2000**, 129, (1-4), 259.
6. Liu, M., Distributions of charged defects in mixed ionic-electronic conductors. I. General equations for homogeneous mixed ionic-electronic conductors. *Journal of the Electrochemical Society* **1997**, 144, (5), 1813.

Electrodes

NANOSTRUCTURED CERAMIC SUSPENSIONS FOR ELECTRODES AND THE BRAZILIAN SOFC NETWORK "REDE PaCOS"

R. C. Cordeiro[1], G. S. Trindade[1], R. N. S. H. Magalhães[2], G. C. Silva[1], P. R. Villalobos[1], M. C. R. S. Varela[2], P. E. V. de Miranda[1].
1 – COPPE/UFRJ, Rio de Janeiro, Brazil
2 – UFBA, Salvador, Brazil

ABSTRACT

Important challenges for the development of electrodes for solid oxide fuel cells nowadays include the easiness of fabrication and industrialization, keeping the lower possible costs and improved electrocatalytic properties. The choice of screen printing processing may satisfy these requirements if special attention is given to structuring the ceramic suspension as a way to improve electrode electrocatalytic properties. In this paper an analysis is made of the ceramic suspensions available for that purpose and results will be shown concerning newly developed suspensions based on nanostructured nickel oxide and commercial YSZ and vehicles processed using planetary ball milling as a quick and simple way to produce nanostructured suspensions. Such colloids were used to obtain electrolyte supported green anodes that were fired and sintered to produce adequate porosity. The structure and morphology of these electrodes were characterized as well as their electrochemical performance. The powder synthesis, processing and characterization work were conducted using the facilities of laboratories that participate as members of the Brazilian network on solid oxide fuel cells, "Rede PaCOS". This network was created in 2004 with the objective to allow a synergistic advancement of this area in Brazil and includes today research groups from several universities, as well as state and private companies.

I – INTRODUCTION

An important experience has been accumulated in scientific developments and industrial production of ceramic components by the use of ceramic suspensions or inks, composed of a colloidal dispersion of thin ceramic powders in a liquid medium to which organic additives are mixed to guarantee important features associated to viscosity, homogeneity and stability. Nanostructured ceramic powders have been found to strongly emphasize good features of ceramic suspensions [1]. However, the smaller the inorganic particles the bigger the reactive surface area and stronger the forces that will appear and induce the particles to agglomerate and form aggregates [2]. These aggregates are detrimental to suspension stability; pose a processing problem that may lead to film inhomogeneities and a poor final microstructure of the electrode. The characteristics of the aggregates will depend on the particles initial size and morphology, as well as on the magnitude of the forces acting on them. For example, van der Waals capillary forces have the tendency to form weak aggregates, whereas chemical reactions, dryings and calcinations will rather give rise to strong aggregates [3]. This calls for the introduction of organic dispersants in the ceramic suspension, which will adsorb to the surface of the inorganic particles, impeding bigger size agglomerates to be formed. In that case, large stabilizing forces, such as electrical double-layer repulsion or steric interactions are applied to offer an energy barrier against aggregation. Various commercial dispersants are made based on texanol, therpineol, ethylcelulose and poly(ether-imide). It was shown that stabilizations of nanoparticles with polyethylenimine as a dispersant for ZrO_2 nanopowders has warranted better rheological results than electrostatic stabilization alone due to steric effects of the polymer [4].

Other organic constituents of the ceramic suspensions include solvents, plasticizers and binders. Several types of solvents may be used, that is, aqueous, low viscosity non-aqueous, with low evaporation rates, which include water, ethanol and therpineol, among others. Binders are added to promote adhesion of the ceramic suspension within the green body and to the substrate after evaporation of the solvent, as a way to impede the formation of cracks and defects, whereas plasticizers help the binders to improve flexibility and workability of the ceramic coating [5]. The ceramic suspensions are applied to the substrate to be subsequently dried, calcined and sintered for several applications[6-8], including solid oxide fuel cells (SOFC) [9, 10].

Several physical and chemical methods were developed for processing ceramic materials with the objectives of producing ceramic coatings[5]. These procedures, thoroughly used in early [11] and present [12] scientific works and industrial trials, require very sophisticated equipments and very specialized control of process parameters for the fabrication of electrolytes and electrodes. However, the easiness associated with making and applying ceramic suspensions have opened the way for simpler and cheaper processes that, apparently, are also more suitable for industrial mass production. Nowadays it is possible to find in the market several companies that commercialize ceramic suspensions ready to be used for the fabrication of functional ceramic coatings, such as electrodes for SOFC, as well as their components all alone, including ink vehicles, dispersants, other organic additives and the inorganic ceramic powders, so that distinctively featured home made ceramic suspensions may be made. Screen printing is a particularly interesting mechanical process that is commonly used for the fabrication of thick films in which a viscous ceramic suspension is pressed against the mesh of a screen to be deposited onto a substrate. This technique was initially developed for the electronic circuitry industry, but became very attractive for the fabrication of electrodes and electrolytes for SOFC, due to its low operation cost and because it enabled the production of homogeneous coatings with thicknesses as thin as 10 μm, depending on the aperture of the screen utilized [9, 13].

Nevertheless, some parameters must be well controlled to produce sound films that will be homogeneous and will have good adhesion to the substrate at the end of the processing procedure. It means that care has to be taken to optimize the characteristics of the ceramic suspension, of the screen, the preparation of the substrate and the rheology of the ceramic suspension [11]. The adequate preparation of the substrate is of utmost importance to improve the wetting ability of the ceramic suspension onto the substrate, which is affected by modification of its surface energy caused by physical aspects, such as roughness, to chemical ones, as the presence of impurities and grease.

Materials used for the fabrication of SOFC cathodes and anodes using ceramic suspensions must present some basic characteristics that are typical of this application. Materials for cathode should present a high catalytic activity for the reduction of oxygen, good compatibility with the electrolyte and high electrical conductivity [14]. Lanthanum strontium-doped manganite (LSM), lanthanum strontium-doped ferrite (LSF), lanthanum strontium and cobalt-doped ferrite (LSCF) and their composites with yttria-stabilized zirconia (YSZ) or gadolinium-doped ceria (GDC) present such properties and are commercially available in powder form or already within ceramic suspensions. Materials for anodes should bear important properties such as good catalytic activity for the fuel oxidation, good electrical conductivity, offering catalytic sites for the fuel reaction with the oxygen ions that transverse the YSZ or GDC electrolyte [15]. Several materials satisfy those requirements if processed into a composite ceramic-metal (cermet). NiO-YSZ and NiO-GDC cermets are the most commonly used and commercialized ones. They are also commercially available in powder form or already incorporated into ceramic suspensions possessing vehicles based on ethyl-cellulose, therpineol, dibutylftalate and buthylcarbitol, among others. Materials for cathode or anode must undergo comminution to be homogenized into the vehicle using ultra-sound or milling.

The present paper will focus on the use of ceramic suspensions for the fabrication and test of electrodes for solid oxide fuel cells, using materials considered to be conventional for this application, but which were fabricated and processed into ceramic suspensions in such a way to become nanostructured and highly homogenized. The work was developed by laboratories belonging to the Brazilian Network on Solid Oxide Fuel Cells, Rede PaCOS, which will also be described in the text.

II – EXPERIMENTAL PROCEDURES

II.1 Fuel Cell Fabrication

SOFC single cells were prepared by the use of ceramic suspensions and printed on 20 mm diameter, 500 µm thick 8%mol YSZ discs produced by Kerafol GmbH.

The NiO powder used to prepare the anode was obtained as follows. A nickel hydroxide sample was prepared by adding a concentrated ammonium hydroxide aqueous solution (28%) to a nickel nitrate ($Ni(NO_3)_2.6H_2O$) aqueous solution (1 M) at room temperature under stirring. The final pH was 9. The resulting solution was then heated at 90°C and kept under stirring at this temperature for 2 h. The sol was centrifuged (2000 rpm, 5 min.) and the gel was rinsed with water to remove the nitrate ions from the starting material. The sol was subsequently centrifuged, testing for the presence of the supernatant nitrate ions, by adding concentrated sulfuric acid and observing the formation of brown rings, until nitrate ions were no longer detected [16]. The steps of rinsing and centrifugation were repeated several times until the nitrate ions were not detected in the supernatant anymore. The gel thus obtained was dried in an oven at 120 °C for 24 h and heated at 10°C min.[-1] under air until 500 °C and kept at this temperature for 2 h.

Nickel oxide was mixed with 4.2 % mol YSZ (Melox5Y MEI Chemicals, mean particle size 0.5µm) and milled in a Retsch PM4 planetary mill at 100 RPM using 99.9% Al_2O_3 grinding bowls and balls (10mm diameter), first in a dry form and then in the presence of a Nextech vehicle to produce the anode ceramic suspension. Typically a 50% bowl load was used with a powder to grinding balls mass ratio of 0.4, the total milling time was 22 minutes. The amounts of NiO and YSZ used were 55%wt and 45%wt respectively (as to achieve 40%vol of Ni).

The ceramic suspension used to produce the cathode was an as-received Nextech Materials ltd. commercial one, containing LSM with a stoichiometry of $La_{0.8}Sr_{0.2}MnO_3$.

Thermal gravimetric analyses of the ceramic suspension vehicle and of the inorganic powders were obtained to design the sequence of heat treatments to be performed for the fabrication of the electrodes.

As usual, the sequence of fabrication of the single cell included first the preparation of the anode (sintering at 1300°C for 120 minutes) and then that of the cathode (sintering at 1100°C for 120 minutes), for then applying gold current collector (ESL Electro-Science 8884-G) and gold wire current conductors in both sides. Both cathode and anode covered an area of approximately 1.0 cm^2 on opposite sides of the 8%mol YSZ electrolyte disc, by the deposition of the respective ceramic suspension reached a thickness of less than 20 µm.

II.2 Raw Materials and Fuel Cell Physical and Morphological Characterization

The structure of raw materials and electrodes was characterized by room temperature X-ray diffraction, using Cu K_α ($\lambda = 0.15418$ nm) in a Shimadzu XRD 6000, with divergence and scattering $1°$ slits and a 0.3mm receiving slit, on a $\theta/2\theta$ step scan mode with a $0.02°$ pitch and 1.2s sampling time.

Morphological characterization was performed by scanning electron microscopy using a Jeol JSM 6460 LV scanning electron microscope to observe electrodes-electrolyte interfaces, particles sizes, electrodes porosity level and level of adherence of the porous electrodes to the dense electrolyte in fractured cells.

Contaminant level analysis was performed by X-ray fluorescence, using Rigaku Rix3100 equipment with an Rh source. Thermo gravimetric analysis was performed in a Rigaku TAS-100/TG 8110 in 13%O_2/N_2 atmosphere with a heating rate of $10°$C.min^{-1}.

II.3 Fuel Cell Testing

Figure 1 presents a schematic drawing of a detail of the experimental test bench showing the fuel cell test set up. The single fuel cell was positioned on top of an alumina tube with the anode facing the interior of the tube and connected to the fuel inlet system. The fuel cell was sealed to the alumina tube using a commercial Electro Science sealing (ESL Electro-Science sealing glass paste 4460), applied at room temperature and dried at 200°C before heating to the test temperature (950°C).

Room temperature water humidified fuel is introduced by a concentric inner alumina tube that gets close to the anode surface. Excess non-reacted fuel and water vapor are purged out through outlet tubing and eventually through a security valve if the internal pressure builds up. The cathode is exposed to the ambient air. Tests were performed at 950°C using hydrogen as a fuel, at a flow rate of 5 L h^{-1}. Upon introducing hydrogen onto the heated anode, the open circuit voltage reached a value greater than 1.0 V in a few minutes, but more than an hour was allowed to guarantee reduction of the anode before obtaining voltage versus current density and power density versus current density curves.

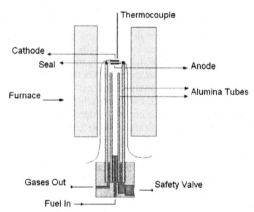

Figure 1 – Schematic representation of the fuel cell test set up.

III – RESULTS AND DISCUSSIONS

III.1 Raw Materials and Fuel Cell Physical and Morphological Characterization

The preparation of the ceramic suspension to fabricate the anode initiated through the precipitation of a nickel oxide powder by alkalinizing a nickel nitrate solution. Figure 2 presents the X-ray diffraction pattern, confirming that the powder so obtained is composed solely of the NiO phase with a cubic structure, as identified by comparison with a standard reference file (PCPDF # 44-1159).

It is well known that nickel oxides with different properties such as various degrees of crystallinity, particles sizes, morphologies and specific surface areas can be obtained by different preparative methods, which strongly affect these properties [17, 18]. During the formation of a precipitate, several factors can affect the final solid, like the metal concentration, pH, temperature, aging time, nature of anions and cations present and other topochemical transformations [19], as well as the temperature and the kind of atmosphere used in the calcination step. Therefore one can get, for example, nickel oxide with particle sizes varying from nanophases to millimeters, or with specific surface areas ranging from values as low as $1 \text{ m}^2 \text{ g}^{-1}$ to values as high as $150 \text{ m}^2 \text{ g}^{-1}$ [18], by controlling the pH, the solution concentration and the calcination temperature.

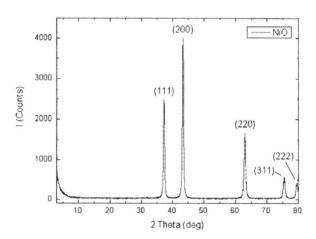

Figure 2 – XRD pattern for nickel powder as obtained by chemical precipitation.

In spite of the large applications of nickel oxide in fuel cells, agglomeration of the nickel oxide powder obtained is often noted, becoming an important drawback of the process. Agglomerates are not broken during compaction and lead to incomplete densification and strength-limiting processing flaws [20]. In this paper a controlled method to prepare weak-agglomerated nanocrystalline nickel oxide was used, overcoming the problem of agglomeration.

Scanning electron microscopy analysis of the NiO powder showed that it is composed of fairly spherical interconnected agglomerates with a diameter of about 7 μm (Figure 3a). However, it was found that each of these agglomerates is made up of very many tiny little rod-like particles with squared cross section, presenting a length of about 400 nm with the side of the squared cross section equal to about 180 nm, as depicted in Figure 3b.

Figure 3 – (a) SEM micrograph of NiO agglomerates produced by chemical precipitation. Original magnification: 12.000 X. (b) SEM micrograph shows that NiO agglomerates produced by chemical precipitation are made up of tiny rod-like particles with squared cross section. Original magnification: 30.000 X.

The aim of the next processing step is to further decrease the nickel oxide particle size and to improve the dispersion of these fine particles in the cermet by co-milling the nickel oxide and yttria stabilized zirconium oxide, first in a dry manner and then in the presence of a dispersant, resulting in a fine and homogeneous ensemble of nanostructured particles, following the procedure presented schematically in Figure 4.

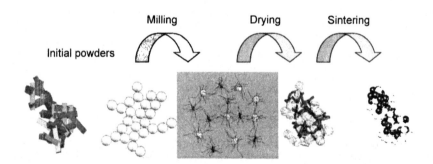

Figure 4 – Schematic representation of the controlled procedure used to prepare weak-agglomerated nanocrystalline nickel oxide – YSZ powder to fabricate ceramic suspensions for SOFC anode.

Figure 5 presents a scanning electron micrograph of the as-processed anode ceramic suspension showing the fine oxide particles supported by the dispersant. The latter is subsequently eliminated by the calcination and sintering routes used to produce the anode electrode.

Figure 5 – SEM micrograph of fine oxide particles supported by the dispersant in the as-processed anode ceramic suspension.

Planetary milling is a high energy procedure capable of transferring a substantial amount of energy to a sample compressed by the collision of hard grinding ball projected against the wall of a grinding bowl by the centrifugal forces created by its planetary movement around an axis [21]. The fragile agglomerates are broken down to the NiO nanoparticles with the resulting material being ejected in the dispersion media. The newly formed surfaces are stabilized in the presence of the dispersant components to produce a nanostructured anode suspension.

An important concern in this procedure is to keep a low level of contaminants as this process may carry away some of the milling bowl or grinding ball materials. Contaminant level was controlled by first conditioning the grinding bowl by exhaustive milling an anode powder preparation. Once the grinding surfaces were covered with the anode materials, the milling operations here described were performed in the mildest regime available in the equipment. X-ray fluorescent analysis showed that the aluminum oxide level present in the so obtained sample was around 0.032% mass/mass.

Figure 6 presents the thermal gravimetric results obtained in oxidizing atmosphere for the Nextech vehicle all alone and for the anode ceramic suspension prepared using the nickel oxide powder produced by chemical precipitation followed by milling in the presence of this vehicle.

These results were important to design the several stages and their heating rates for the heat treatment performed with electrolyte "painted" with the anode ceramic suspension in order to guarantee the formation of a SOFC anode with the desired level of porosity and adherence to the electrolyte.

Characterization of the thermal behavior of the dispersion shows that the mass loss occurs in two main stages, an abrupt initial one associated with the volatilization of the liquid media and consequent film formation, followed by a slower one, associated with the decomposition of the higher molecular mass process assistants, like polymers and binders. The final mass obtained after combustion of the dispersant above 600°C points to a suspension with a solid content of 61.21%.

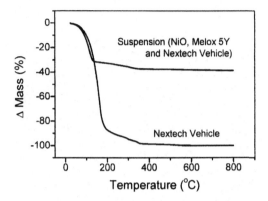

Figure 6 – Thermal gravimetric curves for the Nextech vehicle and anode ceramic suspension.

Figure 7a shows a cross section of an anode – electrolyte interface obtained after sintering the electrode. Figure 7b presents the morphology of a cross section of the cathode – electrolyte interface. In both cases, for the anode and for the cathode, oxide particles are well attached to the electrolyte, offering an adequate adherence for the fuel cell operation.

The comparison of Figures 3a, 3b with Figures 5 and 7a shows that the final morphology changed markedly if compared against the as produced NiO powder. No agglomerates could be found in the resulting electrode structure after sintering indicating that the milling process was effective in breaking up the large agglomerates present in the initial nickel powder.

Figure 7 – (a) SEM micrograph showing a cross section of the anode-electrolyte interface. Original magnification: 10.000 X. (b) SEM micrograph showing a cross section of the cathode-electrolyte interface. Original magnification: 10.000 X.

The squared cross-section rod-like particles found in the as produced NiO powder were fractured during the milling procedure and dispersed throughout the suspensions. Firing results in somewhat globular particles with partial coalescence, compatible with an adequate porous microstructure required for the electrode after the sintering heat treatment.

III.2 Fuel Cell Performance

The single SOFC was tested in humidified hydrogen at 950°C to explore the stability of the open circuit voltage (OCV) and that of the potential under load as function of time, as well as to characterize its general performance under load.

Figure 8 shows the operating performance of the fuel cell presenting the polarization curves. The open circuit voltage was equal to 1.1 V and the maximum current density obtained was equal to 16.38 mA cm^{-2}. A maximum power density equal to 3.32 mW cm^{-2} was reached.

The values for the maximum current density and for the maximum power density are small due to the very thick (500 µm) electrolyte used and also because of the use of a 4% mol YSZ powder to fabricate the anode. If current density values at maximum power are corrected for the effect of thickness, using conventional Fickian formulation [22], new maximum power density values would be obtained equal to 7.2, 32.8 and 424,1 mW.cm^{-2}, for electrolytes 300, 150 and 50 µm thick, respectively. These projections for the maximum power density for fuel cells possessing electrolytes with different thicknesses should still take into consideration the effect of the electric field, to be rigorous [23,24]. The fuel cell performance can be very much influenced by the adherence of the electrodes to the electrolyte and also by the activity at the triple phase boundary. It would be further enhanced by using more conductive zirconia powders in the anode cermet, as compared to the one available for this work, which had a low yttria content, and also by decreasing the electrolyte thickness.

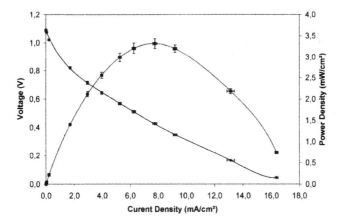

Figure 8 - Fuel cell performance under load, using humidified hydrogen at a flow rate of 5 L h^{-1}.

III.3 Rede PaCOS, the Brazilian Network on Solid Oxide Fuel Cells

The Brazilian Network on Solid Oxide Fuel Cells "Rede Cooperativa Pilha a Combustível de Óxido Sólido – Rede PaCOS" is one of the national networks belonging to the Brazilian Ministry of Science and Technology along with three others, namely "Rede PEM", "Rede Produção de Hidrogênio" and "Rede Engenharia e Integração", respectively, the national networks on polymer electrolyte fuel cells, on hydrogen production and on engineering and integration.

Rede PaCOS was created in March 2004, when its First Seminar was organized by and held at the Hydrogen Laboratory, Coppe-Federal University of Rio de Janeiro, with the purposes of disseminating information and fostering the implementation of cooperative public-private research, development and demonstration projects on solid oxide fuel cells and related themes in Brazil. The first actions were directed to consolidate the research and development groups already established in the country, for bringing about newer actors to the scene from public and private sectors and for publicizing the area.

Nowadays the following universities and research centers belong to Rede PaCOS, as indicated in Figure 9 and listed in Table II.

Figure 9 – Map of Brazil showing some of the institutions members of Rede PaCOS.

Table II – Main Activities of Rede PaCOS Research Groups in the SOFC Area

Research Group	Main Activities			
	raw materials and components	anode research	cathode research	fuel cell performance testing
Hydrogen Laboratory – COPPE/UFRJ http://www.labh2.coppe.ufrj.br Federal University of Rio de Janeiro	X	XXX	XXX	XXX
NUCAT – COPPE/UFRJ http://www.peq.coppe.ufrj.br/areas/nucat/old/ Federal University of Rio de Janeiro	XXX	-	-	-
LaMPaC – UFMG Materials and Fuel Cells Laboratory Federal University of Minas Gerais http://www.qui.ufmg.br/~lampac/	X	X	XXX	X
LAMAV - UENF State University of North Fluminense http://www.uenf.br/Uenf/Pages/CCT/Lamav/	XX	--	XXX	--
DEMA - UFSCar Federal University of São Carlos http://www.dema.ufscar.br	XXX	X	X	XX
Energetic and Nuclear Research Institute - IPEN http://www.ipen.br/cel/002_2.htm	XXX	X	X	X
Faculty of Science University of São Paulo State – Unesp http://www.fc.unesp.br	--	XXX	X	X
Federal University of Bahia – UFBA http://www.geccat.ufba.br http://www.quimica.ufba.br/main_pesquisa.html#labcat	XXX	XX	--	X
Federal University of Rio Grande do Norte – UFRN http://www.dem.ufrn.br http://www.ppgq.ufrn.br/php/index.php	XXX	--	--	--
Federal University of Maranhão – UFMA http://www.ufma.br/graduacao/quimica/index.php	XX	--	--	--

Collaboration was established between Rede PaCOS and SOFCNet, the European Solid Oxide Fuel Cell Network, since its creation, which resulted in the participation of an official representative from SOFCNet at the Rede PaCOS First Seminar. Privileged information exchange, contact between representatives and official staff meetings followed on to the participation of SOFCNet in Rede PaCOS's Second Seminar, on March 2005. New joint projects to be developed by members of the two networks are already under discussion.

An approximation was also done with the Japanese SOFC Society, resulting in a staff meeting held in Japan, the participation of the President of the Japanese SOFC Society at the Second Rede PaCOS Seminar in 2005 in Rio de Janeiro and the proposition for organizing a joint event on solid oxide fuel cells.

Rede PaCOS's actions are subdivided in two branches; one devoted to the diffusion of information about the area of interest, Rede PaCOS$_I$, and the other fostering the development of cooperative joint research projects among the different research groups working in this area in the country, Rede PaCOS$_{P\&D}$. Information is available in: http://www.redepacos.coppe.ufrj.br.

The Ministry of Science and Technology is investing about eight million dollars in 2006 on scientific projects within the above mentioned networks. A two-million-dollar, 30 month long, collaborative project initiated in 2006 is being jointly developed by laboratories belonging to Rede PaCOS, concerning the development and characterization of SOFC raw materials, components and single cells.

The Third Rede PaCOS Seminar was held in Salvador, Bahia, from March 19th to 22nd, 2006. The Fourth Seminar will also be held in the same city, from April 1st to 4th, 2007. Information is available in the network's web site at http://www.redepacos.coppe.ufrj.br.

V – CONCLUSIONS

The solid oxide fuel cell test bench developed to obtain the results presented in this paper has shown to be safe and reliable. The method of synthesis used to produce the nickel oxide powder coupled with the processing in a high energy planetary mill and subsequent heat treatments have been successful in fabricating anodes with adequate microstructure. This fabrication procedure turned out to be rapid and quite simple, opening new possibilities for facilitating the production of electrodes for SOFC. Further experiments are under way to use the procedures herein presented to enhance the electrocatalytic performance of the electrodes.

A solid oxide fuel cell network was created in March 2004, which has contributed to foster the development of this area in Brazil.

VI – ACKNOWLEDGEMENTS

The authors acknowledge Bianca Ferreira for the technical support given to this work, NUCAT – COPPE for performing thermal and x-ray fluorescence analyses, MEI Chemicals for providing YSZ powder, as well as *CNPq* (grants numbers 309174/2003-1. 500125/02-3, 500147/02-7, 400633/04-3 and 504869/04-3) and *FAPERJ* (grant number E-26/152.396/02) for the financial support granted.

VII – REFERENCES

[1]P. Bowen and C. Carry, "From powders to sintered pieces: forming, transformations and sintering of nanostructured ceramic oxides", *Powder Technology*, **128**, 248-55 (2002).

[2]A. J. Moulson and J. M. Herbert, "Processing of ceramics"; pp. 95-134 in *Electroceramics*. John Wiley & Sons Ltd., West Sussex, England. 2003.

[3]R. J. Pugh, "Dispersion and stability of ceramic powders in liquids"; pp. 127-192 in *Surface and colloid chemistry in advanced ceramics processing*. Edited by R. J. Pugh and L. Bergström. Marcel Dekker Inc., New York. NY, 1994.

[4]C. Duran, Y. Jia, Y. Hotta, K. Sato and K. Watari, "Colloidal processing, surface characterization, and sintering of nano ZrO_2 powders", *J. Mater. Res.*, **20** [5] 1348-55 (2005).

[5]H. Altenburg, J. Plewa, G. Plesch and O. Shpotyuk, "Thick films of ceramic, superconducting, and electro-ceramic materials", *Pure Appl. Chem.*, **74** [11] 2083-96 (2002).

[6]N. Ramakrishnan, P. K. Rajesh, P. Ponnambalam and K. Prakasan, "Studies on preparation of ceramic inks and simulation of drop formation and spread in direct ceramic inkjet printing", *Journal of Materials Processing Technology*, **169**, 372-81 (2005).

[7]H. Debeda-Hickel, C. Lucat and F. Menil, "Influence of the densification parameters on screen-printed component properties", *Journal of the European Ceramic Society*, **25** [12] 2115-19 (2005).

[8]K. Yao, X. He, Y. Xu and M. Chen, "Screen-printed piezoelectric ceramic thick films with sintering additives introduced through a liquid-phase approach", *Sensors and Actuators A: Physical*, **118**, 342-48 (2005).

[9]P. V. Dollen and S. Barnett, "A study of screen printed yttria-stabilized zirconia layers for solid xxide fuel cells", *Journal of the American Ceramic Society*, **88** [12] 3361-68 (2005).

[10]M. M. Seabaugh, S. L. Swartz, W. J. Dawson and B. E. McCormick. NexTech Materials Ltd. *Ceramic electrolyte coating methods*. US 6803138B2. 2004.

[11]V. Vej and J. Schoonman, "Thin-film techniques for solid oxide fuel-cells", *Solid State Ionics*, **57** [1-2] 141-145 (1992).

[12]Y. Liu , W. Rauch , S. Zha and M. Liu, "Fabrication of $Sm_{0.5}Sr_{0.5}CoO_{3-\delta}$-$Sm_{0.1}Ce_{0.9}O_{2-\delta}$ cathodes for solid oxide fuel cells using combustion CVD", *Solid State Ionics*, **166**, 261-68 (2004).

[13]D. Rotureau, J. P. Viricelle, C. Pijolat, N. Caillol and M. Pijolat, "Development of a planar SOFC device using screen-printing technology", *Journal of the European Ceramic Society*, **25** [12] 2633-36 (2005).

[14]H. Yokokawa and T. Horita, "Cathodes"; pp.119-147 in *High temperature solid oxide fuel cells: Fundamental, design and applications*. Edited by S. C. Singhal and K. Kendall. Elsevier Ltd., Oxford, UK, 2003.

[15]A. McEvoy, "Anodes"; pp.149-171 in *High temperature solid oxide fuel cells: Fundamental, design and applications*. Edited by S. C. Singhal and K. Kendall. Elsevier Ltd., Oxford, UK, 2003.

[16]F. Feigl and V. Anger, *Spot Tests in Inorganic Analysis*. Elsevier Ltd., Amsterdam, 1972.

[17]Y. C. Kang, S. B. Park and Y. W. Kang, "Preparation of high surface area nanophase particles by low pressure spray pyrolysis", *Nanostructured Materials*, **5** [7-8] 777-791 (1995).

[18]P. J. Anderson and R. F. Horlock, "Thermal decomposition of magnesium hydroxide", *Trans. Faraday Soc.*, **58**, 1993-2004 (1962).

[19]E. Matijevic and P. Sheiner, "Ferric hydrous oxide sols .3. preparation of uniform particles by hydrolysis of Fe(III)-chloride, Fe(III)-nitrate, and Fe(III)-perchlorate solutions", *J. Colloid Interf. Sci.*, **63** [3] 509-24 (1978).

[20]G. Li, X. Huang, Y. Shi and J. Guo, "Preparation and characteristics of nanocrystalline NiO by organic solvent method", *Materials Let.*, **51** [4] 325-30 (2001).

[21]R. Janot and D. Guérard, "Ball-milling in liquid media - applications to the preparation of anodic materials for lithium-ion batteries", *Progress in Materials Science*, **50**, 1–92 (2005).

[22]G. Petot-Ervas and C. Petot, "Experimental procedure for the determination of diffusion coefficients in ionic compounds – application to yttrium-doped zirconia", *Solid State Ionics*, **117**, 27-39 (1999).

[23]E. Siebert, G. Boureau, M. Mokchah, F. Millot and M. Chemla, "Analysis if transport phenomena in yttrium-doped zirconia in an oxygen chemical potential gradient and in short-circuit condition – reply to the discussion from G. Petot-Ervas and C. Petot", *Solid State Ionics*, **143**, 259-62 (2001).

[24]T. Arima, K. Fukuyo, K. Idemitsu and Y. Inagaki, "Molecular dynamics simulation of yttria-stabilized zirconia between 300 and 2000 K", *Journal of Molecular Liquids*, **113**, 67-73 (2004).

MODELING OF MIEC CATHODES: THE EFFECT OF SHEET RESISTANCE

David S. Mebane, Erik Koep and Meilin Liu
School of Materials Science and Engineering, Georgia Institute of Technology
771 Ferst Dr. NW
Atlanta, GA 30332-0245

ABSTRACT

The critical behavior of electrode polarization resistance versus feature height for patterned MIEC cathodes of constant surface area and TPB length suggests a competition between bulk ionic resistance and electronic resistance. A model taking both effects into account is derived and applied to a two-dimensional reduction of the problem. Testing the model using parameters that are realistic for an LSM-YSZ system reveals that the model structure does, in fact, replicate the critical behavior. This lends credence to the hypothesis of electronic-ionic competition as the source for this behavior, and provides a basis for further quantitative investigation of the oxygen reduction process using patterned electrodes.

INTRODUCTION

Patterned electrode experiments are very promising tools for quantitative analysis of reactions occurring at SOFC electrodes. Resistive second phases and variation due to impurities can be eliminated since the test cells are fabricated using low-temperature, atomic-level deposition techniques in a cleanroom environment. Close control of the electrode geometry is also possible, with such critical parameters as overall TPB length and total surface area known to a high degree of accuracy. In fact, it is possible to fabricate a broad range of geometries by closely controlling the pattern shape and feature size.[1, 2] The idea is that it then becomes easier to distinguish between different sources of resistance, such as that due to TPB vs. bulk reduction in cathodes.

In practice, quantitative analysis of some systems may suffer from additional noise introduced in patterns with especially high aspect ratios. Prominent among these is electronic resistance (or sheet resistance) in a patterned MIEC cathode. Sheet resistance becomes a factor in electrochemical measurements on MIECs when the breadth of the electronic transport pathway becomes very small compared to the distance between current collectors.

Normally, sheet resistance is neglected in phenomenological models of bulk MIEC behavior, since the conductivity of electronic species is typically much greater than that of ionic species. The fast movement of electronic species will then squelch any electric field that would influence ionic drift, and ionic transport becomes a process of ambipolar diffusion. Mathematically, the drift term is eliminated by assuming an infinite mobility for electronic species, leading to a zero gradient in electrostatic potential. However, when the transport path for ions becomes very short and broad compared to that for electronic species (as illustrated in Figure 1), the effect of resistance to electronic transport can become considerable. This is the case for the LSM patterns of very low height (< 1 μm) in ref. 2, where critical behavior was observed in interfacial polarization resistance (as determined from impedance spectra taken under open circuit conditions) with respect to decreasing pattern height. We assume that this critical behavior – a minimum in resistance at heights less than 1 μm – results from increasing activation of the bulk pathway at the top surface of the pattern as heights decrease, competing with decreasing activation of all reaction pathways due to sheet resistance at long distances from the current collectors.

If this assumption is correct, then taking advantage of very fine-scale patterned electrode geometries for analysis of oxygen reduction pathways on LSM requires an electrochemical model that does not assume infinite mobility of electronic species. This greatly complicates the mathematics, turning a linear diffusion equation into a coupled, nonlinear set of two second-order differential equations. However, the problem is still tractable from a standpoint of numerical methods of solution. In this paper, we will pose the problem and report on the solution for a two-dimensional version of a patterned cathode with sheet resistance. We will further demonstrate that the structure of the model does, in fact, allow for a drop in potential across the length of the pattern and a critical resistance with respect to pattern height at low heights for LSM.

Figure 1. A pattern with a high aspect ratio offers a greatly reduced transport pathway for vacancies compared to that for electronic species.

MODEL DEVELOPMENT

The geometry used for the model appears in Figure 2. This is essentially a cross-section of the patterned cathode taken along the length of the pattern. The figure demonstrates one of the salient assumptions of the model, which are the lack of a TPB reaction (indeed, there is no TPB in Figure 2) and the corresponding lack of surface diffusion. The system involves three domains: the electrolyte (bottom), the mixed-conductor (top) and a stationary, one-dimensional air-exposed surface domain along part of the top surface of the MIEC. A counter electrode forms the bottom boundary to the electrolyte. The top boundary of the MIEC not exposed to air is a current collector.

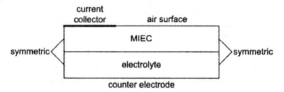

Figure 2. The model geometry.

In addition to these geometric assumptions, the model assumptions are:

1. The MIEC is a dilute solution of vacancies and holes, with an evenly distributed, negative stationary background charge.
2. Activity coefficients are unity.
3. Electroneutrality

4. Reversible counter electrode and current collector
5. Simple surface mechanism: dissociative adsorption and incorporation (specified below)

With these assumptions, the constitutive equations become:

$$N_i = -c_i u_i \nabla \overline{\mu}_i \tag{1}$$

where N is the molar flux, c is the molar concentration, u is the molar mobility and $\overline{\mu}_i$ is the electrochemical potential. The index i may refer to either vacancies or holes. The electrochemical potential is:

$$\overline{\mu}_i = \mu_i^0 + RT \ln c_i + z_i F \varphi \tag{2}$$

where μ_i^0 is the standard chemical potential, z is the valence number and φ is the electrostatic potential. Combining (1) and (2) leads to:

$$N_i = -RTu_i \nabla c_i - z_i Fu_i c_i \nabla \varphi \tag{3}$$

Note that (3) is nonlinear in c and φ. If we assumed infinite hole mobility, then we would be left with the condition that $\nabla \varphi = 0$. Equation (3) then becomes irrelevant for holes and reduces to Fick's 1st law for vacancies.

To define the problem, we have mass continuity:

$$\frac{\partial c_i}{\partial t} = \nabla \cdot N_i \tag{4}$$

and electroneutrality:

$$\sum_i z_i Fc_i + \rho_b = 0 . \tag{5}$$

The latter allows us to solve for the concentration of holes in terms of vacancy concentration and the background charge, ρ_b. For the MIEC, this results in a coupled, nonlinear parabolic-elliptic problem whose field variables are c_v and φ. In the electrolyte, there is only one mobile species. Electroneutrality declares that its concentration is constant throughout, which turns (3) into Ohm's law and (4) into Laplace's equation.

The simple boundary conditions are those at the counter electrode (fixed potential), the current collector (zero vacancy flux, fixed potential) and the symmetrical boundaries (zero flux for all species). More complex are boundaries where electrochemical reactions take place, such as the air surface and the MIEC-electrolyte interface. For the former, we have the simple adsorption-incorporation process referred to in the assumptions:

$$\frac{1}{2}O_2 + s \rightarrow O'_{ads} + h^\bullet \tag{6}$$

$$O'_{ads} + V_O^{\bullet\bullet} \rightarrow O_O^x + h^\bullet + s \tag{7}$$

where s is an adsorption site on the surface. The rate expressions for these reactions are:

$$r_{ads} = k_{ads}^0 \left[\frac{(1-\theta)}{(1-\theta_0)} \exp\left(\frac{-\alpha_{c1} F \Delta\chi}{RT} \right) - \frac{c_h \theta}{c_{h,0}\theta_0} \exp\left(\frac{\alpha_{a1} F \Delta\chi}{RT} \right) \right] \tag{8}$$

for (6), and

$$r_{mc} = k_{mc}^0 \left[\frac{c_v \theta}{c_{v,0}\theta_0} \exp\left(\frac{\alpha_{c2} F \Delta\chi}{RT} \right) - \frac{c_h (1-\theta)}{c_{h,0}(1-\theta_0)} \exp\left(\frac{-\alpha_{a2} F \Delta\chi}{RT} \right) \right] \tag{9}$$

for (7), where k^0 is the exchange rate, θ is the adsorbed site fraction, α terms are transfer coefficients, $\Delta\chi$ is the deviation of surface potential from equilibrium, and the subscript 0 indicates equilibrium values. (These expressions are derived from transition state theory, using Parsons's method of electrochemical potentials, in a recently submitted article.[3]) One may calculate $\Delta\chi$ using a Helmholtz model of the double layer and the concentration of oxide ions on the surface. At the MIEC-electrolyte boundary, we have vacancy transfer, described by a similar expression:

$$r_v = k_v^0 \left[\frac{c_{v,\text{YTE}}}{c_{v,\text{YTE},0}} \exp\left(\frac{2\alpha_{c3} F \Delta\chi}{RT} \right) - \frac{c_{v,\text{MIEC}}}{c_{v,\text{MIEC},0}} \exp\left(\frac{-2\alpha_{a2} F \Delta\chi}{RT} \right) \right] \tag{10}$$

For the moment, we are only interested in the steady-state (or periodic steady state). We could take the additional step of replacing time-dependent terms (or decaying transients in frequency space) with zero, but instead we chose to use a time-dependent method of solution. The principal reason for this is that it enables the use of a linear solver. The details will be presented in a subsequent paper; here it suffices to say that the method used for discretization and solution was a fully implicit finite volume method, using a line-by-line iterative solver. The method features second-order accuracy in space and is stable.

Values for the parameters were adapted from ranges given by Svensson for LSM.[4] Since Svensson derived her ranges from a survey of the literature on direct measurement techniques, we expect these values to be realistic, if not exactly accurate. The one parameter not taken from Svensson's list was the equilibrium concentration of holes, which was loosely based on the defect equilibria for LSM. More importantly, this value reflects the fact that holes are much more numerous in the MIEC than vacancies. Table I lists the parameters and values used in the study.

RESULTS

The vacancy distribution for the MIEC in a 1 μm^2 test case is shown in Figure 3. Note the gentle gradient moving toward the air surface. Note also that the concentration of vacancies is nearly one order of magnitude higher than the equilibrium value throughout the MIEC. In the absence of any surface path. this reduction of the MIEC bulk is required to drive the reaction at the air surface. The potential distribution is not shown, for at this geometry it is a nearly uniform -0.1 V.

Table I. Parameter values.

Parameter	Value
$c_{v,0}$	2.0×10^{-4} mol/m^2
$c_{h,0}$	30.0 mol/m^2
$\rho_{b, EL}$	-6.4×10^7 C/m^2
k^0_v	3.0×10^{-2} mol/m·s
k^0_{mc}	1.0×10^{-6} mol/m·s
k^0_{ads}	1.0×10^{-2} mol/m·s
θ_0	2.0×10^{-3}
T	1173 K
$u_{v, MIEC}$	1.0×10^{-12} mol·m^2/J·s
$u_{v, EL}$	1.0×10^{-12} mol·m^2/J·s
$u_{h, MIEC}$	1.0×10^{-11} mol·m^2/J·s
surface site concentration	1.0×10^{-3} mol/m
α	0.5

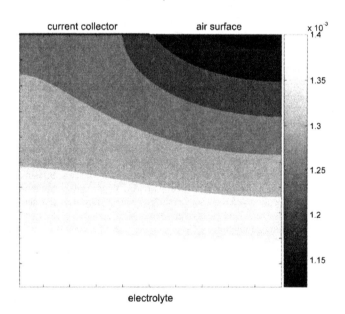

Figure 3. Vacancy distribution, in mol/m^2, for a 1 μm^2 MIEC under -0.1 V polarization.

At higher aspect ratios, the potential becomes nonuniform with distance from the current collector. Figure 4 illustrates that this is a sheet resistance phenomenon, as the potential flattens out asymptotically with increasing hole mobility. Potentials shown in the figure were measured across the top surface of an MIEC 1 μm high and 200 μm long.

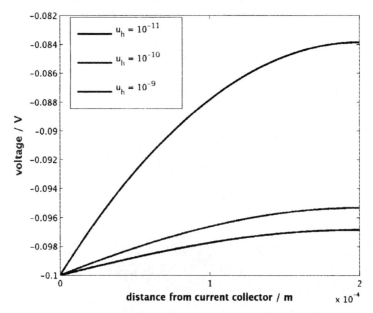

Figure 4. Potential versus distance from the current collector for a 1 μm high, 200 μm long pattern and three different values for molar mobility of holes.

The model output for different MIEC heights is displayed in Figure 5, with 5(a) the potential and (b) the vacancy distribution along the top surface. As the height decreases from 3 μm to 60 nm, the potential increases along the length of the pattern. The vacancy distribution, on the other hand, rises dramatically across the board between 3 and 0.6 μm, but then becomes very uneven at 60 nm. Since the overall activity of the cathode is proportional to the area under the vacancy distribution curve, this indicates critical behavior. Figure 6 – a plot of current versus height for several 200 μm long simulations – shows that the overall current does move through a maximum at heights below 0.6 μm.

CONCLUSION

A competition between sheet resistance and ionic resistance in patterned cathodes was simulated using a novel electrochemical model and finite volume discretization. Using model parameters realistic to LSM cathodes, it was demonstrated that the structure of the model does allow for the critical behavior seen in experimental data.

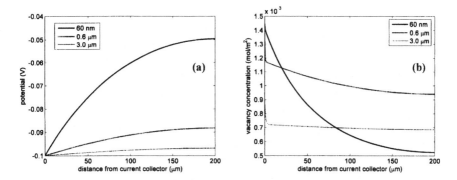

Figure 5. (a) Potential and (b) vacancy distributions for the top surfaces of 200 μm long patterns with different heights.

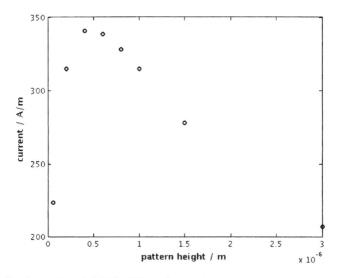

Figure 6. Total current vs. height for 200 μm long patterns.

ACKNOWLEDGEMENTS

The authors wish to thank the Department of Energy for support through NETL's SECA program, Grant no. DE-FC26-02NC41572. Dr. Yingjie Liu contributed to the development of the finite volume solver. The assistance of Rupak Das is also appreciated.

REFERENCES

[1]Koep E, Compson C, Liu ML, Zhou ZP, "A Photolithographic Process for Investigation of Electrode Reaction Sites in Solid Oxide Fuel Cells," *Solid State Ionics,* **176**, 1-8 (2005).

[2]Koep E, Mebane DS, Das R, Compson C, Liu ML, "The Characteristic Thickness for a Dense $La_{0.8}Sr_{0.2}MnO_3$ Electrode," *Electrochemical and Solid State Letters,* **8**, A592-5 (2005).

[3]Mebane DS, Liu, ML, "Classical, Phenomenological Analysis of the Kinetics of Reactions at the Gas-Exposed Surface of Mixed Ionic Electronic Conductors," *Journal Of Solid State Electrochemistry,* in press.

[4]Svensson AM, Sunde S, Nisancioglu K, "Mathematical Modeling of Oxygen Exchange and Transport in Air-Perovskite-Ysz Interface Regions.1. Reduction of Intermediately Adsorbed Oxygen," *Journal Of The Electrochemical Society,* **144**, 2719-32 (1997).

CATHODE THERMAL DELAMINATION STUDY FOR A PLANAR SOLID OXIDE
FUEL CELL WITH FUNCTIONAL GRADED PROPERTIES: EXPERIMENTAL
INVESTIGATION AND NUMERICAL RESULTS

Gang Ju and Kenneth Reifsnider
Department of Mechanical Engineering,
University of Connecticut, Storrs, CT, 06269

Jeong-Ho Kim
Department of Civil & Environmental Engineering,
University of Connecticut, Storrs, CT, 06269

ABSTRACT
        Solid oxide fuel cell in nature acts as a functional ceramic material for power
conversions. Three-layer formation of electrolyte, cathode and anode system, or five- layer
formation that is consisted of two more barrier layers between electrolyte and electrodes,
usually exhibits bimaterial properties with discontinuities in elastic modulus, coefficient of
thermal expansion (CTE), and thermal conductivity. Due to relatively lower sintering
temperature, cathode can delaminate during the in-service operation. 10 mol% scandia
stabilized zirconia (10ScSZ) and 10 mol% scandia stabilized zirconia (6ScSZ) electrolytes
were investigated for elastic properties, hardness and fracture toughness by nanoindentation
together with microindentation techniques. Interfacial energy release rate for lanthanum
strontium ferrite (LSF) cathode: 6ScSZ electrolyte was determined experimentally. 2-D
Thermal fracture of mechanically functional graded material (FGM) system including cathode
and electrolyte is studied numerically with considering spatially continuous and temperature
dependent material properties. The thermal stress intensity factors, and strain energy release
rate were compared for both conventional bimaterial system and functional graded material
experiencing uniform temperature change, as well as locally exponential varying temperature
change.

INTRODUCTION
        Solid oxide fuel cell (SOFC) acts as a candidate for power conversion tool under
active investigations. The success of ceramic components of it may arise from residual
stresses from manufacturing, thermal expansion coefficients, local temperature gradient, and
operation environments[1]. Functional graded material design for cathode is developed to avoid
mismatch of thermal expansion coefficients between cathode and electrolyte that could cause
delamination during thermal cycling, and to assist the electrochemical reaction for lowering
cathodic polarization[2-4]. Selçuk and Atkinson studied porosity dependence of elastic
properties for ceramic oxides in SOFC[5]. Sørensen and Primdahl investigated the failure
mechanisms for yttria stabilized zirconia (YSZ):NiO/YSZ multilayer[6]. They found increase in
the interfacial fracture energy with increasing sintering temperature. Sørensen later studied
numerically of thermally induced delamination for symmetrically graded multilayers[7]. He
calculated the steady-state energy release rate for symmetrically graded ceramic multilayers
and suggested that for asymmetric multilayers, delamination can be suppressed by inserting a
few additional interlayers. Fracture mechanics of FGM has been developed recently and
applied successfully to the thermal barrier coatings for gas turbines[8 10-12]. In this paper, elastic

properties of electrolytes (10ScSZ and 6ScSZ) and cathode (LSF) were measured via nanoindentation and microindentation. Fracture toughness of 10ScSZ and 6ScSZ was determined. Interfacial energy release rate of LSF:6ScSZ was also calculated. Finally, the strain energy release rate and stress intensity factor were calculated for bimaterial and FGM systems for uniform temperature change.

EXPERIMENTAL SETUP
The 10% mol scandia stabilized zirconia powder with the average particle size ~0.8 μm was hydrostatic pressed at room temperature to 90 Mpa. The pellets were then fired into three batches: 1400 ^{0}C, 1450 ^{0}C, and 1550^{0}C for two hours. The sintered pellets have geometry of 2.5 mm thick and 25mm in diameter. The symmetric cathode half cell under study has an electrolyte (6ScSZ) and two cathodes (LSF) on each side, where cathode layer and cathode reference of approximate 15~20 µm thick and 25 mm in diameter was coated on the 100 µm thick electrolyte substrate. Nanoindentation tests (MTS Nanoindenter XP) were performed to obtain the elastic modulus and toughness of the specimens. Microindentation tests were conducted for fracture toughness. All samples were mounted in epoxy and polished (Fig.1.a).

NUMERICAL MODELING
ABAQUS6.5 was used to study thermal fracture of mechanically functional graded material (FGM) system including cathode and electrolyte with considering spatially continuous and temperature dependent material properties. Experimental results were used as the material property inputs in the model. The thermal stress intensity factors, and strain energy release rate were compared for both conventional bimaterial system and functional graded material experiencing uniform temperature change, as well as locally exponential varying temperature change.

EXPERIMENTAL RESULTS
Fig.1.b shows the cathode delamination after cyclic thermal loading of 6ScSZ:LSF. Although elastic modulus is not strongly grain size dependent, it is porosity sensitive. Porosity dependence of elastic modulus is studied against porosity of specimens. Fig.2 shows the microstructure of 10ScSZ sintered at difference temperatures and 6ScSZ sintered at 1300^{0}C. Some portion of pores are present both intra-grain and inter-grain for 10ScSZ specimens. There are only inter-grain pores existed for 6ScSZ.

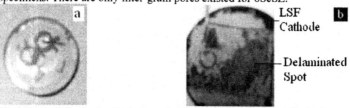

Fig.1. Mounted samples for indentation (a) and delaminated cathode (b).

specimens. There are only inter-grain pores existed for 6ScSZ. Image analysis software, ImageJ, is used to compute the elastic modulus based on composite sphere model (CSM)[5]:

$$E = E_0 \frac{(1-\varepsilon)^2}{(1+b_E\varepsilon)}$$

$$v = 0.25 \frac{4v_0 + 3\varepsilon - 7v_0\varepsilon}{1 + 2\varepsilon - 3v_0\varepsilon} \tag{1}$$

Where, E is elastic modulus, E_0 is fully dense specimen modulus, here is approximated by 10ScSZ1550, ε is porosity, v is Poisson ratio, and $b_E = 2 - 3v_0$. 10ScSZ specimens moduli are fitted well by equation (1), with Young's modulus at room temperature of 208 ± 12 GPa, 214 ± 10 GPa, and 222 ± 15 GPa for 10ScSZ1400, 10ScSZ1450 and 10ScSZ1550, respectively. Young's modulus of 6ScSZ is 195 ± 20 GPa. Young's modulus for porous LSF cathode is 45 ± 2.4 GPa (Table.I).

Fig.3 is indent resulting from nanoindentation. Elastic modulus and hardness were obtained from the load-displacement curve (Fig.4) upon unloading. Since the indent is relatively small (Fig.3), crack was not emanated from it. Fig.5.a and Fig.5.b are microindentation results. We can clearly see that the crack is evident from microindentation. By measuring the crack length, fracture toughness is obtained[13]:

$$K_{IC} = \alpha \left(\frac{E}{H} \right)^{1/2} \left(\frac{P}{a^{3/2}} \right) \tag{2}$$

where α is an empirical constant according to the geometry of the indenter, for Vickers indenter α is 0.016. E is the elastic modulus, H is the hardness, a is the crack length and P is the indentation load. The average value for 10ScSZ1400, 10ScSZ1450, 10ScSZ1550, and 6ScSZ are 1.56 ± 0.04 MPa m$^{1/2}$, 1.38 ± 0.035 MPa m$^{1/2}$, 1.3 ± 0.042 MPa $m^{1/2}$, and 0.4 ± 0.056 MPa m$^{1/2}$, respectively (Table I). Combined with Fig.2, Table I also demonstrates porosity dependence of Young's modulus, and grain size dependence of fracture toughness.

Table I Tested fracture toughness and hardness for 10ScSZ and 6ScSZ

	Young's Modulus, E, (Gpa)	Fracture Toughness, K_{IC}, (MPa m$^{1/2}$)
10ScSZ1400	208 ± 12	1.56 ± 0.04
10ScSZ1450	214 ± 10	1.38 ± 0.035
10ScSZ1550	222 ± 15	1.3 ± 0.042
6ScSZ	195 ± 20	0.4 ± 0.056

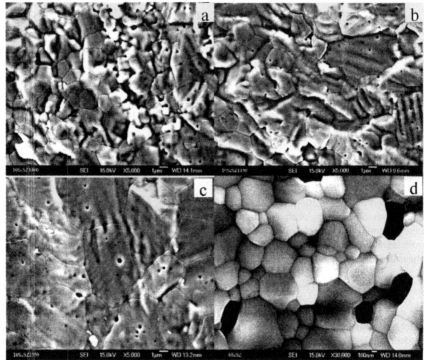

Fig.2. FESEM micrograph of the 10ScSZ sintered at 1400 ^{0}C (a), 1450 ^{0}C (b), 1550 ^{0}C (c), and 6ScSZ.

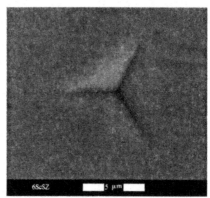

Fig.3. ESEM micrograph of 6ScSZ nanoindent.

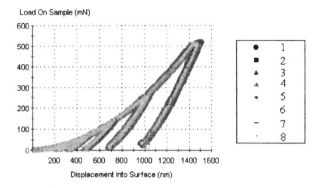

Fig.4. Nanoindentation load-displacement curve.

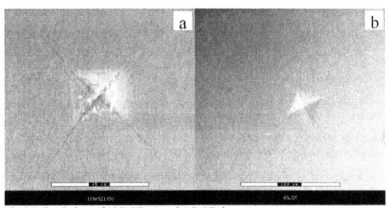

Fig.5. Microindent of 10ScSZ (a) and 6 ScSZ (b).

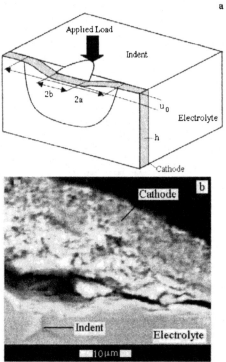

Fig.6. Schematic of cross-sectional nanoindentation (CSN) test configuration (a) and ESEM micrograph of cross section view of CSN test results.

Cross-sectional nanoindentation (CSN) technique is applied to obtain interfacial energy release rate. According to it, a Berkovich indenter normal to th··· ··nd at a distance d to the interface of interest is impressed in to the electrolyte. Cross section delamination for layered structure is achieve by indenting the cross section of the electrolyte up to a load at which indentation cracks form and propagate to the interface of cathode and electrolyte (Fig.6). For ceramic-ceramic interface. an analytical model was developed to calculate interface energy release rate, G_{ic}, in terms of maximum film deflection, u_0, geometry, Young's modulus, and Poisson's ratio of the thin film[14-15],

$$G_{ic} = \frac{D(1-v^2)u_0^2}{(a-b)^4}(1-\lambda)^4(2F+F') \qquad (3)$$

$$F(\lambda) = \frac{2\ln\lambda + \frac{1+\lambda}{1-\lambda}\ln^2\lambda}{[(1+\lambda)\ln\lambda + 2(1-\lambda)]^2} \text{ , with } F' = dF/d\lambda \qquad (4)$$

where, D is bending stiffness of the thin film plate, $D = Eh^3 \Big/ 12(1-v^2)$, a and b denote the radii of delamination and contact, respectively. $\lambda = b/a$. Calculated interfacial energy release rate for 6ScSZ:LSF is 1.84 J/m^2.

MODELING RESULTS

Model Setup and Verification
   In this study, interfacial edge crack subject to a uniform temperature change is considered (Fig.7). Models for bimaterial and FGM of uncracked semi-infinite strips are built and verifications against Erdogan and Wu's analytical results are made[8,16]. Fig.8 and Fig.9 are the meshes and simulated results of tensile loading and thermal loading compared to analytical results with following material properties following exponentially graded relation:
$E_2 / E_1 = 2, 5, 10, 20$, $\alpha_2 / \alpha_1 = 2$. $k_2 / k_1 = 10$. $T(x=0) = T_1 = 0.05T_0$, and
$T(x=W) = T_2 = 0.5T_0$ (Fig.8). The tensile stress is normalized against σ_t, applied tensile stress. Normalizing thermal stress is $\sigma_0 = E_1 \alpha_1 T_0 /(1-v)$ that is a conventional quantity for thermal loading under plane strain.

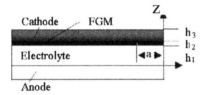

Fig.7. Schematic of edge crack at the cathode and electrolyte interface with a FGM layer.

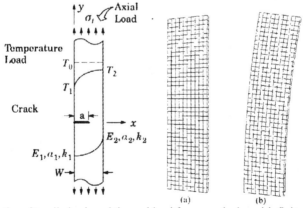

Fig.8. Schematics of tensile loads and thermal load for uncracked semi-infinite strip. finite element mesh (a) and tensile loading (b).

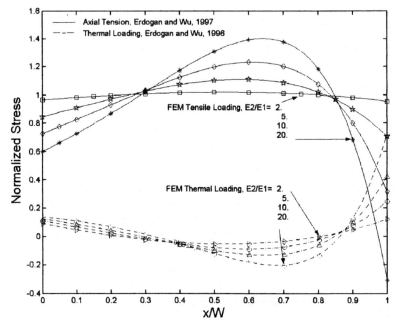

Fig.9. Erdogan and Wu analytical results (solid and dashed lines) for $\sigma_{yy}(x)$ and finite elements results (symbols)[16].

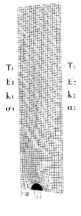

Fig.10. FGM specimen under axial tension with $a/W = 0.4$, W is width, and $E_2/E_1 = 10$ and under thermal loading with $a/W = 0.5$, $E_2/E_1 = 10$, $\alpha_2/\alpha_1 = 2$, $k_2/k_1 = 10$, $T(x = 0) = T_1 = 0.05T_0$, $T(x = W) = T_2 = 0.5T_0$.

Fig.10 shows mesh generated for cracked specimen under axial tension loading and thermal loading. Path independent domain integrals (J-values) at crack tip is computed from ABAQUS. Stress intensity factors K_I-values are converted from J as:

$$K_I(s) = (J(s)E^*(s))^{1/2} \tag{5}$$

Where $E^*(s) = E(s)/(1-v^2)$ for plane strain, $E^*(s) = E(s)$ for plane stress, and $E(s)$ is local Young's modulus value at crack front. For ease of comparison to Erdogan and Wu's analytical solution, K_I-values are normalized:

$$K_{In} = \frac{K_I}{\sigma_t \sqrt{\pi a}} \tag{6}$$

Where σ_t is the applied tensile stress, and a is crack length. From FEM calculation, K_{In} is 1.582 that is -0.37% different from analytical result.

For thermal loading, a common normalization of K_I-values is:

$$K_{In\_T} = \frac{K_I}{E_1' \alpha_1 T_0 \sqrt{\pi a}} \tag{7}$$

Where a is crack length, T_0 is initial temperature, α_1 is coefficient of thermal expansion at crack front, $E_1' = E_1/(1-v)$ for plane strain, and $E_1' = E_1$ for plane stress, where E_1 is crack front Young's modulus. The FEM obtained stress intensity factor K_{In_T} in this study with material properties: $a/W = 0.5$, $E_2/E_1 = 10$, $\alpha_2/\alpha_1 = 2$, $k_2/k_1 = 10$, $T(x=0) = T_1 = 0.05T_0$, $T(x=W) = T_2 = 0.5T_0$ is 0.0407 that is -0.7% different from analytical results.

Numerical Results

Thermal stress intensity factors K_I and K_{II}, and strain energy release rate G_{tot} are studied for bimaterial systems and FGM systems with following materials properties in Table1.The following relations are then applied to FGM layer to represent the thickness variation of the thermoelastic parameters:

Table.II. Material constants at 25^{0}C and 1000^{0}C

Material	E(GPa)	$\alpha\ (\times 10^{-6}\ K^{-1})$
Electrolyte: 6ScSZ[a,17,18]	188	10.8
	165	12.8
Cathode: LSF[a,17]	45	12.3
	25	17.6
Anode: Ni|YSZ[17,18]	50	12.8
	30	15.3

[a]Table 1

$$E(z,T) = \begin{cases} E_s(T), for(0 \leq z \leq h_1) \\ E_c(T), for(h_1 + h_2 \leq z \leq h_1 + h_2 + h_3) \\ E_c(T)+(E_s(T)-E_c(T))(\dfrac{h_1+h_2-z}{h_2})^{n_1}, for(h_1 \leq z \leq h_1 + h_2) \end{cases} \quad (9)$$

$$\alpha(z,T) = \begin{cases} \alpha_s(T), for(0 \leq z \leq h_1) \\ \alpha_c(T), for(h_1 + h_2 \leq z \leq h_1 + h_2 + h_3) \\ \alpha_c(T)+(\alpha_s(T)-\alpha_c(T))(\dfrac{h_1+h_2-z}{h_2})^{n_2}, for(h_1 \leq z \leq h_1 + h_2) \end{cases} \quad (10)$$

where, $n_1 = n_2 = 3$

Fig.11 is the calculated Thermal stress intensity factors K_I and K_{II}, and strain energy release rate G_{tot} with the normalization:

$$K_0 = E_s(T)\alpha_s(T)T_0\sqrt{\pi a} \quad (11)$$

$$G_0 = (1-v_s^2(T_0))\frac{K_0^2}{E_s(T_0)} \quad (12)$$

Where, E_s, α_s, v_s are elastic modulus, thermal expansion coefficient, and Poisson' ratio of electrolyte, and a is crack length. From Fig.11 with temperature independent material properties, one can see that by replacing homogenous coatings by FGM layers would both enhance the bonding strength and reduce the driving forces at the crack tip. Also, Because of self-equilibrating stress state, the stress intensity factors K_I's, and strain energy release rate G_{tot} tends to be zero at r=0 and r=W. In Fig.12. temperature dependent material properties are studied. It suggested in Fig.12 that compared to temperature independent case, the influence of the temperature dependence is significant for bimaterial case. Also due to self-equilibrating stress state, the stress intensity factors K_I's, and strain energy release rate Gtot tends to be zero at r=0 and r=W. Finally, the mode mixity for FGM is changed.

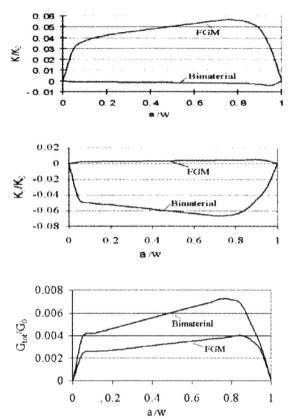

Fig.11. Normalized stress intensity factors and strain energy release rate for interface cracks of bimaterial and FGM systems under uniform temperature change with temperature independent material properties: $E = E(z), \alpha = \alpha(z)$.

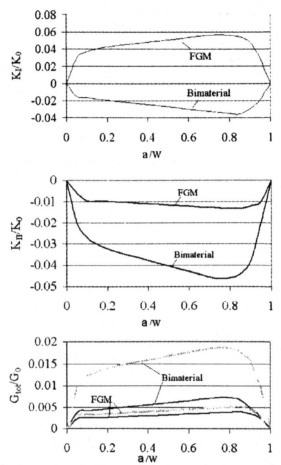

Fig.12. Normalized stress intensity factors and strain energy release rate for interface cracks of bimaterial and FGM systems under uniform temperature change with temperature independent material properties: $E = E(z,T), \alpha = \alpha(z,T)$.

CONCLUSIONS

ScSZ electrolytes sintered at different temperatures exhibit porosity dependence of Young's modulus, and grain size dependence of fracture toughness. Larger grain size has lower fracture toughness. Cross-sectional nanoindentation technique is applied to obtain interfacial energy release rate of 6ScSZ:LSF.Numerical modeling results for bimaterial and FGM systems show that temperature dependence material properties can alter strain energy release rate significantly. To make interface as graded material can both enhance the bonding strength and reduce the driving forces at the crack tip.

REFERENCES

[1]A.Atkinson and A.Selçuk, "Mechanical Behavior of Ceramic Oxygen Ion-conducting Membranes", *Solid State Ionics*, 134, 59-66 (2000).

[2]C.Xia, W.Rauch, et al., "Functionally Grades Cathodes for Honeycomb Solid Oxide Fuel Cells", *Electrochemical and Solid-State Letters*, 5, A217-A220 (2002).

[3]Y.Liu, C.Compson, and M.Liu, "Nanostructured and functionally graded cathodes for intermediate temperature solid oxide fuel cells ", *J. Pow. Sources*, 138, 194-198 (2004).

[4]S.Zha, Y.Zhang, and M.Liu, "Functionally graded cathodes fabricated by sol-gel/slurry coating for honeycomb SOFCs ", *Solid State Ionics*, 176, 25-31 (2005).

[5]A. Selçuk and A. Atkinson, "Elastic Properties of Ceramic Oxides Used in Solid Oxide Fuel Cells (SOFC) ", *J. Euro. Ceram. Soc.*, 17, 1523-1532 (1997).

[6]B. Sørensen and S. Primdahl, "Relationship between Strength and Failure Mode of Ceramic Multilayers", *J. Mater. Sci.*, 33, 5291-5300 (1998).

[7]B. Sørensen, "Thermally Induced Delamination of Symmetrically Graded Multilayers", *J.Am.Ceram.Soc.*, 85, 858-864 (2002).

[8]F. Erdogan, and B.H.Wu, "Crack Problems in FGM Layers Under Thermal Stresses", *J. Therm. Stresses*, 19, 237-265 (1996).

[9]J.H. Kim, and G.H. Paulino, "T-stress, Mixed-mode Stress Intensity Factors, and Crack Initiation Angles in Functionally Graded Materials: A Unified Approach Using the Interaction Integral Method", *Comput. Methods Appl. Mech Engrg.*, 192, 1463-1494 (2003).

[10]N. Konda, and F.Erdogan, "The Mixed-mode Crack Problem in a Nonhomogeneous Elastic Medium", *Engrg. Fracture Mech.*, 47, 533-545 (1994).

[11]G. W. Schulze and F.Erdogan, "Periodic Cracking of Elastic Coatings ", *Intl. J. Solids & Struct.*, 35, 3615-3634 (1998).

[12]T. Chiu and Fazil Erdogan, "Debonding of Graded Coatings Under In-plane Compression ", *Intl. J. Solids & Struct.*, 40, 7155-7179 (2003).

[13] B. R.Lawn, A. G. Evans, and D. B. Mashall, "Elastic/plastic Indentation Damage in Ceramics: the Median/radial Crack System" .*J. the Am. Ceram. Soc.*, 63, 574-589. (1980).

[14]M.R.Elizalde, et al., "Interfacial Fracture induced by Cross-sectional Nanoindentation in Metal-Ceramic Thin Film Structures", *Acta Mater.*,51, 4295-4305 (2003)

[15]J. M. Sánchez, et al., "Cross-sectional Nanoindentation: A New Technique for Thin Film Interfacial Adhesion Characterization ", *Acta Mater.*,47, 4405-4413 (1999)

[16]F. Erdogan, and B.H.Wu, "The Surface Crack Problems for a Plate with Functionally Graded Properties", *ASME J. Appl. Mech.*,64, 449-456 (1997)

[17]D.Lee, et al., "Characterization of Scandia Stabilized Zirconia Prepared by Glycine Nitrate Process and its Performance as the Electrolyte for IT-SOFC", *Solid State Ionics*, 176, 1021-1025 (2005)

[18]C.Y.Park, and A.J.Jacobson, "Thermal and Chemical Expansion Properties of LSFTi" *Solid State Ionics*, 176, 2671-2676 (2005)

ELECTROCHEMICAL CHARACTERISTICS OF Ni/Gd-DOPED CERIA AND Ni/Sm-DOPED CERIA ANODES FOR SOFC USING DRY METHANE FUEL

Caroline Levy, Shinichi Hasegawa, Shiko Nakamura, Manabu Ihara[1]
[1] Research Center for Carbon Recycling and Energy, Tokyo Institute of Technology,
2-12-1 Ookayama, Meguro-ku,
Tokyo, 152-8852, Japan.

Keiji Yamahara[2]
[2] Mitsubishi Chemical Corporation,
Dai-ichi Tamachi Building, 33-8, Shiba 5-chome, Minato-ku,
Tokyo, 108-0014, Japan

ABSTRACT

This research was carried out on the study of SOFCs working with hydrogen (H_2) and dry methane (CH_4) fuels with a cermet anode of nickel (Ni)/fluorite type material as Gadolinium doped ceria (GDC) and Samarium doped ceria (SDC) to compare with Ni/Yttria-stabilized zirconia (YSZ) anode at 900°C. We observed that the performances with Ni/GDC and Ni/SDC anodes obtained with H_2 and dry CH_4 were better than those obtained with Ni/YSZ anode. After degradation tests, the SOFC with Ni/GDC anode conserved high performances, while Ni/YSZ anode was degraded. The degradation of the SOFCs in dry CH_4 fuel was accelerated by the interruption of the current. Thus, Ni/ doped ceria anodes had similar behaviour and could be suitable candidates as anodes in SOFC with dry CH_4 fuel.

INTRODUCTION

Hydrogen (H_2) is usually used as a convenient fuel in solid oxide fuel cells (SOFCs) [1]. The technological drawbacks related to the hazardous character of H_2 storage, oriented some research towards cells functioning with other possible fuels, i.e. hydrocarbons such as methane (CH_4) [2,3]. The reaction of H_2 production issued from methane takes place into the anodic compartment. The catalytic performances for the transformation of the CH_4 into H_2 thus depend on anodic material. The steam reforming of methane is the classical route for H_2 production [4], but considering the technological difficulty of this process we therefore focused our research on the use of dry methane as a possible fuel. However, cokage, i.e. carbon deposition, of the anode under CH_4 conditions is the essential chemical problem that occurs and it leads to the final degradation of the catalytic activity of the anode. Ni/YSZ anodes are usually described in SOFCs applications [5], but they can not stably operated because of the carbon deposition. It is then also necessary to find new anode materials which allow operators to work under dry methane fuel conditions with the minimum of degradation [6]. This paper reports recent studies of a SOFC working with dry methane fuel versus various anode materials. The used anodes are a cermet of nickel (Ni) and a ceria based oxide, doped with Gadolinium (GDC) or with Samarium (SDC). The performances of the cells will be compared with those obtained with the conventional Ni / yttria-stabilized zirconia (YSZ) anodes. The durability of Ni/GDC and Ni/YSZ anodes were studied using adapted degradation tests.

EXPERIMENTAL

Fabrication of the cells

The cell is composed by an electrolyte disc support of 8-mol%-Y_2O_3-ZrO_2 (YSZ, 20mm diameter, 0.3mm thickness), a porous film of $La_{0.85}Sr_{0.15}MnO_3$ (LSM) as the cathode (Figure 1b) and a cermet anode of nickel (Ni)/fluorite type material as doped ceria (Gadolinium and Samarium 20mol%-doped Ceria (GDC, SDC)) to compare with Ni/yttria-stabilized zirconia (YSZ) (Figure 1a). The powder of SDC was synthesized by the glycine-nitrate process (GNP), a self-combustion method using glycine as fuel and nitrates of the metal components as oxidant. The GNP leads to obtain small particles. The anode paste was made by mixing NiO powder and SDC, GDC or YSZ powders with α-terpineol solvent by ball milling. Ethylcellulose powder was then added in the paste to increase its viscosity and the porosity of the anode. The electrodes were deposited by coating with a mask on both faces of the electrolyte with the paste of synthesized materials. The anodes were treated at 1300°C and the cathode at 1200°C for 4h. The thickness of the electrodes are 35~50μm for the anodes and ~15μm for the cathode respectively.

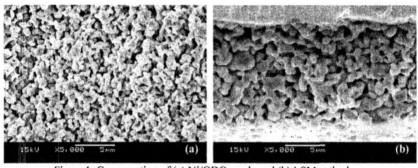

Figure 1: Cross-section of (a) Ni/GDC anode and (b) LSM cathode.

Power generation experiments of SOFC using Ni/YSZ, Ni/SDC and Ni/GDC anodes

The power generation experiments were carried out at 900°C under dry CH_4 at a feed rate of 50ccm. To compare the power generation at 50ccm of dry CH_4, the SOFCs had been also tested with 200ccm of H_2. Pure O_2 was supplied in the cathodic compartment at a feed rate of 60ccm. The characteristics of the anodes have been evaluated, using the current interruption method. The durability of the cells with Ni/GDC and Ni/YSZ anodes was evaluated by the degradation tests. As first degradation test applied to the cells with Ni/GDC and Ni/YSZ anodes, one cycle for the test was the current interruption during 10 minutes and the successive I-V measurements. 20 cycles were carried out. As second degradation test applied to the cell with Ni/GDC anode, the terminal voltage was measured during the interrupting of current for 30 minutes and the power generation for 1h at current density of 300mA/cm². The operation was repeatedly carried out 5 times.

RESULTS AND DISCUSSION

The initial performances of Ni/GDC and Ni/SDC cells were better than those of Ni/YSZ cell as reported on the I-V curves (Figure 2). The performances obtained with H_2 were found better than those obtained with dry CH_4 for SOFC with Ni/YSZ and Ni/GDC and Ni/SDC anodes. The anodic overpotential with dry methane was higher than that with hydrogen fuel whatever the anode material.

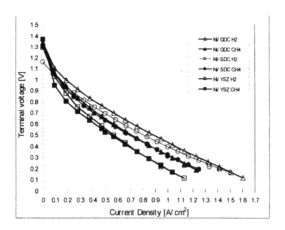

Figure 2: I-V characteristics of SOFC with Ni/YSZ (60/40wt%), Ni/SDC (55/45wt%) and Ni/GDC (55/45wt%) anodes with H_2 and dry CH_4 at same fuel utilization.

The relationship between the current density (i) and the oxygen activity (a_O), calculated from the ohmic-resistance free anode potential vs. 1atm O_2 (E_a') as $a_O = exp(2FE_a'/RT)$, was shown in a logarithmic graph (Figure 3). The slope of the logarithmic curves of i= f(a_O), was varied as function of the anode type. For Ni/YSZ anode, the slope was equal to about 1. In contrast, the slopes for Ni/GDC and Ni/SDC anodes were about 0.5 and 0.6 respectively whatever the gas fuel. We assumed that the reaction was a typical Langmuir reaction. A model, representing the competitive adsorption mechanism on Ni/YSZ anode in H_2 was previously established [7]. From this model, we could qualitatively estimate the coverage of oxygen at the triple-phase boundary (TPB) on the anode. The slope of ~1 for Ni/YSZ indicates that the reaction rate of the rate determining reaction on the anode is proportional to the a_O. The smaller slope than 1 indicated that the coverage of oxygen at the TPB was possibly higher than that on Ni/YSZ anode Therefore, the coverage of oxygen at the TPB on Ni/GDC and Ni/SDC anode was higher than that on Ni/YSZ anode. This could be attributed to the kind of the oxide material. Indeed, SDC and GDC oxides are mixed ionic-electronic conductors by applying a difference of oxygen partial pressure, as it is on the anodic compartment, whereas YSZ oxide is only ionic conductor. Thus, the surface exchange for the gas on the anode for Ni/SDC and Ni/GDC is carried out on the extended area of the surface of the oxide grains around TPB, while it is only at the TPB, which is at the contact between Ni, YSZ and the gas for the Ni/YSZ anode. Moreover, the difference between the lines in H_2 and dry CH_4 with the Ni/YSZ anode in the figure 3 was larger than those with the Ni/GDC or Ni/SDC anodes. It means that the small difference of I-V curves for Ni/YSZ (as shown in figure 2) was caused by the change of the other voltage loss than the anodic overpotential, such as that from contact resistance. Thus, Ni/GDC and Ni/SDC anodes were found more suitable to be used in dry methane fuel than Ni/YSZ anode.

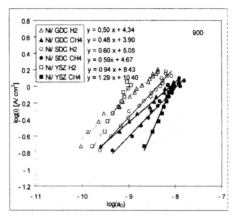

Figure 3: Relationship of the current density as function of oxygen activity of TPB with Ni/YSZ (60/40wt%), Ni/SDC (55/45wt%) and Ni/GDC (55/45wt%) with H_2 and dry CH_4 at same fuel utilization.

The current density of the SOFC with Ni/YSZ, SDC and GDC anodes in dry CH_4 at a constant a_O was smaller than that in H_2, when the Ni content exceeded the oxide content (Figure 3 and Figure 4). In contrast, the current density of some cells with the same amount of Ni/SDC or Ni/GDC was higher in dry CH_4 than that in H_2 (Figure 4). The dependence on the Ni content seems to be related to some factors such as the electronic conductivities and the microstructures of the anodes.

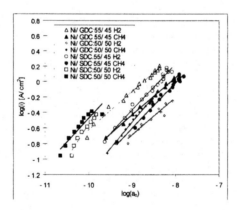

Figure 4: Comparison of the current density versus the oxygen activity of cells with Ni/SDC and Ni/GDC anodes with 50/50 wt% and 55/45wt% of Ni/oxide content with H_2 and dry CH_4 at same fuel utilization.

We reported that the SOFC with Ni/GDC anode in dry CH₄ fuel was not degraded after 120h °. To investigate the durability in more detail, the accelerated degradation tests had been performed under pure dry CH₄ (50ccm). According to the Figure 5, after shutting off the current during 10min, the terminal voltage decreased. This experiment had been realized 19 times. As shown in Figure 5, the degradations were observed just after the beginning of the cycles, however the degradation was almost saturated after 10 cycles. During these cycles, the interruption of the current due to the separation of the overpotentials at a I-V measurement, had been done for each measurement. The interruptions for the measurements also seemed to enhance the degradation. As shown in Figure 5, the interruptions during the many times of I-V measurements caused only slight degradation (the difference between the initial curve and that before the first degradation test). The decrease of performances was due to the carbon deposition until to attend saturation where the performances remain constant. According to the figure 6, the performances of the Ni/GDC anode all along the cycles after the cycle 10 tended to recover to the performances obtained after the first shutting off.

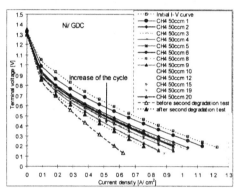

Figure 5: I-V characteristics of SOFC with Ni/GDC (55/45wt%) anode after each shutting off the current for 10min (first degradation test). (And the I-V characteristics before and after the second degradation test shown in the figure 10).

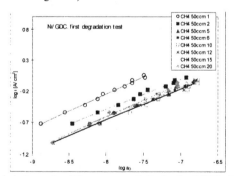

Figure 6: Relationship between i and the oxygen activity for the SOFC with Ni/GDC (55/45wt%) anode for different cycles of the first degradation test.

The same degradation test was carried out on the cell with Ni/YSZ anode (Figure 7). Contrary to the cell with Ni/GDC, the degradation was serious for Ni/YSZ. As shown in Figure 7, the interruptions during the many times of I-V measurements caused the degradation. The maximum of current density fell down to about 60% after the 20 cycles for Ni/YSZ anode whereas it decreased to only 20% for Ni/GDC anode. However, like Ni/GDC anode degradation, the degradation occurred essentially during the 10 first cycles and after cycle 10 there was a saturation of the degradation. Thus, the degradation of the cell with Ni/YSZ anode showed a fast kinetic. As shown in the figure 8, the performances of Ni/YSZ anode decreased considerably in the first cycles. After the 10th cycle, the carbon deposition formed a bridge between the working and the reference electrodes, and the anode potential could not be measure correctly. Thus, the i=f(a$_o$) curves after the 10th cycle were not be plotted.

Therefore, Ni/GDC anode is more suitable to be used in a SOFC comparing to Ni/YSZ anode, which is totally degraded after the first degradation test.

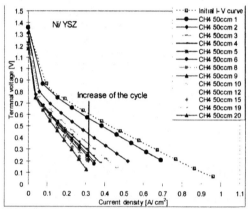

Figure 7: I-V characteristics of SOFC with Ni/YSZ (60/40wt%) anode after each shutting off the current for 10min (first degradation test).

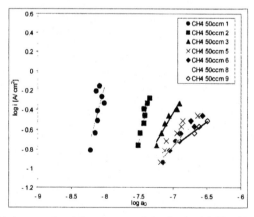

Figure 8: Relationship between i and the oxygen activity for the SOFC with Ni/YSZ (60/40wt%) anode for the nine first cycles of the degradation test.

Because of the few degradation of the Ni/GDC anode after the first degradation test, a second degradation test had been performed. After 12h in pure Ar, the performances decreased as represented on the Figure 5 ("before second degradation test" curve). Then, to further accelerate the degradation, the current kept shutting off for 30min and the power generated for 1h at 300mA/cm². The cycle was carried out 5 times (Figure 9). We observed the increase of the performances during the second degradation test. At this time, the voltage between the anode and the reference-electrode in cathode side (Wa-Rc) increased significantly, whereas the voltage between the cathode and the reference-electrode in cathode side increased slightly due to the increase of the cathode overpotential (Figure 10). Thus, the increase of the performance was attributed to the anode.

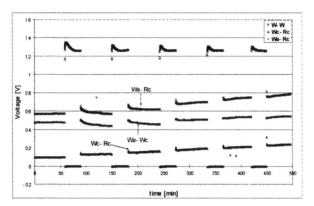

Figure 9: Degradation test of SOFC with Ni/GDC (55/45wt%) by shutting off the current during 30min at 300mA/cm². (Wa-Wc: terminal voltage, Wa-Rc: voltage between anode and reference-electrode in cathode side, Wc-Rc: voltage between cathode and reference-electrode in cathode side).

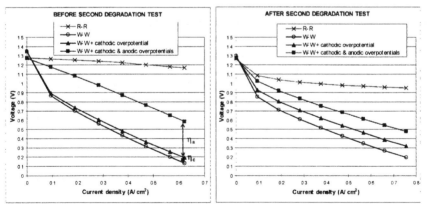

Figure 10: I-V characteristics of the Ni/GDC (55/45wt%) cell before and after the second degradation test.

We measured the I-V characteristics after the second degradation test. The performance recovered to that after the first degradation test during the second degradation test (as shown in Figure 5: "after second degradation test" curve). Therefore, after the initial degradation, the

Ni/GDC anodes conserved good performances even after drastic conditions where carbon deposition occurred. The high tolerance of Ni/GDC to carbon deposition is possibly related to the reaction between carbon and oxide ion [9] and also to the cleaning effect due to the presence of the water formed during the power generation.

CONCLUSION

We demonstrated that the SOFCs with Ni/SDC and Ni/GDC anodes presented higher performances in dry CH_4 and in H_2 than those with Ni/YSZ anode. Ni/GDC and Ni/SDC anodes had similar performances. At least, the stable operation of the SOFC with Ni/GDC anode in dry CH_4 have been confirmed even after strong conditions, whereas Ni/YSZ anode was degraded in the same conditions. The degradation of the SOFCs in dry CH_4 fuel was accelerated by the interruption of the current. Therefore, Ni/doped ceria materials are promising candidate as anode in SOFC in dry CH_4 fuel.

REFERENCES

[1] T. Ishihara, T. Shibayama, H. Nishiguchi, and Y. Takita, "Nickel-Gd-doped CeO_2 cermet anode for intermediate temperature operating solid oxide fuel cells using $LaGaO_3$-based perovskite electrolyte," Solid State Ionics **132** (3-4), 209-216 (2000).

[2] K. Eguchi, H. Kojo, T. Takeguchi, R. Kikuchi, and K. Sasaki, "Fuel flexibility in power generation by solid oxide fuel cells," Solid State Ionics **152-153**, 411-416 (2002).

[3] Y. Lin, Z. Zhan, J. Liu, and S. A. Barnett, "Direct operation of solid oxide fuel cells with methane fuel," Solid State Ionics **176** (23-24), 1827-1835 (2005).

[4] S. J. A. Livermore, J. W. Cotton, and R. M. Ormerod, "Fuel reforming and electrical performance studies in intermediate temperature ceria-gadolinia-based SOFCs," Journal of Power Sources **86** (1-2), 411-416 (2000).

[5] J.-H. Koh, Y.-S. Yoo, J.-W. Park, and H. C. Lim, "Carbon deposition and cell performance of Ni-YSZ anode support SOFC with methane fuel," Solid State Ionics **149** (3-4), 157-166 (2002).

[6] N. Laosiripojana and S. Assabumrungrat, "Catalytic dry reforming of methane over high surface area ceria," Applied Catalysis B: Environmental **60** (1-2), 107-116 (2005).

[7] M. Ihara, T. Kusano, and C. Yokoyama, "Competitive adsorption reaction mechanism of Ni/Yttria-stabilized zirconia cermet anodes in H_2-H_2O solid oxide fuel cells," Journal of the Electrochemical Society **148** (3) (2001).

[8] M. Ihara, K. Matsuda, H. Sato, and C. Yokoyama, presented at the 5th European SOFC forum, Switzerland, 2002.

[9] M. Ihara, K. Matsuda, H. Sato, and C. Yokoyama, "Solid state fuel storage and utilization through reversible carbon deposition on an SOFC anode," Solid State Ionics **175** (1-4), 51-54 (2004).

CONTROL OF MICROSTRUCTURE OF NiO-SDC COMPOSITE PARTICLES FOR DEVELOPMENT OF HIGH PERFORMANCE SOFC ANODES

Koichi Kawahara, Seiichi Suda, Seiji Takahashi
Japan Fine Ceramics Center
2-4-1 Mutsuno, Atsuta-ku
Nagoya, 456-8587, Japan

Mitsunobu Kawano, Hiroyuki Yoshida, Toru Inagaki
The Kansai Electric Power Co., Inc.
11-20 Nakoji, 3-chome
Amagasaki 661-0974, JAPAN

ABSTRACT

Three different types of nickel oxide (NiO)-samarium doped ceria (SDC) composite particles were synthesized using the spray pyrolysis technique. The starting conditions were controlled to produce (1) capsule-type particles in which NiO is enveloped with SDC, (2) matrix-type in which SDC is finely dispersed within a matrix of NiO and (3) hollow-type, with a shell of less than 100nm in thickness. The capsule- and matrix-type composite particles were spherical, whereas the hollow-type particles were observed to be a mixture of hollow spheres and flake-like pieces which were probably broken particles. Measurements of the performance of single cells with $La_{0.9}Sr_{0.1}Ga_{0.8}Mg_{0.2}O_{3-\delta}$ as the electrolyte, $La_{0.6}Sr_{0.4}CoO_{3-\delta}$ as the cathode and each of the three particle types as the anode were made at $750^{\circ}C$. The cells fabricated from the matrix-type particles showed higher power density than those made from the capsule- and hollow-type particles. The better performance of the matrix-type composite was attributed to a lower ohmic loss. It is therefore considered that the matrix type of particle is likely to be best suited to the development of a fine and well-connected network structure which avoids aggregation of nickel phase during cell fabrication, resulting in a lower ohmic loss.

INTRODUCTION

It is well known that the performance of solid oxide fuel cells (SOFC) depends critically on the microstructure of the electrode as well as on materials properties such as conductivity and catalytic activity. The electrochemical reaction is considered to occur at the triple-phase boundary[1] (TPB) where the fuel, electrode and electrolyte meet. The TPB length in the anode must therefore be one of the most important factors in the development of high-performance SOFC anodes. The TPB density can be modified by changing the microstructure of the anode, for example the cermet composition or grain size. Sunde made Monte Carlo simulations of the polarization resistance of composite electrodes as a function of the microstructural characteristics[2]. One of his results was that, for a given anode cermet material and a given composition, the polarization resistance (activity of the anode) depended on the ratio between the grain size of the metal phase and that of the ceramic. This indicates that the anode microstructure should be optimized for the development of high performance SOFC anodes.

The evolution of the electrode microstructure during anode fabrication will be affected by the microstructure of the starting powder. Recently, we have synthesized nickel oxide (NiO)-samarium doped ceria (SDC) composite particles for the SOFC anode using the spray pyrolysis

183

technique[3]. This is a solution process, and the synthesized particles are spherical and around 1μm in size, with a narrow size distribution. Each is composed of both NiO and SDC primary grains; this is advantageous in the evolution of a desirable anode network structure during electrode fabrication, without excessive densification and/or aggregation. Suda et al. reported that SOFC single cells with an anode fabricated from composite particles synthesized by spray pyrolysis showed lower overpotential loss than those fabricated from NiO/SDC mixed powder[4]. This is attributed to the higher density of TPBs in the anode. Composite particles synthesized by spray pyrolysis should therefore have an advantage in the development of a fine anode network structure.

Since spray pyrolysis consists of mist generation, drying and pyrolysis processes occurring in turn, it should be possible to control the microstructure of the synthesized particles by controlling the conditions of these three processes. Suda et al. showed that the width of the particle size distribution was increased by increasing the temperature during the mist generation stage[5]. However, the average particle size varied less with temperature. The cell performance depended on the width of the particle size distribution when the resulting composite particles were used as an anode. The dependency was not monotonic, but went through a maximum at an intermediate particle size distribution, suggesting the existence of an optimum width for this distribution.

Kawano et al. investigated the effect of the addition of nitric acid on the morphology of composite particles synthesized by spray pyrolysis, and discussed the relation between the morphology of these particles and the cell performance[6]. The surface morphology and the specific surface area depended strongly on the amount of nitric acid added to the starting solution. A higher specific surface area was obtained using a solution of lower pH. Cells with anodes fabricated from these particles with higher specific surface area showed a better and more consistent cell performance.

In this study, careful attention has been paid to the effect of the starting solution conditions, particularly the addition of ethylene glycol (EG), on the microstructure of the synthesized composite particles. The role of this microstructure in determining the performance of single cells in which the resulting composite is used as an anode has been investigated.

EXPERIMENTAL
Synthesis of NiO-SDC Composite Particles

NiO-SDC composite particles with different microstructures were synthesized by spray pyrolysis. The details of this technique are described elsewhere[4]. To prepare the solutions, samarium oxide was first dissolved into conc. nitric acid, then cerium nitrate hexahydrate and nickel acetate tetrahydrate were added. The composition of NiO-SDC composite particles was set to be 5 : 2 by volume; this is equivalent to an Ni-SDC cermet (Ni:SDC) of composition 6:4 by volume. The ratio of cerium (Ce) and samarium (Sm) in SDC (Ce:Sm) was 4:1 (atomic ratio).

Three different starting solutions for spray pyrolysis were prepared. In two of these solutions, a molar amount of ethylene glycol (EG) equal to five times that of all cations was added. One of these two solutions was then heat-treated at 80°C for 5h in order to allow chelation. This solution was designated 5EGH. The second solution, denoted 5EGN, did not undergo heat-treatment. The third and final solution did not have EG added; this will be referred to as 0EG. Prior to spray pyrolysis, all solutions were diluted with water to give an

SDC concentration of 0.1mol/l in SDC. Spray pyrolysis was conducted in four series furnaces independently controlled to be 200°C, 400°C, 800°C and 1000°C, respectively. The as-synthesized composite particles were calcined at 1000°C for 24h in air prior to cell fabrication.

SEM observation
The surface morphology of the synthesized NiO-SDC composite particles was observed using scanning electron microscopy (SEM). A HITACHI S-4500 was used at an acceleration voltage of 20kV. To avoid charging-up during SEM observation, samples were coated with a thin (10~20nm) layer of Au.

TEM/EDX analysis
Transmission electron microscope (TEM) observations were conducted to observe the microstructures of the composite particles synthesized from the different starting solutions. A TOPCON EM 002B was used at an acceleration voltage of 200kV. Samples suitable for TEM observation were prepared using copper grids (#200) of 3mm in diameter with a collodion film. Samples were analyzed using energy dispersive X-ray spectroscopy (EDX) in order to clarify distribution of chemical elements (Ni, Ce, Sm) within the composite particles.

Cell fabrication
SOFC single cells were fabricated from the synthesized NiO-SDC composite particles as the anode, doped lanthanum gallate ($La_{0.9}Sr_{0.1}Ga_{0.8}Mg_{0.2}O_{3-\delta}$ (LSGM)) of 200μm in thickness as the electrolyte and a commercially available $La_{0.6}Sr_{0.4}CoO_{3-\delta}$ (LSC) powder as the cathode. The paste of a mixture of polyethyleneglycol (PEG) and NiO-SDC composite particles or LSC powder was screen-printed onto the LSGM surface with the electrode area of 0.283cm^2. The anode was fired in air at 1250°C for 2h and the cathode at 1000°C for 4h.

Electrochemical measurements
The performance of the SOFC single cells with the experimental anodes was evaluated electrochemically by the current interruption method; measurements were conducted at 750°C. The flow rates of the fuel (dry hydrogen) and the oxidant (air) were fixed to be 50cm^3/min.

RESULTS AND DISCUSSION
SEM observations of composite particles
Figure 1 shows SEM images of NiO-SDC composite particles as synthesized from solutions 0EG, 5EGH and 5EGN. None of the X-ray diffraction patterns from the particles showed any other peaks than those expected from the NiO and SDC phases; indicated that no unexpected phase was present. The high-magnification images revealed that the particles were composed of primary grains as small as about 10nm, which were probably NiO and SDC grains. No noticeable difference in the primary grain size between particle types was observed. The morphology of the composite particles strongly depended on the starting solution conditions. Particles synthesized from a solution of 5EGN were hollow in appearance, with shells less than 100nm in thickness. Some of these had broken into flake-like pieces. Both of the other types of particle (0EG and 5EGH) were spherical in shape but the surfaces of 5EGH particles were rougher than those of 0EG particles.

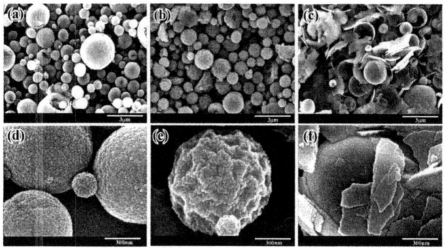

Figure 1   SEM images of NiO-SDC composite particles synthesized from the solutions 0EG ((a) and (d)). 5EGH ((b) and (e)) and 5EGN ((c) and (f)).

Figures 2 (a) and (b) show the particle size distributions, measured from SEM images of particles synthesized from the solutions 0EG and 5EGH, respectively. Because of their flake-like shapes, such a distribution could not be measured from the 5EGN particles. It can be seen that the particle distribution widths from both 0EG and 5EGH solutions are rather narrow and that the average particle sizes are about 0.75 and 0.76µm, respectively. Given the width of the particle size distributions and the experimental error in size measurement, it was considered that the sizes from solutions 0EG and 5EGH were about the same.

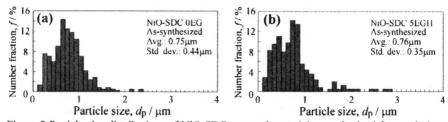

Figure 2 Particle size distributions of NiO-SDC composite particles synthesized from solutions (a) 0EG and (b) 5EGH.

Figure 3 shows SEM images of the NiO-SDC particles after they had been calcined at 1000°C for 24h. This treatment did not appear to modify the size of any of the three types of particles. However, the calcination led to remarkable growth of the primary grains. as shown in the high-magnification image, in all three particle types. In addition, the flake-like pieces observed in Fig.1 (c) and (f) were further broken into debris composed of several grains. Figure 4 shows the primary grain size distributions of each type of composite particle, as measured from SEM images. In all cases, these grain sizes were found to be 10 or more times greater than those measured before calcination. It is notable that the average primary grain size of the composite particles from the 0EG solution was larger than those from 5EGH and 5EGN. It is known that grain growth phenomena are greatly affected by the composition in multi-phase materials. However, in this case, the same composition was used for each of the three particle types, so any difference in primary grain size after calcination must result from the microstructure of the as-synthesized particles.

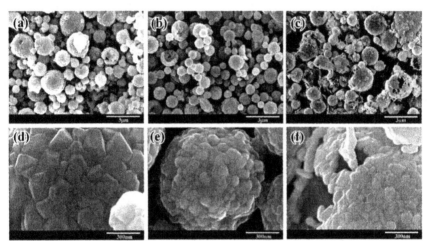

Figure 3 SEM images of NiO-SDC composite particles calcined at 1000°C for 24h from the solutions 0EG ((a) and (d)), 5EGH ((b) and (e)) and 5EGN ((c) and (f)).

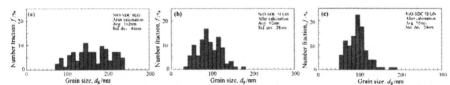

Figure 4 Primary grain size distributions of composite particles after calcination at 1000°C for 24h synthesized from solutions   (a) 0EG, (b) 5EGH and (c) 5EGN.

TEM/EDX analyses of composite particles

Figure 5 shows TEM/EDX analyses of particles synthesized from the solutions 0EG and 5EGH after the calcination treatment described above. In all cases, the distribution of samarium was the same as that of cerium. Nickel and cerium were rather homogeneously dispersed in the particles synthesized from the 5EGH solution, indicating that this type of particle had a matrix-type structure. By contrast, in the particles fabricated from 0EG, cerium was preferentially distributed at the surface region. indicating that these particles had a capsule-type structure in which the NiO phase was enveloped with SDC. The mechanism for the formation of this capsule-like structure from 0EG has not yet been elucidated. However, the fact that the matrix-type structure was obtained using EG and a subsequent heat treatment (chelation treatment) indicates that the microstructure of the composite particles is strongly affected by the presence of the chelating agent (EG). The primary grain size of calcined composite particles synthesized from the solution of 5EGH would be, therefore, smaller than that from the solution of 0EG because of its microstructure of finely dispersed NiO and SDC phases. Moreover, since the primary grain size of the calcined composite particles synthesized from the solution of 5EGN was as small as that from the solution of 5EGH, it was likely that the microstructure within the thin shell and pieces of the composite particles synthesized from the solution of 5EGN had a similar microstructure (matrix-type).

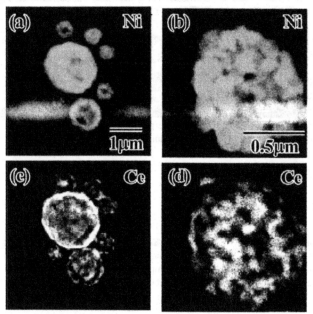

Figure 5 TEM/EDX analyses of the composite particles synthesized from solutions (a), (c) 0EG and (b), (d) 5EGH. The nickel distribution is shown in micrographs (a) and (b), and the cerium distribution in micrographs (c) and (d).

Cell performance evaluation

Figure 6 shows the cell voltage and power density measured as a function of the current density at 750°C. The I-V curves for cells fabricated from all particle types showed similar characteristics. In the region of lower current density, the cell voltage rapidly decreased with increasing current density, while in the higher current density region, the voltage decreased almost linearly with increasing current density, resulting in the convex downward shape seen in (a). The open-circuit voltages of all the cells were similar (about 1.23V). However, the voltage loss at a given current density depended on the particle type. Hence, the cell performance strongly depended on the microstructure of the particles used to make the anode. The best cell performance was obtained from the anode fabricated using 5EGH (matrix-type microstructure).

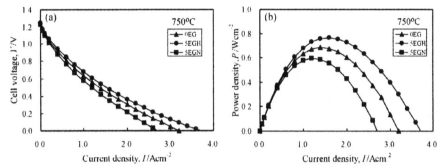

Figure 6   Performance of cells with anodes fabricated from different types of composite particles: (a) cell voltage and (b) power density as a function of current density.

Figure 7 shows ohmic loss and overpotential loss as a function of current density evaluated by the current interruption method. It can be seen that the ohmic loss depends on the particle type, whereas the overpotential loss does not show such a dependence. It is therefore concluded that the differences in the cell performance seen in Fig. 6 can be attributed to the ohmic loss. The measured ohmic loss contained contributions from the electrolyte, cathode and anode. However, all the components of the cells, apart from the anodes, were prepared from the same materials with the same dimensions. It is therefore reasonable to consider that the differences in ohmic loss observed are a result of differences in the microstructure of the anode material. The main electron path contributing to the decrease in ohmic loss in the anode is probably a nickel skeleton. Since the particles synthesized using 5EGH had a matrix-type microstructure, it appeared that this type of particle was most likely to present an advantage in the development of fine- and well-connected nickel network structures without aggregation. It is worth noting that although the composite synthesized from 5EGN (which consisted of hollow, fragmented particles) probably had a similar microstructure to those fabricated using 5EGH (matrix-type), the former (5EGN) showed the lowest cell performance of all. This suggests that the anode performance depends not only on the distributions of NiO and SDC within the particles, but also on the shape of the particles themselves, i.e. whether they are hollow, fragmented, solid, etc.

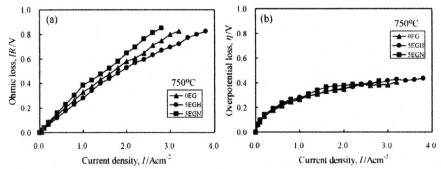

Figure 7   Current interruption evaluations showing (a) ohmic loss and (b) overpotential loss as a function of current density.

The reaction between NiO-SDC anode and LSGM electrolyte has been reported by Huang et al.[7] to form high-resistive compounds.   The buffer layer preventing the reaction between anode material and LSGM sometimes used to avoid this effect[7-9].   Although we did not use any buffer layer, the electrochemical performance seemed not to be affected so much.   The reactivity between NiO-SDC composite particles and LSGM, and its effect on the electrochemical performance of SOFC should be clarified in the future work.

SUMMARY
The microstructure of composite particles synthesized by spray pyrolysis depended strongly on the starting solution conditions.   The addition of ethylene glycol into the solution resulted in a matrix-type microstructure in which the SDC phase was finely dispersed within a matrix of NiO, whereas particles synthesized without the use of ethylene glycol had a capsule-type microstructure in which the NiO was enveloped with SDC.   Anodes fabricated from matrix-type composite cgave the highest performance among those evaluated in this study.   It is clear that control of the microstructure of the starting composite particles is of great importance for the development of high-performance SOFC anodes.

REFERENCES
[1]J.Mizusaki, T. Tagawa, T. Saito, K. Kamitani, T. Yamamura, K. Hirano, S. Ehara, T. Takagi, T. Hikita, M. Ippommatsu, S. Nakagawa and K. Hashimoto. "Preparation of Nickel Pattern Electrodes on YSZ and Their Electrochemical Properties in H_2-H_2O Atmospheres", *J. Electrochem. Soc.*, **141**. 2129-34 (1994).
[2]S. Sunde, "Monte Carlo Simulations of Polarization Resintance of Composite Electrodes for Solid Oxide Fuel Cells", *J. Electrochem. Soc.*, **143**, 1930-39 (1996).
[3]T. Inagaki, H. Yoshida, K. Miura, S. Ohara, R. Maric, X. Zhang, K. Mukai and T. Fukui, "Intermediate Temperature Solid Oxide Fuel Cells with Doped Lanthanum Gallate Electrolyte, La(Sr)CoO₃ Cathode, and Ni-SDC Cermet Anode", *Electrochemical Proceedings (SOFC VII)*, **2001-16**, 963-72 (2001).

[4]S. Suda, M. Itagaki, E. Node, S. Takahashi, M. Kawano, H. Yoshida and T. Inagaki, "Preparation of SOFC anode composites by spray pyrolysis", *J. Euro. Ceram. Soc.*, **26**, 593-97 (2006).

[5]S. Suda, S. Takahashi, M. Kawano, H. Yoshida and T. Inagaki, "Effects of atomizing conditions on morphology and SOFC anode performance of spray pyrolyzed $NiO-Sm_{0.2}Ce_{0.8}O_{1.9}$ composite particles", submitted to *Solid State Ionics*.

[6]M. Kawano, K. Hashino, H. Yoshida, H. Ijichi, S. Takahashi, S. Suda and T. Inagaki, "Synthesis and characterizations of composite particles for solid oxide fuel cell anodes by spray pyrolysis and intermediate temperature cell performance", *J. Power Sources*, **152**, 196-99 (2005).

[7] K. Huang, Jen-Hau Wan and J.B. Goodenough, "Increasing Power Density of LSGM-Based Solid Oxide Fuel Cells Using New Anode Materials", *J. Electrochem. Soc.*, **148**, A788 - A794 (2001).

[8]W. Gong, S. Gopalan and U.B. Pal, "Materials System for Intermediate-Temperature (600-800°C) SOFCs Based on Doped Lanthanum-Gallate Electrolyte", *J. Electrochem. Soc.*, **152**, A1890 - A1895 (2005).

[9]J. Yan, H. Matsumoto, M. Enoki and T. Ishihara, " High-Power SOFC Using $La_{0.9}Sr_{0.1}Ga_{0.8}Mg_{0.2}O_{3-\delta}/Ce_{0.8}Sm_{0.2}O_{2-\delta}$ Composite Film", *Electrochem. Solid-State Lett.*, **8**, A389 - A391(2005).

ELECTROCHEMICAL CHARACTARIZATION AND IDENTIFICATION OF REACTION
SITES IN OXIDE ANODES

T. Nakamura, K. Yashiro, A. Kaimai, T. Otake, K. Sato, G.J. Park, T. Kawada, J. Mizusaki
Institute of Multidisciplinary Research for Advanced Materials, Tohoku University
2-1-1, Katahira, Aoba-ku, Sendai 980-8577, Japan

ABSTRACT
    Anode performance and reaction sites on $LaCrO_3$ based oxides, $SrZrO_3$ based oxides, and
ceria based oxides were evaluated by means of AC impedance and steady state polarization
measurements in H_2-H_2O-Ar gas mixtures. All oxide anodes tended to show poorer performance
than Ni/YSZ cermet anode which is a popular anode for solid oxide fuel cells. Ceria based oxide
showed relatively high performance among these oxides. Extraordinary large capacitance was
observed in the $LaCrO_3$ based oxide and ceria based oxide anodes. The measured large
capacitance is considered to be a kind of chemical capacitance due to oxygen nonstoichiometry
of the electrode material. It means that the electrode reaction proceeds via O^{2-} diffusion through
the electrode particles, which is mixed ionic and electronic conductor. Further semi-quantitative
analysis indicated that O^{2-} diffuses through almost all particles in the electrode layer. The
reaction site of hydrogen oxidation could be either distributed on the whole surface of the
electrode particles or located at the interface of the current collector and the electrode layer. They
also imply that Au current collector might act as a catalyst and is not appropriate for evaluation
of oxide anodes.

INTRODUCTION
    Solid oxide fuel cell (SOFC) is expected as one of the advanced energy conversion
systems. In principle, they achieve higher performance than other types of fuel cells, and
hydrocarbons can be used directly. Now, most of anode materials contain metal Ni as a catalyst
and as electronic paths. Although Ni containing anodes show high performance, it has some
drawbacks caused by Ni presence. For example, Ni particles gradually aggregate at high
temperatures, which cause degradation of the performance in a long term operation. Carbon
deposits on the Ni surface when hydrocarbon is directly introduced to anode. It reduces cell
performance drastically. In addition, Ni is easily oxidized to NiO in oxidizing atmosphere.
Volume change of Ni to NiO breaks anode microstructure[1-3]. One of the candidates of Ni free
anode is a mixed conducting oxide. Oxides are stable in oxidizing atmospheres, and it is hard to
coarsen. It may prevent carbon deposition because oxide ion and electron can move in the
electrode and react with the deposited carbon. Moreover, it may enlarge reaction area to
electrode/gas two phase boundary. Under a SOFC system with mixed conducting oxide anode,
one can quickly shut down the system without introducing any protective gas to anode, and
startup/shutdown frequently. One can remove a reformer from a SOFC system and do not have
to control water content in a hydrocarbon fuel. Mixed conducting oxide anode will contribute to
realization of more flexible and convenient SOFC system.
    Because of such advantages, some mixed conducting oxides have been evaluated their
electrode performance, catalytic activity, and chemical stability. For example, perovskite type
oxide, fluorite type oxide, pyrochlore, tungsten bronze and so on has been studied[4-8]. However
there have been few reports about reaction kinetics or reaction sites of the oxide anodes.
Knowledge about how and where electrode reaction occurs in oxide anodes makes it easier to

improve the electrode performance of the oxide anode. The aim of this study is to make clear the reaction sites of an oxide anode.

EXPERIMENTAL

$La_{0.9}Ca_{0.1}Cr_{1-x}Al_xO_3$ (x=0.2: LCCA9182, x=0.8: LCCA9128) was prepared via combustion of mixtures of nitrate solution and citric acid. $SrZr_{0.825}Y_{0.1}Ru_{0.075}O_3$ (SZYR) was prepared by solid-state reaction method. Niobium doped ceria was prepared by co-precipitation method, and commercial $Ce_{0.9}Gd_{0.1}O_{1.95}$ (GDC) powder was used. They were confirmed to be single phase by XRD. Anode inks were prepared by mixing as prepared oxide powders and 20 vol% ethyl cellulose with an organic solvent. Anode inks were screen printed on YSZ substrate in a circle of 6 mm in diameter. Pt paste was symmetrically printed in 6 mm on the opposite surface as a counter electrode. Pt wire was wound at the side of the substrate and fixed with Pt paste as a reference electrode. The both electrodes were sintered at 1173 K for 3 h. Overview of cell is shown in Fig.1. For the electrochemical measurement, 3-terminal measurement was applied. Au mesh and Pt mesh were used as current collectors to anode and cathode, respectively. Figure 2 is a schematic view of the experimental set up. H_2-H_2O-Ar gas mixture was introduced as fuel. Air was fed to the counter and the reference electrode. Using this experimental set up, AC impedance measurement and steady state polarization measurement were carried out.

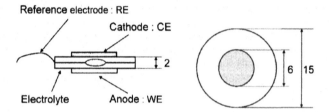

Fig. 1 Cell overview

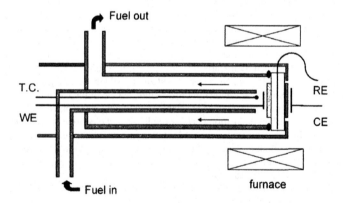

Fig. 2 Schematic view of the experimental set up

RESULTS AND DISCCUSION
Results of polarization measurement of oxide anodes

Figure 3 shows the results of polarization measurement of oxide anodes and Ni/YSZ which was prepared by the same manner as the oxide anode. Ceria based oxide anodes showed relatively high performance among the oxide anodes. However all oxide anodes showed poorer performance than that of Ni/YSZ. In oxide anodes, liming current behavior was observed under anodic polarization

In order to check the effect of Au-mesh current collector on the electrode reaction, a porous YSZ layer was prepared in stead of an oxide electrode layer in the same manner as an oxide anode was prepared. In that case, YSZ works as a part of the electrolyte and the Au mesh works as an electrode. Electrode reaction takes place at the triple phase boundary (Au/YSZ/gas). It is called "porous YSZ/Au electrode" in the following discussion. Porous YSZ/Au electrode showed comparable performance to an oxide anode. It indicates that Au has some extent of catalytic activity. In this case, perovskite type oxide anodes showed poorer performance than porous YSZ/Au electrode. It indicates that Au played major role to electrode reaction in perovskite type oxide anodes. On the contrary, ceria based anodes showed better performance than porous YSZ/Au. The effect of Au is considered to be minor in ceria based anodes.

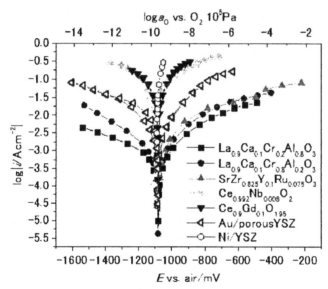

Fig. 3 Results of polarization measurement at 1173 K in H_2-2.3%H_2O

Capacitance due to oxygen nonstoichiometry

Extraordinary large capacitance was observed in ceria based and $LaCrO_3$ based anodes. The values of capacitance were varied by electrode materials. They are in the range of 10^2 to 10^6

μFcm^2. Such large capacitance is considered as a kind of chemical capacitance reported by Jamnik and Maier[9]. In case of mixed conducting oxide, chemical capacitance due to oxygen nonstoichiometry has been already reported in Sr doped $LaCoO_3$ cathode[10] and ceria based electrolyte in reducing atmosphere[11, 12].

Oxygen potential in the electrode is equal to that in the gas phase at open circuit condition (dashed line in Fig. 4). When small DC voltage ΔE is applied to the electrode/electrolyte interface, oxygen potential in the electrode distributes like shown in Fig. 4 (solid line in Fig.4). It is assumed that electron transfer through electrode/electrolyte interface is fast enough to be in equilibrium. Then, the oxygen potential at the interface can be defined as

$$\mu_{O,int} = \mu_{O,gas} + 2F\Delta E \tag{1}$$

$\mu_{O,gas}$ is the equilibrium oxygen potential in the gas phase. F and ΔE are faraday constant and applied voltage, respectively. Abrupt change of oxygen potential is built up where rate determining step exist. Oxygen nonstoichiometry δ of electrode and the electrolyte around the interface reaches to an equilibrium value corresponding to the new oxygen potential distribution. Under AC voltage, oxygen nonstoichiometry changes continuously. Incorporation and release of oxygen is measured as capacitance. Larger capacitance is observed when surface reaction determine reaction rate predominantly. If relation between oxygen nonstoichiometry δ and $P(O_2)$ is known preliminarily, capacitance due to oxygen nonstoichiometry can be calculated as

$$C = \frac{dQ}{dE} = \frac{d\int(-2F)\delta dV}{dE} \tag{2}$$

$$= -\frac{8F^2}{RTV_m} \frac{d\int\delta dV}{d\ln P(O_2)_{eff}} \tag{3}$$

R, T, V_m, and $P(O_2)_{eff}$, are gas constant, temperature, molar volume and effective oxygen partial pressure at the interface corresponding to $\mu_{O,int}$. Detail introduction of the equation is written in ref. 10.

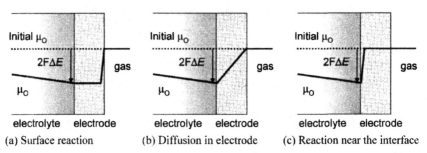

(a) Surface reaction   (b) Diffusion in electrode   (c) Reaction near the interface

Fig. 4 Oxygen potential profile around electrode under small DC voltage ΔE in case that rate determining step is (a) surface reaction (b) diffusion in electrode (c) reaction near electrode/electrolyte interface

In this paper, the values of capacitance were calculated under two assumptions. First, electrode reaction takes place at the whole surface of electrode particles. It means oxide ion and electron can migrate throughout the electrode layer. Second, surface reaction is rate determining step. Second assumption is considered to be reasonable since electrode resistances from I-V curve shown in Fig. 3 are far larger than those estimated from the resistance of ion migration through the electrode particles. It also means that diffusion controlled situation shown in Fig.4-(b) could not be realized. Whole electrode contributes to capacitance under these assumptions.

Estimated value from eq. (3) and measured capacitance from impedance measurement are listed in table 1. Literature data were used in estimation[13-14]. Capacitance observed in porous YSZ/Au and Ni/YSZ seems to be typical capacitance which comes from electrode/electrolyte interface[15-16]. SZYR also show small capacitance among oxide anode. Possible reasons are, reaction sites of SZYR anode may be quite different from that of other oxide anodes, or SZYR show small oxygen nonstoichiometry under experimental condition. Estimated values from eq. (3) show relatively good agreement to the measured capacitance. Then the two assumptions, i.e. electrode reaction takes place at the surface of electrode and surface reaction is rate determining step, seems to be valid.

Table 1 Capacitance in H_2-2.3%H_2O at 1173 K

Sample	Measured capacitance $[\mu Fcm^{-2}]$	Estimated value $[\mu Fcm^{-2}]$
$La_{0.9}Ca_{0.1}Cr_{0.2}Al_{0.8}O_3$	7×10^2	4×10^3
$La_{0.9}Ca_{0.1}Cr_{0.8}Al_{0.2}O_3$	2×10^4	8×10^4
$Ce_{0.992}Nb_{0.008}O_2$	6×10^5	-
$Ce_{0.9}Gd_{0.1}O_{1.95}$	9×10^5	3×10^6
$SrZr_{0.825}Y_{0.1}Ru_{0.075}O_3$	1	-
Porous YSZ/Au	7	-
Ni/YSZ	4×10	-

SUMMARY

Ceria based oxide anode show relatively high performance among oxide anode. However, its performance is far poorer than that of Ni/YSZ. Electrode surface seems to play important role to electrode reaction in some oxide anodes. There still remains one big question. That is where is the true "surface" we mention here. Is it really electrode/gas two phase boundary? Or Au/electrode interface? Porous YSZ/Au electrode show decent performance and electrode reaction of oxide anodes seems to occur at the top of the electrode. Such facts are indirect evidences which indicate that Au works as catalytically active site and electrode reaction takes place at the Au/electrode/gas interface that is the very top of the electrode.

REFERENCES

[1]Caine M. Finnerty, Neil J. Coe, Robert H. Cunningham, R. Mark Ormerod, "Carbon formation on and deactivation of nickel-based/zirconia anodes in solid oxide fuel cells running on methane" *Catalysis Today*, **46**, 137-145 (1998)

[2]D. Simwonis, F. Tietz, D. Stöver, "Nickel coarsening in annealed Ni/8YSZ anode substrates for solid oxide fuel cells" *Solid State Ionics*, **132**, 241-251 (2000)

[3]D. Waldbillig, A. Wood, D.G. Ivey, "Electrochemical and microstructural characterization of the redox tolerance of solid oxide fuel cell" *J. Power Source*, **145**, 113-124 (1996)

[4]A.L. Sauvet, J.T.S. Irvine, "Catalytic activity for steam methane reforming and physical characterization of $La_{1-x}Sr_xCr_{1-y}Ni_yO_{3-\delta}$" *Solid State Ionics*, **167**, 1-8 (2004)

[5]Olga A. Marina, Nathan L. Canfield, Jeff W. Stevenson, "Thermal, electrical, and electrocatalytical properties of lanthanum-doped strontium titanete" *Solid State Ionics*, **149**, 21-28 (2002)

[6]Olga A. Marina, Carsten Bagger, Søren Primdahl, Mogens Mogensen, "A solid oxide fuel cell with a gadolinia-doped ceria anode: preparation and performance" *Solid State Ionics*, **123**, 199-208 (1999)

[7]E. Ramirez-Cabrera, A Atkinson, D. Chadwick, " Catalytic steam reforming of methane over $Ce_{0.9}Gd_{0.1}O_{2-x}$" *App. Catal. B: Envi.*, **47**, 127-131 (2004)

[8]A. Kaiser, J.L. Bradley, P.R. Slater, J.T.S. Irvine, "Tetragonal tungsten bronze type phases $(Sr_{1-x}Ba_x)_{0.6}Ti_{0.2}Nb_{0.8}O_{3-\delta}$: Material characterization and performance as SOFC anodes" *Solid State Ionics*, **135**, 519-524 (2001)

[9]J. Jamnik, J. Maier, "Generalised equivalent circuit ofr mass and charge transport: chemical capacitance and its implications" *Phys. Chem. Chem. Phys.*, **3**, 1668-78 (2001)

[10]T. Kawada, J. Suzuki, M. Sase, A. Kaimai, K. Yashiro, Y. Nigara, J. Mizusaki, K. Kawamura, H. Yugami, "Determination of Oxygen Vacancy Concentration in a Thin Film of $La_{0.6}Sr_{0.4}CoO_{3-\delta}$ by an Electrochemical Method" *J. Electrochem. Soc.*, **149**, E252-259 (2002)

[11]Alan Atkinson, Sylvia A. Baron, Nigel P. Brandon, "AC Impedance Spectra Arising from Mixed Ionic Electronic Solid Electrolyte" *J. Electrochem. Soc.*, **151**, E186-193 (2004)

[12]Wei Lai, Sossina M. Haile, "Impedance Spectroscopy as Tool for Chemical and Electrochemical Analysis of Mixed Conductors: A Case Study of Ceria" *J. Am. Ceram. Soc.*, **88**, 2979-97 (2005)

[13]K. Yashiro, S. Onuma, A. Kaimai, Y. Nigara, T. Kawada, J. Mizusaki, K. Kawamura, T. Horita, H. Yokokawa, "Mass transport properties of $Ce_{0.9}Gd_{0.1}O_{2-\delta}$ at the surface and in the bulk" *Solid State Ionics*, **152-153**, 469-476 (2002)

[14]Now pressing, M. Hasegawa, M. Oishi, A. Kaimai, K. Yashiro, H. Matsumoto, T. Kawada, J. Mizusaki, "Defect structure and electrical conduction of $La_{0.9}Ca_{0.1}Cr_{1-y}Al_yO_{3-\delta}$ ($y = 0$ – 0.9)" *Ionic and Mixed Conducting Ceramics V*, T. Ramanarayanan, et al. editors

[15]J. Mizusaki, H. Tagawa, K. Isobe, M. Tajika, I. Koshiro, H. Maruyama, K. Hirano, "Kinetics of the Electrode Reaction at the H_2-H_2O Porous Pt/Stabilized Zirconia Interface" *J. Electrochem. Soc.*, **141**, 1675-83 (1994)

[16]M. Brown, S. Prindahl, M. Mogensen, "Structure/Performance Relations for Ni/Yttria-Stabilized Zirconia Anodes for Solid Oxide Fuel Cell" *J. Electrochem. Soc.*, **147** 475-485 (2000)

Interconnects and Protective Coatings

CORROSION PERFORMANCE OF FERRITIC STEEL FOR SOFC INTERCONNECT APPLICATIONS

M. Ziomek-Moroz, G.R. Holcomb, B.S. Covino, Jr., S.J. Bullard, P.D Jablonski, D.E. Alman
U.S. Department of Energy, National Energy Technology Laboratory
1450 Queen Avenue, SW
Albany, OREGON, 97321

ABSTRACT

Ferritic stainless steels have been identified as potential candidates for interconnects in planar-type solid oxide fuel cells (SOFC) operating below 800°C. Crofer 22 APU was selected for this study. It was studied under simulated SOFC-interconnect dual environment conditions with humidified air on one side of the sample and humidified hydrogen on the other side at 750°C.

The surfaces of the oxidized samples were studied by scanning electron microscopy (SEM) equipped with microanalytical capabilities. X-ray diffraction (XRD) analysis was also used in this study.

INTRODUCTION

Planar solid oxide fuel cells (PSOFCs) are an emerging power generation technology that produces electricity and heat by electrochemically combining a gaseous fuel and oxidizing gas via an ion-conducting electrolyte.[1] The gaseous fuel could be in the form of H_2, CH_4, or CO/H_2 and the oxidizing gas could be in the form of oxygen or air. During operation, oxygen ions (O^{2-}) formed during the reduction reaction at the cathode move through an ion-conducting electrolyte to the anode to participate in the oxidation reaction along with hydrogen to form water. During the oxidation reaction, electrons are released and used in an outer circuit. An example of a SOFC single unit[2] is shown in Figure 1.

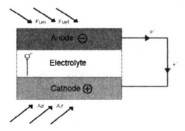

Figure. 1. Schematic diagram of a single unit of a solid oxide fuel cell showing the generation of useful power.

The open circuit voltage of this type of the cell is approximately 1 V. To generate higher voltages, these cells are connected through a conductive interconnect or a bipolar separator as shown in Figure 2.

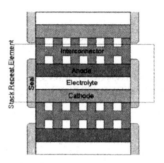

Figure. 2. Schematic diagram of a planar solid oxide fuel cell (PSOFC) stack.

The interconnect serves as a gas separator to prevent mixing of the fuel (anode) and air (cathode) and provides electrical connection between individual cells. Therefore, it must be durable and functional in the anodic and cathodic gases, while providing electrical connection in series with the fuel cell components of a PSOFC stack.

Significant progress has been made in reducing the operating temperature of the PSOFC stack from ~1000°C to below 800°C.[3] This decrease in operating temperature allows the use of metallic materials for the interconnect components. There are several advantages of using metals over currently used ceramic materials based on doped LaCrO₃: 1) achievement of gas tightness between fuel and air gases, 2) ease of handling, which lowers fabrication cost, and 3) high electronic and thermal conductivity, which increases the cell performance.[4]

During operation at 800°C, the p_{O2} is usually at 0.21 atm at the cathode and approximately 10^{-21} atm at the anode for the hydrogen fuel containing 3% H_2O. Therefore, a potential metallic candidate undergoes oxidation in both the anodic (fuel) and cathodic (air) environments. Oxide scale formation takes place on the metallic material surface as a result of the material reacting with the fuel and atmospheric gases.[5,6]

Chromium sesquioxide-forming metallic materials appear to be the most promising candidates since they show relatively low electrical resistance, high corrosion resistance, and suitable thermal expansion behavior.[7,8]. High chromium ferritic steels appear to be promising candidates to fulfill the technical and economical requirements among commercial Cr_2O_3 - forming alloys.

This paper reports the oxidation performance of the commercial ferritic stainless steel Crofer 22 APU in moist hydrogen and moist air atmospheres and compares the different oxidative effects of those two atmospheres.

EXPERIMENTAL

All experiments were carried out on commercial low coefficient of thermal expansion (CTE) commercial ferritic stainless, Crofer 22 APU, developed for SOFC applications by

Forschungszentrum Julich and commercialized by ThyssenKrupp VDM. Chemical compositions of the Crofer 22 APU samples used in this research study are shown in Table 1.

Table 1: Chemical Composition of Crofer 22 APU

Chemical I Composition (wt %)								
Fe	Cr	Mn	Ti	La	Al	Si	Cu	Ce
Bal.	22.3	0.53	0.055	0.10	0.0056	0.10	0.0046.	0.0009

   Corrosion experiments were carried out on flat samples in the shape of squares or discs. The square samples (25.4 mm x 25.4 mm x 0.5 mm) were exposed separately to fuel or moist air under isothermal conditions, i.e. a single environment. The discs (25.4 mm in diameter and 1 mm in thickness) were simultaneously exposed to fuel on one side and moist air on the other side, i.e., a dual environment. The simulated fuel was a mixture of $H_2+3\%$ H_2O and moist air was a mixture of air+3% H_2O.
   The square samples were oxidized isothermally within the temperature range 750-800°C. Before each test, the samples were polished with 600 grit SiC paper and ultrasonically cleaned in acetone. In each single environment experiment, the samples were removed from the furnace, cooled, weighted, and reinserted into the furnace. The heating and cooling cycles were repeated several times during the exposure.
   In the dual environment experiment, the samples were mounted onto a dual environment fixture (Figure 3) and installed inside the furnace.

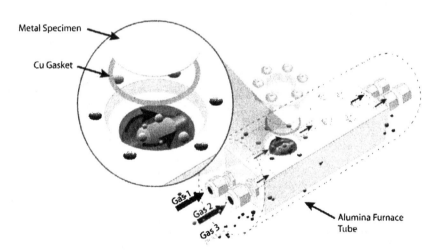

Figure 3. Experimental setup for flat samples in dual environments.

The samples were simultaneously exposed to the $H_2+3\%$ $H_2O/$ air+3% H_2O dual environment, with $H_2+3\%$ H_2O on one side and air+3% H_2O air on the other side (Figure 1: Gas 1 and Gas 3) for 200 h at 750°C. Also, in the same experiment a set of samples were simultaneously exposed to a single environment consisting of air+3%H_2O on both sides of each sample. (Figure 1: Gas 2 and Gas 3).

The post-oxidation surface investigations involved the use of the X-ray diffraction (XRD) to identify possible phases present in the scale, scanning electron microscopy (SEM) to determine morphology of the oxide scales, and x-ray dispersive energy spectroscopy (EDS) to generate concentration profiles of the mounted and polished cross sections of the investigated materials.

RESULTS AND DISCUSSION

The results of the gravimetric experiments for Crofer 22 APU in $H_2+3\%$ H_2O at 750°C and in air+3% H_2O 800°C, plotted as (mass change/area)^2 versus oxidation time, are shown in Figure 4.

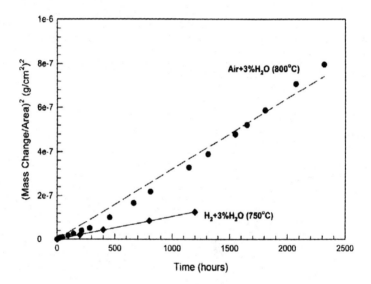

Figure 4: (Mass change/area)^2 versus oxidation time for Crofer 22APU oxidized in air+3% H_2O 800°C and $H_2+3\%$ H_2O at 750°C.

The plots are linear for the material studied in both environments, which are in agreement with the parabolic rate law of oxidation

$$(\Delta m/A)^2 = k_g t$$

where, Δm is mass change, A is the sample surface area, k_g is the parabolic rate constant determined from the gravimetric measurements, and t is the oxidation time. The values of k_g are shown in Table 2.

Table 2. Parabolic rate constants for Crofer 22 APU in single environments of air+3% H_2O and H_2+3% H_2O.

Environment	k_g (g^2/cm^4 h)
Air + 3% H_2O (800°C)	3.40E-10
H_2 + 3%H_2O (750°C)	1.07E-10

Crofer in H_2+3% H_2O has a lower k_g value than in air+3% H_2O. This indicates that Crofer 22 APU has faster oxidation kinetics in air+3% H2O than in H_2+3% H_2O.

The XRD results obtained for the material in humidified hydrogen and humidified air revealed the presence of a Cr_2O_3-like phase and a (Mn, Cr) spinel-like phase. An example of the XRD pattern for Crofer in air+3% H_2O after 2000 h exposure is shown in Figure 5.

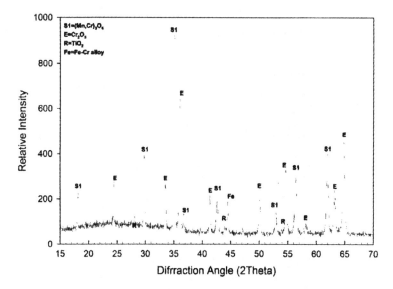

Figure 5. XRD pattern for Crofer 22 APU exposed to air+3% H_2O at 800 °C for 2000h.

Also, literature data indicate formation of a thicker oxide scale on Crofer in air than in H_2-H_2O. This difference in scale thickness is caused by different p_{O2} in air (0.2 atm) and H_2-H_2O (10^{-21} atm).[9]

A significant difference in scale thickness was observed for Crofer exposed simultaneously to air+3% H_2O on one side and H_2+3%H_2O on the other side (dual environment). Figure 6 shows SEM cross sections of the scale formed on the Crofer sample in the dual environment with one surface oxidized in H_2+3% H_2O and the other surface in air+3% H_2O.

H_2 + 3% H_2O side

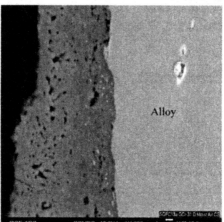

Air + 3% H_2O side

Figure 6. SEM (backscattered electron) cross-section micrographs of Crofer 22 APU after exposure to dual H_2+3% H_2O/ air+3% H_2O environment for 200h at 750°C.

Much thicker scale formed on the air side of the sample than on the hydrogen side. This indicates that the surface exposed to humidified air oxidized faster than the surface exposed to humidified hydrogen.

Figure 7 shows the elemental distribution of O, Cr, Mn, and Fe inside the scale formed in $H_2+3\%$ H_2O as determined by EDS.

10 μm

Figure 7. EDS maps of oxygen, chromium, manganese, and iron for scale formed on Crofer 22 APU during exposure to $H_2+3\%$ H_2O in dual $H_2+3\%$ $H_2O/$ air$+3\%$ H_2O environment for 200h at 750°C.

Oxygen was found uniformly distributed inside the scale. The highest concentration of Cr was detected in the inner layer, and Mn along with Cr was detected in the outer layer. Fe was not detected in this scale, however it was found in scale formed on the other side of the sample, which was exposed to air+3%H$_2$O, Figure 8.

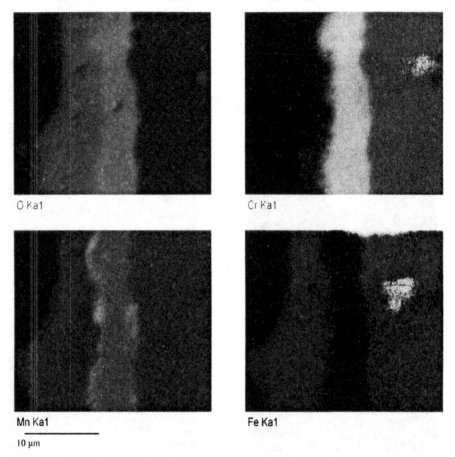

Figure 8. EDS maps of oxygen, chromium, manganese, and iron for scale formed on Crofer 22 APU during exposure to air+3%H$_2$O in dual H$_2$+3% H$_2$O/ air+3% H$_2$O environment for 200h at 750°C.

From Figure 8, it appears that oxygen is distributed uniformly inside the scale, chromium is present in the inner layer, iron is present in the outer layer, and Mn was detected between the Cr inner layer and the Fe outer layer. The presence of Fe in the scale as detected by XRD as a hematite like-phase on the air side in the dual environment indicates an atypical effect of moist hydrogen on the scale formation on the air side for ferritic steels which is in good agreement with literature data.[10-13]

CONCLUSIONS/ SUMMARY

Based on the kinetic results, Crofer 22 APU obeys the parabolic rate law of oxidation in H_2+3% H_2O at 750 °C and in air + 3% H_2O at 800°C. A higher parabolic rate constant was determined for Crofer in air + 3% H_2O and lower in H_2+3% H_2O indicating its higher oxidation rate in air.

The XRD results obtained for Crofer 22 APU in H_2+3% H_2O and air+3% H_2O revealed the presence of Cr_2O_3-like phase and (Mn, Cr) spinel-like phase.

In the H_2+3% H_2O/ air+3% H_2O dual environment, much thicker scale formed on the air+3% H_2O side of the sample than on the H_2+3% H_2O side. This indicates that the surface exposed to humidified air oxidized faster than the surface exposed to humidified hydrogen.

A significant concentration of iron was detected only in the scale formed on the air side in the H_2+3% H_2O/ air+3% H_2O dual environment.

ACKNOWLEDGEMENTS

The authors would like to thank Mr. Dan Davis for engineering support; Mr. Keith Collins, Mr. Steve Matthes, Mr. Richard Chinn, and Mr. David Smith for their analytical support, and Mr. Charles Norris of the U.S. Department of Energy, National Energy Technology Laboratory (formerly: Albany Research Center), for technical support.

REFERENCES

[1]Fuel Cell Handbook, fifth edition, U.S. Department of Energy, Office of Fossil Energy, National Energy Technology Laboratory.

[2]H. L. Tuller, Electroceramics – Materials, Processing and Characterization: Application to Solid Oxide Fuel Cells, NETL Workshop, Morgantown, WV, June 16-17, 2003

[3]BC. Steele, A. Heinzel, *Nature, 414 (2001) 345*

[4]T. Horrita, Y. Xiong, K. Yamaji, N. Sakai, H. Yokokawa, *J. Power Sources 118 35-43 (2003).*

[5]D.M. England and A.V. Virkar, *J. Elechem Soc. 146 (9) 3196-3202 (1999).*

[6]D.M. England and A.V. Virkar, *J. Elechem Soc. 148 (4) A330.*

[7]Z.G Yang, K.S. Weil, D.M. Paxton, and J.W. Stevenson "Selection and Evaluation of Heat-Resistant Alloys for Planar SOFC Interconnect Applications," *Abstracts of 2002 Fuel Cell Seminar.*

[8]S. de Souza, S.J. Visco, L.C. De Jonghe, *J. Electrochem. Soc., 144 (1997) L35.*

[9] P. Huczkowski, S. Ertyl, N. Christiansen, T. Hoefler, F-J. Wetzel, E. Wessel, V. Shemet, L. Singheiser, W.J. Quadakkers, *Abstracts of 2005 Fuel Cell Seminar.*

[10]Z. Yang, M. S. Walker, P. Singh, J.W. Stevenson, *Electrochemical and Solid-State Letters, 6 (10) 1-3 (2003).*

[11]P. Singh, Z. Yang, "Thermo-chemical Analysis of Oxidation Processes in High Temperature Fuel Cells," 131st TMS Annual Meeting, Seattle, February 18, 2004

[12]Z. Yang, G. Xia, P. Singh, J. Stevenson, "Advanced Metallic Interconnect Development," SECA Annual Workshop and Core Technology Peer Review, Boston, May 11-13, 2004.

[13] Z. Yang, G. Xia, P. Singh, J. Stevenson, *Solid State Ionics 176 (2005) 1495.*

HIGH TEMPERATURE CORROSION BEHAVIOR OF OXIDATION RESISTANT ALLOYS UNDER SOFC INTERCONNECT DUAL EXPOSURES

Zhenguo Yang, Greg W. Coffey, Joseph P. Rice, Prabhakar Singh, Jeffry W. Stevenson, and Guan-Guang Xia

Pacific Northwest National Laboratory,
902 Battelle Blvd,
Richland, WA 99352

ABSTRACT

Metals and alloys are considered as promising candidates to construct interconnects in intermediate temperature (600-800°C) solid oxide fuel cell (SOFC) stacks. During SOFC operation, the interconnects are working in a dual environment, i.e. simultaneously exposed to air at cathode side and a fuel (e.g. hydrogen) at the anode side. Our recent studies found the oxidation behavior of metals and alloys in the dual environment can be significantly different from that in a single oxidizing or reducing atmosphere exposure. The anomalous oxidization is attributed to hydrogen diffusion flux from the fuel side to the air side under the influence of a hydrogen gradient across the stainless steel interconnects. This paper summarizes our study on a number of selected metals and alloys under the dual environment.

INTRODUCTION

Advances in materials and cell fabrication have led to a steady reduction in SOFC operating temperatures to an intermediate range of 600-800°C. Consequently, it has become feasible to supplant lanthanum chromite with high temperature oxidation resistant (HTOR) alloys as the stack interconnect materials [1-3]. Compared to doped lanthanum chromite, the alloys potentially offer advantages, including significantly lower raw material and fabrication costs. However, the SOFC operating conditions present significant challenges for the metallic interconnects. To be durable and reliable, the alloys must demonstrate excellent oxidation and corrosion resistance in a very challenging "dual" environment, as they are simultaneous exposed to a fuel (hydrogen or hydrogen–rich) on the anode side and an oxidant (air) on the cathode side during SOFC operation. Depending on the stack design, both Fe-Cr base stainless steels and Ni-Cr-base heat resistant alloys can be potential candidates for interconnect applications [3-7].

The oxidation behavior (scale composition, structure, and growth) of HTOR alloys has been widely investigated for a myriad of applications [5, 8-17], including recent work on the SOFC interconnect applications [5, 13-17]. These studies, however, were usually carried out using single atmosphere exposure conditions (i.e., either an oxidizing [5,13-16] or reducing environment [17]), and were presumably based on the implicit assumption that oxidation behavior as measured in either an oxidizing or reducing environment will be essentially identical to the scale growth and corrosion occurring on the air or fuel side of the material when it experiences dual atmosphere conditions.

However, our recent investigation of various metals and alloys under dual atmosphere conditions found that the oxidation and corrosion behavior on the air side differs significantly from the behavior observed when the steels are exposed to air only [18-21]. The effects of dual exposures on the oxidation behavior of metals and alloys have also been observed by others [22-

24]. Consequently, an extensive study has been carried out at PNNL on interconnect candidate alloys under the SOFC interconnect dual exposures. The alloys studied include ferritic stainless steels and Ni-base alloys. For purpose of comparison and mechanistic understanding, elemental metals such pure Ni were also investigated. This paper reports and summarizes the experimental procedure and key results of this research.

EXPERIMENTAL

The dual atmosphere study was carried out in a specifically designed apparatus that allows for simultaneous testing of separate alloy coupons in 3 different atmospheres (air, fuel and dual atmosphere) [18,19]. The materials under investigation are three selected ferritic stainless steels, AISI430 (17%Cr), Crofer22 APU (22%Cr), and E-brite (26%Cr), and three Ni-base alloys, Haynes 242, Hastelloy S, and Haynes 230. The chemical compositions of the ferritic stainless steels and Ni-base alloys are listed in Table I and II, respectively.

Table I. Chemical compositions of ferritic stainless steels.

Ferritic stainless steels	Nominal composition, wt%								
	Fe	Cr	Mn	Si	C	Ti	P	S	La
AISI430*	Bal	17.0	0.50	0.51	0.10	--	0.03	0.03	--
Crofer22 APU*	Bal	22.8	0.45	--	0.005	0.08	0.016	0.002	0.06
E-brite*	Bal	27.0	0.01	0.025	0.001	--	0.02	0.02	--

*AISI430 and E-brite were provided by Alleghany Ludlum; Crofer22 APU by ThyssenKrupp VDM.

Table II. Chemical compositions of Ni-Cr base alloys.

Alloys	Nominal composition, wt%											
	Ni	Cr	Fe	Co	C	Mn	Si	Mo	W	Al	B	Others
Haynes 230*	Bal	22.0	3.0^m	5.0^m	0.10^m	0.5	0.4	2.0	14.0	0.30	0.015^m	0.5Cu^m
Hastelloy S*	Bal	16.0	3.0^m	2.0^m	0.02^m	0.5	0.4	15.0	1.0^m	0.25	0.015^m	0.02La
Haynes 242*	Bal	12.0	2.0^m	2.5^m	0.03^m	0.8^m	0.8^m	25.0	--	0.5^m	0.015^m	

* Haynes, Hastelloy, 230, S and 242 are registered trademarks of Haynes International; m: maximum.

The dual atmosphere specimen was prepared by sealing a 0.5 mm thick, 25 mm diameter circular disk of a stainless steel to the end of an E-brite tube using BNi-2 braze, a Ni-based braze. After brazing, a helium leak test was performed to assure that the metal disk was hermetically sealed to the E-brite tube. After assurance of hermetic sealing, the E-brite tube and the ferritic stainless steel disk were placed in a test-stand, along with two coupons that would be exposed to air only or fuel only. Moist hydrogen (with ~3% moisture) as the fuel was introduced into the furnace by flowing hydrogen through a water bubbler at room temperature. Ambient air or moist air that was prepared as the same way as for the moist hydrogen was flowed into the furnace using a Tetratec AP200 pump. The furnace was then heated to 800°C and kept at temperature for 300 hours in an isothermal mode or cycled for three times with each cycle lasting 100 hours.

In addition, pure Ni (99.9%) metal was also investigated under the dual exposure conditions in a double tube setup. Due to the availability of thin wall Ni tubes, the test was carried out in an apparatus that is different from the previous coupon tests. In this setup, two identical Ni tubes were arranged vertically in a vertical tube furnace with air flowing through one tube, and moist hydrogen through the other.

The surfaces of the tested samples were first analyzed on a Philips XRG-3100 X-ray Generator with Cu K_α radiation and then under a JEOL scanning electron microscope (model 5900LV) equipped with energy-dispersive spectroscopic X-ray (EDS) capability at an operating voltage of 20 kV. After surface analysis, the coupon samples were subsequently cross-sectioned for further SEM analyses. In addition, JEOL 2010 high resolution transmission electron (TEM) was also used to study the scales grown on Crofer22 APU under different conditions.

RESULTS and DISCUSION

Ferritic Stainless Steels

For Crofer22 APU, with 22-23%Cr, it was found that, on the moist hydrogen side of coupons exposed to the dual atmosphere, the scale crystal structure, microstructure, and composition were similar to scales grown on coupons that were exposed to moist hydrogen on both sides. In contrast, the scale grown on the air side of the dual atmosphere coupons was significantly different from scales on coupons that were exposed to air on both sides. SEM/EDS and TEM/EDS analyses all indicated that the scale grown on the air side of the coupon held at 800°C for 300 hours under dual atmospheres consisted of an iron-rich spinel (i.e. $(Mn,Cr,Fe)_3O_4$) top layer, while little iron was observed in the spinel top layer on the scale grown on the coupon exposed to air only. Figure 1 shows the TEM bright field image (Figure 1(a)) of the scale grown on the sample that was exposed to moist air at both sides and that (Figure 1(b)) of the scale grown at the airside of dual exposure samples after 300 hours at 800°C.

Besides Crofer22 APU, AISI430 and E-brite were also investigated under single and dual exposure conditions. It was found that, for AISI430 with 17%Cr, hematite (α-Fe_2O_3) rich nodules formed on the air side even under isothermal oxidation, as shown in Figure 2 (c) and (d). In addition to this localized attack, the air side scale was also found to be more prone to defects, such as porosity, than the scale grown on the coupon that was exposed to air on both sides. In contrast, a uniform scale was grown on the AISI430 coupon when exposed to air at both sides. For E-brite with 26%Cr, under the current test scheme no iron-rich spinel layer or hematite nodules were observed in the scale grown at the air side of the dual atmosphere coupon. However, the scale grown on the coupon that was exposed to air on both sides appeared to be denser that the scale grown on the airside of the dual atmosphere coupon. Again, no anomalous scale growth or corrosion was observed on the moist hydrogen side.

In addition to the alloy composition, water vapor content and temperature were found to be other important factors affecting the anomalous oxidation behavior of the ferritic stainless steels [21].

Ni-Base Alloys

In addition to the ferritic stainless steels, three Ni-base alloys, Haynes 242 (9%Cr), Hastelloy S (17%Cr), and Haynes 230 (22%Cr) were also investigated under the SOFC interconnect dual exposure conditions.

For Haynes 230 and Hastelloy S, study indicated that the dual exposures led to a different oxidation behavior at the air side of the samples that were exposed to hydrogen fuel at the other side. But unlike the ferritic stainless steels, no localized attack via accelerated growth of iron oxide nodules was observed at the airside of the Ni-base alloy samples. Instead, a uniform scale was observed on the alloy surface after 300 hours at 800°C under the dual exposure conditions. As shown in Figure 3, the Ni-based Hastelloy S was characterized by formation of a uniform, fine chromia dominated scale at the airside (see Figure 3(b)), which also contained a small amount of spinels (see Figure 3 (a)). In comparison, the scale grown on the Haynes S alloy that was exposed to moist air at both sides appeared less uniform and NiO was observed in the scale (see 3 (c)). Thus the oxidation behavior of the Ni-base alloy at the air side of the dual exposure samples was different from that when exposed to air or moist air at both sides. In contrast, the scale grown on the fuel side of the Hastelloy S samples that were exposed to moist air at the other side was similar to the scale growth on the sample that was exposed to fuel at both sides. A similar oxidation behavior was also observed on Haynes 230 when exposed to the dual environment.

Similar to Hastelloy S and Haynes 230, no substantial difference was observed between the scale grown on the fuel side of Haynes 242 that was exposed to moist air at the other side, and the scale grown on the sample that was exposed to a fuel at both sides. Both of these two scales appeared to be uniform and contained chromia and spinel phases. In contrast, X-ray diffraction analysis, as shown in Figure 4 (a)), indicated that the scale grown on the sample that was exposed to moist air at both sides appeared less uniform and contained more NiO in the scale. The difference was however not obvious during SEM examination, as shown in Figure 4 (b) and (c). Both the scale grown on the airside of the dual exposure sample and the scale grown on the sample exposed to moist air at both sides were comprised of a thick NiO top layer over a chromia and spinel rich sub-layer.

Overall it appears that, for Ni-based alloys with enough chromium, such as Haynes 230 and Hastelloy S, dual exposure tended to facilitate the formation of a uniform chromia dominated scale. This effect was however not apparent in Haynes 242, an alloy with only 9%Cr.

Pure Elemental Ni

As previously mentioned, the oxidation behavior of ferritic stainless steels and Ni-based alloys at the airside of dual exposure samples can be significantly different from that observed on samples exposed to air at both sides. To gain further insight into the anomalous oxidation behavior, Ni was chosen as an example of a pure elemental metal system and its oxidation behavior under the dual exposures was studied. Figure 5 shows microstructures of the surfaces and cross-sections of scales that were grown on the outside surfaces of the Ni tubes (exposed to air) during high temperature exposure at 800°C for 300 hours. The surface microstructure of the scale (see Figure 5(b)) grown on the tube with flowing moist hydrogen inside appeared to be similar to that of the scale (see Figure 5(a)) grown on the tube with flowing air inside, except for pin holes observed in some NiO crystals in the first scale (Figure 5(b)). Nevertheless, significant differences were clearly found during the SEM examination on the cross-sections. While porosity was observed at the metal/scale interface of the tube with flowing air inside, no porosity was observed at the same interface for the tube through which moist hydrogen was flowed. Also, the scale grown on the tube that with moist hydrogen flowing inside was thicker than that grown on the tube with flowing air inside. Thus it appears that the dual exposure conditions alter the NiO scale growth on Ni due to a hydrogen flux across the tube wall thickness.

CONCLUSIONS

In summary, when metals and alloys were simultaneously exposed to air on one side and moist hydrogen at the other, the scale grown on the airside differed from the scale observed during exposure to a single air atmosphere only. In contrast, no substantial difference was observed between the scales grown on the fuel side and scales grown on coupons exposed to moist hydrogen only. Also, some differences in scale microstructure and composition were observed in the scales grown in air and in moist hydrogen. The anomalous scale growth on the airside of dual exposure coupons was dependent on the composition of the alloys.

The ferritic stainless steels, in particular those with a relative low Cr%, tended to exhibit a localized attack at the airside via inhomogeneous scale growth. For example, AISI430, with 17% Cr, formed Fe_2O_3 hematite nodules on the airside even during isothermal heating at 800°C, while the spinel top layer of the scale on the airside of Crofer22 APU (23%Cr) was merely enriched in iron under similar conditions. For E-brite (27% Cr), no unusual composition or phase was found in the airside scale after isothermal heating at 800°C for 300 hours, but it wasobserved that the scale was less dense and appeared to be more prone to defects than scales grown in air only.

For Ni-based alloys with sufficient chromium, e.g. Haynes 230 and Hastelloy S, dual exposure tended to facilitate the formation of a uniform chromia-dominated scale; This effect was however not obvious for Haynes 242, an alloy with only 9%Cr. Effects of dual atmosphere exposure on oxidation behavior were further observed on pure elemental Ni.

The anomalous oxidation behavior of metals or alloys on the airside of dual exposure samples was presumably related to the transport of hydrogen through the steel (from the fuel side to the airside) and its subsequent presence at the metal/oxide scale interface and in the scale. Further work is needed to achieve a mechanistic understanding of this behavior.

ACKNOWLEDGEMENTS
The authors would like to also thank Nat Saenz and Shelly Carlson for their assistance in metallographic and SEM sample preparation, Jim Coleman for SEM analysis, and Chong-Min Wang for TEM samples preparation and analysis. The work summarized in this paper was funded as part of the Solid-State Energy Conversion Alliance (SECA) Core Technology Program by the U.S. Department of Energy's National Energy Technology Laboratory (NETL). PNNL is operated by Battelle Memorial Institute for the U.S. Department of Energy under Contract DE-AC06-76RLO 1830.

REFERENCES
1. N.Q. Minh, J. Am. Ceram. Soc., 76, 563 (1994).
2. B.C.H. Steele, Nature (London), 414, 345 (2001)
3. W.J. Quadakkers, J. Piron-Abellan, V. Shemet, and L. Singheiser, Materials at High Temp., 20, 115 (2003).
4. Z. Yang, K.S. Weil, D.M. Paxton, and J.W. Stevenson, J. Electrochem. Soc., 150, A1188 (2003).
5. Tietz F. in High-Temperature Solid Oxide Fuel Cells: Fundamentals, Designs and Applications, Ed. Singhal SC, Kendal K, Elsevier Science, The Netherlands, 2004, p. 173
6. P. Kofstad and R. Bredesen, Solid State Ionics, 52, 69 (1992).
7. W.Z. Zhu and S.C. Deevi, Mater. Res. Bull., 38, 957 (2003).
8. D.J. Young and S. Watson, Oxid. Met., 44, 239 (1985).
9. F. Gesmundo, and B. Gleeson, Oxid. Met., 44, 211 (1985).
10. P. Kofstad, *High-Tempearture Corrosion*, Elsevier, 1988.
11. E.A. Gulbranssen and T.P. Copen, Nature, 186, 959 (1960).

12. M.P. Ryan, D.E. Williams, R.J. Chater, B.M. Hutton, and D.S. McPhail, Nature, 415, 770 (2002).
13. W.J. Quadakkers, T. Malkow, J. Piron-Abellan, U. Flesch, V. Shemet, and L. Singheiser, in Proc. 4th European Solid Oxide Fuel Cell Forum, A. McEvoy, Editor, p. 827, European SOFC Forum, Switzerland (2000).
14. K. Huang, P.Y. Hou, and J.B. Goodenough, Solid State Ionics, 129, 237 (2000).
15. T. Malkow, U.V.D. Crone, A.M. Laptev, T. Koppitz, U. Breuer, and W.J. Quadakkers, in Proc. 5th Int. Symp. Solid Oxide Fuel Cells, U. Stimming, S.C. Singal, and H. Tagawa, Editors, PV 97-40, p. 1244, The Electrochemical Society, Pennington, NJ (1997).
16. D.M. England and A.N. Virkar, J. Electrochem. Soc., 146, 3196 (1999).
17. D.M. England and A.N. Virkar, J. Electrochem. Soc., 148, A330 (2001).
18. Z. Yang, M.S. Walker, P. Singh, J.W. Stevenson, *Electrochem. & Solid State Lett.*, 6, B35-37 (2003).
19. Z. Yang, M.S. Walker, P. Singh, J.W. Stevenson, T. Norby, *J. Electrochem. Soc.*, 151, B669-678 (2004).
20. P. Sing, Z. Yang, V. Viswanathan, and J.W. Stevenson, J Mater. Perform. Eng., 13, 287(2004).
21. Z. Yang, G-G. Xia, P. Singh, J.W. Stevenson, *Solid State Ionics*, 176, 1495-1503 (2005).
22. M. Ziomek-Moroz, B.S. Covino, S.D. Cramer, G.R. Holcomb, S.J. Bullard, P. Singh, in: Corrosion 2004, NACE International, Houston, TX, 2004, p.1.
23. K. Nakagawa, Y. Matsunaga, T. Yanagisawa, Mater. High Temp., 20 (2003) 67.
24. G.R. Holcomb, M. Ziomek-Moroz, S.D. Cramer, B.S. Covino, Jr., and S.J. Bullard, J. Mater. Perform. Eng., in press.
25. S.K. Yen and Y.C. Tsai, J. Electrochem. Soc., 143, 2736 (1996).
26. P. Kofstad, Oxid. Met., 44, 3 (1995).
27. B. Tveten, G. Hultquist, and T. Norby, Oxid. Met., 51, 221 (1999).

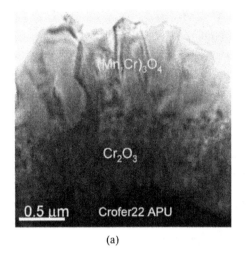

(a)

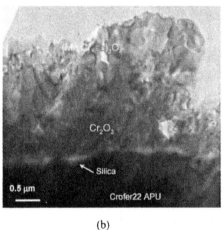

(b)

Figure 1. TEM bright field images of scales grown on (a) a Crofer22 APU coupon that was exposed to air on both sides, and (b) the airside of a coupon that was exposed to moist hydrogen (3% H_2O) on the other side. Both of these coupons were isothermally treated at 800°C for 300 hours.

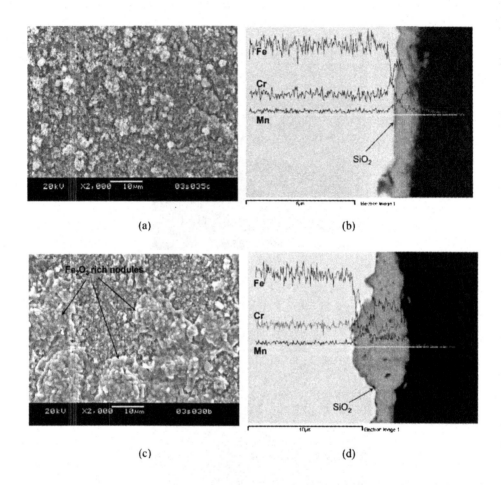

(a)

(b)

(c)

(d)

Figure 2. SEM observation of the scale on AISI430 after heat-treatment at 800°C for 300 hours: (a) surface and (b) cross-section microstructures of the scale grown on a coupon that was exposed to air on both sides; (c) surface and (d) cross-section microstructures of the air side scale on the coupon that was simultaneously exposed to air on one side and moist hydrogen on the other. Results of the EDX linear analysis on cross-sections are also included.

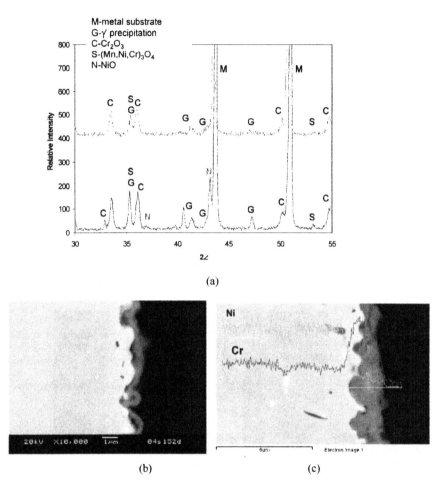

Figure 3. Crystal structure and microstructural analyses on Hastelloy S after heat-treatment at 800°C for 300 hours: (a) X-ray diffraction pattern of the scale grown on the moist air side of Hastelloy S that was heat-treated at 800°C for 300 hours under dual atmospheres, in comparison with that from the scale on a coupon that was subjected to the same heat treatment, but with both sides exposed to moist air, and SEM cross-section microstructures of the scale grown on (b) the coupon that was exposed to moist air on both sides and (c) on the airside of the coupon that was simultaneously exposed to moist air on one side and moist hydrogen on the other.

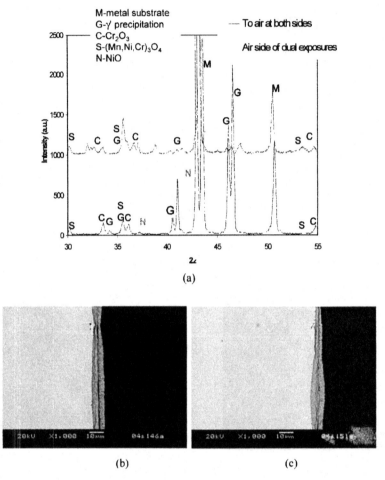

Figure 4. Crystal structure and microstructural analyses on Haynes242 after heat-treatment at 800°C for 300 hours: (a) X-ray diffraction pattern of the scale grown on the moist air side of Haynes242 that was heat-treated at 800°C for 300 hours under dual atmospheres. in comparison with that from the scale on a coupon that was subjected to the same heat treatment, but with both sides exposed to moist air, and SEM cross-section microstructures of the scale grown on (b) the coupon that was exposed to moist air on both sides and (c) on the air side of the coupon that was simultaneously exposed to moist air on one side and moist hydrogen on the other.

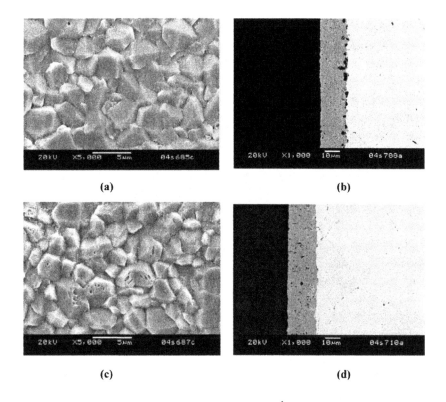

(a)

(b)

(c)

(d)

Figure 5. SEM observation of Ni tubes after being oxidized for 300 hours at 800°C under different exposure conditions: (a) microstructure of outside surface and (b) that of cross-section of the tube that was exposed to air both inside and outside: (c) microstructure of outside surface and (c) that of cross-section of the tube with flowing moist hydrogen inside and ambient air outside.

ELECTRO-DEPOSITED PROTECTIVE COATINGS FOR PLANAR SOLID OXIDE FUEL CELL INTERCONNECTS

Christopher Johnson, Chad Schaeffer, Heidi Barron, Randall Gemmen
US Department of Energy
National Energy Technology Center
Morgantown, WV 26505-0880

ABSTRACT

Chromium containing high temperature alloys coated with stable and conductive oxides are being considered as possible interconnects for planar solid oxide fuel cells. The coatings are necessary to increase oxidation resistance and reduce chromium evaporation. Here we report the use of electro-deposition methods for applying Mn/Co alloy coatings as precursor for the production of Mn/Co spinel protective layers. Characterization of the films is done by SEM and EDS analysis.

INTRODUCTION

Solid oxide fuel cells (SOFC) hold great potential for future electrical energy production (1). This is not only because SOFC are more efficient than current commercial methods of electricity production, but also because they emit lower levels of pollutants than these alternative methods. In addition, because of their high operating temperature relative to other types of fuel cells, they are ideal for combined cycle SOFC/ turbine applications, where efficiency may approach 70-80%. At typical operating temperatures (700-1000°C), internal reforming of hydrocarbon fuels using their excess heat is also a possibility. Despite these very attractive attributes, SOFCs have yet to be widely commercialized. The impediments to commercialization are related to a number of cell and stack materials issues. For instance, in planar type SOFC, reliable and effective seals that can be thermally cycled, are stable relative to other SOFC components, and will last the expected 40,000 hours, have yet to be developed. Another example is the typical Ni/YSZ anode materials which, if not properly designed, degrades due to Ni particle sintering, or may build up carbon deposits if the anode gas mixture is not correctly chosen. In this paper, work designed to solve another material issue, the development of low cost and effective interconnects for planar SOFC, is discussed.

Until recent years the operating temperatures of the typically SOFC was 1000°C. This temperature was required to achieve high ionic conductivity of the YSZ electrolyte. However, with the advent of higher performing electrolytes, operating temperatures of 700-800°C suffice to give equivalent conductance to the thicker electrolyte SOFC. The lowering of the temperature needed for cell operation meant that a number of lower cost materials became potential replacements for previously used materials. One such opportunity was the replacement of the $LaCrO_3$ based ceramic interconnects with low cost high temperature alloys(2). Besides lower cost, the high temperature alloys also have the added benefit of higher mechanical strength, and higher thermal conductivity. These are particularly useful for planar designs where interconnects provide mechanical support, and the higher thermal conductivity gives a more uniform temperature distribution in the fuel cell stack. The best candidate high temperature alloys are chromia formers. This is because chromia

223

has the lowest electronic resistivity relative to other native oxides of high temperature alloys. However, chromia scales are not as resistant to oxide growth at elevated temperatures as the alumina formers, and thus over the long term the oxide scales continue to grow and increase resistance. In addition, in the severe SOFC cathode environment, volatile Cr species can migrate and contaminate the cathode/electrolyte interface, causing degradation of cell performance. Thus, while the high temperature chromia forming alloys are the best replacement candidates for $LaCrO_3$ based SOFC interconnects, additional improvement in corrosion resistance and a solution to the chromia volatility issue remain to be addressed.

A potential method of mitigating the problem with chromia volatility is by changing the native oxide that forms on a high temperature interconnect alloy to one that is both conductive and is more thermodynamically stable than the Cr_2O_3 scale. One such example of this is when small amounts of Mn are added to the alloy. Addition of the correct amount of Mn leads to the formation of $MnCr_2O_4$ spinel as the outer most surface oxide, after some period of time at typical SOFC operating temperatures. This spinel layer is a relatively good electronic conductor and does reduce the Cr evaporation. However, degradation of cell performance continues to be a problem (3), despite the lowered Cr evaporation rate. The problem may be that the Cr preferentially deposits in the electrochemically active region near the cathode electrolyte interface during passage of current through the cell (4). This non-equilibrium deposition of Cr at the worst possible place in the cell (triple phase boundary) means that much lower Cr species vapor pressures must be achieved.

A number of research groups have investigated the use of coatings to improve conductivity of scale and reduce chromia volatility for high temperature alloy interconnects (5,6). Typically, perovskites such as calcium or strontium doped lanthanum chromites have been investigated. The perovskites based on lanthanum chromite however still contain chromium as a component, and while the chromite phases are more thermodynamically stable than other chromium containing oxides such as Cr_2O_3 and MnCr spinels there still is a finite amount of chromia evaporation from the surface. More recently, none chromia containing Mn/Co spinel coatings applied by slurry spraying of the spinel powder have been investigated as a protective coating (7). The authors report good adhesion of the coatings, good area specific resistance (ASR) values, and low Cr levels in the coating after thermal cycling.

In this work we report our initial efforts to obtain Mn/Co spinel coatings via electroplating methods. Electroplating of alloys or individual metals followed by controlled oxidation and/or reaction to the desired spinel may be an effective method for the formation of dense spinel coatings. This method may also be more amenable for the deposition of coatings on interconnects with complex geometries or even deposition on internal pores of foam materials. This report primarily looks at the deposition of a Mn/Co alloy from a single deposition solution on to Crofer APU 22 alloy substrates. Characterization of the deposits is by SEM and EDS analysis.

Experimental

The electrolytes used in this experiment were prepared from $CoSO_4\bullet 7H_2O$, $MnSO_4\bullet H_2O$, saccharin, and disodium EDTA. The specific concentrations used in this work were 0.008M $CoSO_4$, 0.008M EDTA, 0.075M Saccharin, 0.032M $MnSO_4$. Dilute sulfuric acid (H_2SO_4) and sodium hydroxide (NaOH) were used to adjust pH. All solutions were made with deionized water. The deposition cell used the work piece (crofer APU 22 alloy) as the cathode, and two sheets of platinum mesh were used as the anodes. The fixture was suspended in a beaker of the solution and this was placed on a heating and stirring mantle, and a Teflon coated stir bar used to provide agitation. These experiments were done without the use of a reference electrode. All experiments were carried out at 50 °C. The morphology and composition of the deposited samples were analyzed using a JEOL JEOL JSM 6300 FE-SEM equipped with a Thermo Electron EDS system.

Results and Discussion

The effect of the saccharin additive was investigated by depositing single metal Co and Mn electrodeposited from simple single component ($CoSO_4 \bullet 7H_2O$ or $MnSO_4 \bullet H_2O$) solutions with and without saccharin additive. For the cobalt deposition the concentration of cobalt sulphate was 0.1M and saccharin was 0.0025M. The cell was operated potentiostatically at an applied voltage of 4V. Figure 1 shows the SEM images of the results found for the Co deposition. The upper images (A and B) show the morphology of the cobalt deposit when saccharin additive is not present. The lower images (C and D) show the effect of addition saccharin to the deposition solution. In the upper images the large clusters are the cobalt deposits, while the surround dark area is still the uncoated substrate. In the lower images the cobalt has deposited relatively uniformly over the surface of the substrate with only an occasional cobalt cluster present (the light particles intermittently dotting the surface). The effect of saccharin addition on the deposition of Mn is qualitatively similar. The morphology of the Mn is significantly smoother when saccharin is added.

The deposition of alloys can typically be accomplished when the deposition potential for each of the components are approximately the same. This makes the deposition of Mn/Co alloys from simple solutions very difficult. The standard reduction potentials for Mn^{2+} and Co^{2+} are -1.180V and -0.277V respectively. If these were closer in value then it may be possible to simply use concentration differences to control the deposition potentials. However, estimation of the concentrations necessary to bring the deposition potentials to equal levels for Mn^{2+} and Co^{2+} done using the Nernst equation and assuming dilute solutions (so that concentration can be used instead of activity) shows that the Co^{2+} concentration would need to be approximately 30 orders of magnitude more dilute than the Mn^{2+}. Therefore alternative methods of controlling the deposition potentials need to be found.

One method of changing the deposition potential for a given metal is to change the species from which the deposition is made. This can be done by formation of a complex or species with a higher or lower standard reduction potential. For instance, EDTA has been used in Zn-Mn alloy electroplating to change the deposition potential of the Zn^{2+}

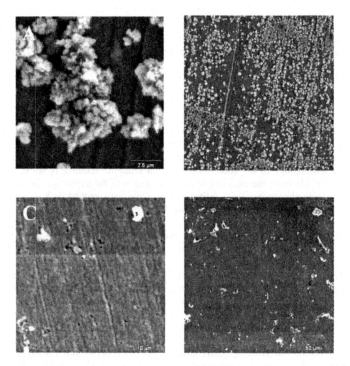

Figure 1. Upper SEM images (A and B) show the morphology of the cobalt deposit without saccharin. Lower SEM images show the cobalt deposit with 0.0025M saccharin.

species(8). The Zn^{2+} was reacted with stoichiometric amounts of EDTA to form an EDTA-Zn complex before the remaining Mn species was added to the solution. The result is an increase in the deposition potential of the Zn species at a given concentration. For the purposes of this study EDTA may be use to complex the Co^{2+} cation and raise its

Reaction 1. $Mn^{2+} + 2H_2O \rightleftharpoons HMnO_2^- + 3H^+$

Reaction 2. $HMnO_2^- + 3H^+ + 2e^- \rightleftharpoons Mn(s) + 2H_2O$ $E° = -0.163$

deposition potential relative to Mn^{2+}. To lower the deposition potential of the Mn^{2+} species we can take advantage of the effect of pH on the presence of a more easily reducible aqueous manganese species (shown in reaction 1) created when Mn^{2+} ions are dissolved in
water. The standard reduction potential of $HMnO_2^-$ (shown along with the deposition reaction, reaction 2) is -0.163V, much lower than -1.180V for the Mn^{2+} species. The K_{sp} of the reaction shown below at 50°C is 6.59×10^{-19}, so that under typical conditions (when the $MnSO_4$ is simply

added to solution) the $HMnO_2^-$ concentration is very low and despite the lower $E°$ value the deposition potential remains high because of the low concentration. However, we can shift the concentration of the deposition species by changing the pH, that is, by consuming the H^+ ions available in solution we can shift the equilibrium for reaction 1 to the right and increase the amount of $HMnO_2^-$ available for deposition.

The concentration of $HMnO_2^-$ species can be calculated at a number of pH values by use of the equilibrium equation below.

$$K = \frac{[HMnO_2^-][H^+]^3}{[Mn^{2+}]} = 1.6 X 10^{-32}$$

Then using the Nernst equation the deposition potential for $HMnO_2^-$ at a number of pH values can be estimated for a given manganese sulfate concentration. Figure 2 shows the calculated deposition potential for $HMnO_2^-$ to Mn(s) as in reaction 2 over a range of pH values. Again, it is assumed that concentrations can be substituted for activities in the Nernst equation. Note that at very high pHs the value of the deposition potential approaches the deposition potential of Co^{2+} (- 0.277V), which would suggest that simply increasing the pH near 9 would lead to alloy deposition without complexation of the Co^{2+} with EDTA. Unfortunately, at pH levels approaching 8, precipitation of hydroxide occurs in the solution, therefore it is still necessary to increase the deposition potential of the Co^{2+}.

Given the above information we determined that we would first dissolve the $CoSO_4$ and react this with the EDTA to form the Co-EDTA complex. The saccharin and $MnSO_4$ were added once the EDTA was completely dissolved. The pH was the controlled using 10% solution of H_2SO4 and NaOH. Figure 3 shows the Mn/Co atomic ratio of the deposited films as a function of pH. In this case the cell was operated galvanostatically at 50mA. It should also be pointed out that at the pH of 8 a small amount of precipitate was present in the solution. Nevertheless, from this graph, we can see that to obtain a 50/50 mixture of Co and Mn in the deposited alloy from this particular solution, the pH should be near 7.5. Comparison of the calculated deposition potential of the Mn species at pH 7.5 (-0.52855) to the deposition potential calculated for Co2+ at 0.008M concentration (-0.34419) allows an estimate of the increase in deposition potential attributable to the complexation of the Co2+ by EDTA. That increase is -0.18436V. Thus if the Co2+ had not been complexed with EDTA, the pH of the solution would need to be higher in order to bring the Mn species deposition potential down to -0.34419 (approximately a pH of 8.5) at which point a large amount of hydroxide material would have precipitated. Care must be taken when operating the cell above a voltage that can dissociate water as the half reaction at the cathode for water electrolysis also creates hydroxide ions. Thus localized increases in pH can occur and produce significant amounts of hydroxide precipitate on the work piece.

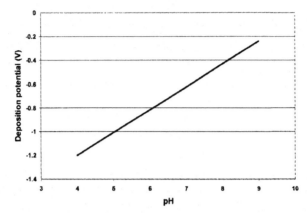

Figure 2. The Calculated deposition potential as a function of pH for deposition of Mn(s) from HMnO$_2^-$ containing solutions. using the Nernst equation and assuming dilute solutions.

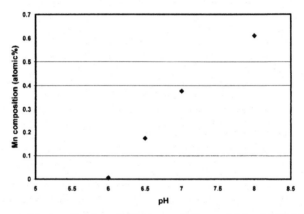

Figure 3. Mn composition as a function of pH using a solution containing 0.008M CoSO4, 0.008M EDTA, 0.075M Saccharin, 0.032M MnSO$_4$. In this case the cell was operated galvanostatically at 50mA.

SUMMARY

The work presented here has focused on developing a solution composition for electrodeposition of Mn/Co alloys. It was found that saccharin addition to the deposition solution acts a leveling agent. leading to much smoother films. It is also necessary to adjust the deposition potentials of the two metals by complexation of the Co2+ ion with EDTA and control of the Mn deposition species by adjusting the pH of the solution. Since precipitation of hydroxides can occur if solution pHs are too high, care must be taken not drive the water splitting reaction so fast that localized pH levels cause hydroxide formation on the work piece.

In future work we will address the oxidation of the electrodeposited alloys and evaluate the effectiveness of the oxide coatings as protective layers for ferritic stainless steels used as interconnect in SOFC.

1. S.C. Singhal, *Solid State Ionics*, 2000, 135, 305-313

2. W.Z. Zhu, S.C. Deevi, *Materials Science and Engineering,* 2003, A348, 227-243

3. K. Fujita, K. Ogasawara, Y. Matsuzaki, T. Sakurai, *J. of Power Sources*, 2004, 131, 261-269

4. S. Taniguchi, M. Kadowaki, H. Kawamura, T. Yasuo, Y. Akiyama, Y. Miyake, T. Saitoh, *J. of Power Sources*, 1995, 55, 73-79

5. S. Linderoth, Controlled reactions between chromia and coatings on alloy surface, *Surface and Coating Technology*, 1996, 80, 185-189

6. J. Kim, R, Song, S. Hyun, *Solid State Ionics*, 2004, in press, available at www.sciencedirect.com

7. X. Chen, P. Hou, C. Jacobson, S. Visco, L. De Jonghe, *Solid State Ionics* 2005, 176, 425-433

8 C. Muller, M.Sarret, and T Andreu, *J. of the Electrochemical Society* (2002), 149 (11) C600-C606

PROPERTIES OF (Mn,Co)$_3$O$_4$ SPINEL PROTECTION LAYERS FOR SOFC INTERCONNECTS

Zhenguo Yang, Xiao-Hong Li, Gary D. Maupin, Prabhakar Singh, Steve P. Simner, Jeffry W. Stevenson, Guan-Guang Xia, and Xiaodong Zhou

Pacific Northwest National Laboratory,
902 Battelle Blvd,
Richland, WA 99352

ABSTRACT
    Ferritic stainless steels are promising candidates for interconnect applications in low- and mid-temperature solid oxide fuel cells (SOFCs). A couple of issues however remain for this particular application, including chromium poisoning due to chromia evaporation, and long-term surface and electrical stability of the scale grown on these steels. Application of a manganese colbaltite spinel protection layer on the steels appears to be an effective approach to solve the issues. For an optimized performance, properties of Mn$_{1+x}$Co$_{2-x}$O$_4$ (-0.5≤x≤1.5) spinels relevant to the protective coating application were investigated. It was found that the spinels with x around 0.5 demonstrated a good CTE match to ceramic cell components, high electrical conductivity, and thermal stability up to 1,250°C. The material suitability was confirmed by a long-term test on a Mn$_{1.5}$Co$_{1.5}$O$_4$ protection layer that was thermally grown on Crofer22 APU, indicating the spinel protection layer not only significantly decreased the contact resistance between a LSF cathode and a stainless steel interconnect, but also inhibited the sub-scale growth on the stainless steel.

INTRODUCTION
    Due to their electrically semi-conducting oxide scale, appropriate thermal expansion behavior, and low cost, chromia-forming ferritic stainless steels are considered among the most promising candidate materials for interconnect applications in intermediate or low temperature SOFCs [1-5]. One issue that potentially hinders their application, however, is the migration of chromium via chromia scale evaporation into SOFC cathodes [6-8], which can lead to a severe degradation in cell electrochemical performance [8-11]. Newly developed alloys such as Crofer22 APU, which is protected at elevated temperatures via formation of a unique scale that is comprised of a (Mn,Cr)$_3$O$_4$ spinel top layer and chromia or chromia-rich sub-layer [12-14], may offer some improvement in this regard due to the lower volatility of Cr from spinel than from chromia. However, volatility measurements at PNNL indicate that the chromium volatility from the spinel may be only a factor of ~3 lower than that from chromia (at 850°C), so Cr volatility from the spinel as well as any exposed chromia that is not covered by the spinel layer, particularly during the early stages of oxidation, may still result in an unacceptable degradation in cell performance [8,14,15]. Also, it appears that a further improvement in long-term scale stability is needed, particularly for SOFC stacks with an operating temperature above 700°C [14,16,17].
    As an alternative approach to the bulk modification of alloy composition, the interconnect can be surface-modified via application of an overlay coating of conductive oxide(s) (e.g. perovskites) on the cathode-side. These overlay coatings may help lower the interfacial

231

contact resistance, but cell performance may still be degraded by chromium migration from either chromium-containing perovskites (e.g. (La,Sr)CrO$_3$) or non-chromium-containing compositions via chromium cation diffusion through the coatings. Other potential challenges associated with perovskite overlay coatings include the thermomechanical stability during thermal cycling due to poor adhesion and the subscale growth from the oxygen anion inward diffusion through coatings. Spinel protective layers have also been investigated. Previous work of Larring and Norby [18] on Plansee Ducrolloy (Cr-5%Fe-1%Y$_2$O$_3$), an interconnect alloy for high temperature (900-1,000°C) SOFCs, indicated that a (Mn,Co)$_3$O$_4$ spinel layer could be a promising barrier to chromium migration. Recently, Yang et al. [19-21] investigated thermal growth of (Mn,Co)$_3$O$_4$ spinel layers, with a nominal composition of Mn$_{1.5}$Co$_{1.5}$O$_4$, onto a number of ferritic stainless steels for interconnect applications in intermediate temperature SOFCs. Zahid, et al. [22] fabricated MnCo$_2$O$_4$ layers by the surface-coating Co$_3$O$_4$ on Crofer22 APU. After 1000 hours of heating in air, the interface between the MnCo$_2$O$_4$ layer and MnCr$_2$O$_4$ sublayer in the subscale on Crofer22 APU remained stable. Chen et al. [23] reported MnCo$_2$O$_4$ coatings on the ferritic stainless steel 430 via slurry coating followed by mechanical compaction and air-heating. Overall it appears that the (Mn,Co)$_3$O$_4$ spinels are promising coating materials to improve the surface stability of ferritic stainless steel interconnects, minimize contact resistance, and seal off chromium in the metal substrates.

To optimize the performance of coated metallic interconnects, the (Mn,Co)$_3$O$_4$ spinels were systematically investigated in terms of properties relevant for the coating application. This paper briefly summarizes the properties of the (Mn,Co)$_3$O$_4$ spinels and the long-term performance of Mn$_{1.5}$Co$_{1.5}$O$_4$ coatings on Crofer22 APU.

EXPERIMENTAL

The spinel powder with varied stoichiometries was synthesized via a glycine-nitrate combustion synthesis process (GNP) [24], followed by calcination in air at 800°C for 4 hours. The calcined spinel powder was then die-pressed into bars and sintered in air at 1250°C for electrical conductivity and thermal expansion measurements. The four-point method was employed to determine the electrical conductivity of the spinels. The thermal expansion behavior was measured using a Linseis L75 dual push-rod dilatometer.

The spinel coatings were fabricated by slurry coating of the calcined spinel powder onto stainless steel coupons. Details about preparation of the spinel powder and slurries for the coating can be found in a recent publication [20]. A spinel slurry was applied by spraying or screen-printing onto Crofer22 APU coupons. After being dried in an oven at 80°C for 1~2 hours, the coated stainless steel coupons were heat-treated in an Ar/3%H$_2$O/2.75%H$_2$ environment at 800°C for 2 hours. Following the reducing heat-treatment, the spinel layer was finally developed during subsequent oxidation in air at elevated temperatures or during an evaluation test under SOFC operating conditions.

The stainless steel samples with the thermally grown spinel protection layers were evaluated in a configuration that simulates the interconnect/cathode structure in SOFC stacks (details can be found in ref. 20). In the simulated structure, La$_{0.8}$Sr$_{0.2}$FeO$_3$ (LSF) was chosen as the cathode composition and La$_{0.8}$Sr$_{0.2}$Co$_{0.5}$Mn$_{0.5}$O$_3$ (LSCM) was used as an electrical contact paste. A six-month long thermal cyclic (from room temperature to 800°C at a rate of 5°C/min) test was performed. The synthesized spinel powder and thermally-grown spinel protection layers on the stainless steels were analyzed by X-ray diffraction (XRD) and SEM with EDX capability.

RESULTS and DISCUSION

Crystal Structure of $Mn_{1+x}Co_{2-x}O_4$ (-0.5≤x≤1.5) Spinels

To understand their crystal structure, $Mn_{1+x}Co_{2-x}O_4$ compositions with x=-0.5, 0, 0.5, 1.0, 1.5 were prepared and analyzed by X-ray diffraction, as shown in Figure 1. Not all the compositions exhibited a single-phase crystal structure; for example, Figure 2 shows the XRD pattern of the calcined GNP powder with a nominal composition of $Mn_{1.5}Co_{1.5}O_4$ (i.e. x=0.5), indicating the presence of $MnCo_2O_4$ (x=0) and Mn_2CoO_4 (x=1.0), as reported in JCPDS as #18-1237 [25] and #23-0408 [26], respectively. $MnCo_2O_4$ is a normal cubic spinel [27-29], with Mn sitting on octahedral interstitial sites and Co on both tetrahedral and octahedral interstitial sites in the face-centered cubic oxygen ion lattice; while Mn_2CoO_4 possesses an intermediate tetragonal spinel structure between the cubic $MnCo_2O_4$ and tetragonal Mn_3O_4 [26]. The cubic and tetragonal crystal structures were further confirmed by the X-ray analyses (see Figure 1) on the $MnCo_2O_4$ and Mn_2CoO_4 powder samples prepared in this work, indicating a good match with JCPDS #18-1237 and #23-0408, respectively. Similarly, the X-ray analysis further found that $Mn_{2.5}Co_{0.5}O_4$ (x=1.5) and $Mn_{0.5}Co_{2.5}O_4$ (x=-0.5) exhibited a tetragonal and cubic structure, respectively, at room temperature. The X-ray analyses on the various spinels are consistent with previously published work [27] that indicated a cubic structure for x<0.3, a tetragonal structure for x>0.9, and a dual phase structure for 0.3≤x≤0.9. A similar combination of spinels was also observed by Aukrust and Muan during their phase-diagram study of the cobalt oxide – manganese oxide system in air [30].

Thermal Expansion and Electrical Conductivity

Figure 3 shows the thermal expansion behavior of $Mn_{1+x}Co_{2-x}O_4$ spinels with x=-0.5, 0, 0.5, 1.0, 1.5. The measurements were carried out in air from room temperature to 1,250°C. Among the five spinels studied, only $Mn_{1.5}Co_{1.5}O_4$ maintained a good linearity up to 1,250°C and exhibited a good thermal expansion match to ferritic stainless steels such as Crofer22 APU and AISI430, as well as perovskite cathode compositions such as $La_{0.8}Sr_{0.2}MnO_3$. The average thermal expansion coefficient (CTE) of $Mn_{1.5}Co_{1.5}O_4$ from room temperature to 800°C is about 11.4×10^{-6} K^{-1}, which is similar to that of Crofer22 APU at 12.6×10^{-6} K^{-1}. The spinel with a nominal composition of $MnCo_2O_4$ also demonstrated good linearity and good CTE match to the ceramic cell components for up to about 850°C, beyond which however a deviation from the linearity occurred due to precipitation of CoO. The other three compositions with x=-0.5, 1.0, or 1.5 appeared to be unstable and/or poorly matched to the other stack components due to phase transformations at relative low temperatures.

Electrical conductivity tests were carried out on $Mn_{1+x}Co_{2-x}O_4$ spinels with x=-0.5, 0, 0.25, 0.5, 0.75, 1.0, 1.5. Figure 4 shows the electrical conductivity of the varied spinels at 800°C in air as a function of stoichiometry. Among the seven tested compositions, $Mn_{1.5}Co_{1.5}O_4$ demonstrated the highest electrical conductivity at 61.6 S.cm⁻¹, while those with a stoichiometry closer to the binary oxides (Mn_3O_4 and Co_3O_4) tended to exhibit a decreased conductivity. Thus it appears the $Mn_{1+x}Co_{2-x}O_4$ spinels with x around 0.5 possess relatively high electrical conductivity, which is approximately 3~4 orders of magnitude higher than Cr_2O_3 [31], and 2~3 orders higher than $MnCr_2O_4$ [32,33], the two dominant phases in the top layer of scales grown on Crofer22 APU. Due to the high conductivity and good CTE match to cell components, the $Mn_{1+x}Co_{2-x}O_4$ spinels with x around 0.5 are therefore good candidates for coating application on ferritic SOFC interconnects.

Electrical Performance and Thermo-Mechanical Stability

To examine its long-term electrical performance and thermomechanical stability, Crofer22 APU with the thermally grown $Mn_{1.5}Co_{1.5}O_4$ spinel protection layer was tested for a period of six months under a total of 125 thermal cycles. During the test (see Figure 5), the contact ASR between a LSF cathode and the coated metallic interconnect at 800°C steadily decreased slightly from the starting value of 15.0 mohm.cm^2 to 14.3 mohm.cm^2, demonstrating excellent stability. SEM analysis on the cross-section of the tested sample indicated good thermomechanical stability of the thermally-grown protection layers. No spallation or chipping was observed at areas with the spinel protection layer, and the sub-scale only grew to a thickness about 2.0 µm. In contrast, the scale on the portion of the Crofer22 APU without a protection layer grew to a thickness up to ~10 µm, and spallation was observed in some areas. Thus, the spinel protection layer on Crofer22 APU not only drastically reduced the interfacial ASR, but also inhibited the scale growth on the ferritic stainless steel by limiting oxygen ion diffusion inward through the spinel layer. The excellent thermomechanical stability and stable electrical performance are mainly attributed to the inhibited sub-scale growth and a good thermal expansion match between the spinel and the metal substrate.

Additionally, the spinel protection layer acted as an effective mass transport barrier in stopping chromium migration from the metal. EDS analysis (see Figure 6) revealed a sharp Cr profile across the interface between the sub-scale and the spinel protection layer, with no chromium detectable in the spinel protection layer at its surface after the six months thermal cycling test. No Cr was found in the LSCM contact material and the LSF cathode. As verified by the long-term test, the manganese cobaltite spinel protection layers thermally grown on ferritic stainless steel interconnects appear to be very effective in improving the surface stability and electrical conductivity of these metallic interconnect materials, and in preventing outward chromium cation diffusion to the interconnect surface.

CONCLUSIONS

At room temperature, $Mn_{1+x}Co_{2-x}O_4$ spinels exhibit a cubic or tetragonal structure (or a dual phase mixture of both), depending on their cation stoichiometry. Among the manganese cobaltite spinels, those with x around 0.5 demonstrated a good CTE match to other stack components, relatively high electrical conductivity, and good thermal stability up to 1,250°C. Thus the spinels with x around 0.5 appear to be good materials candidates for a protection layer on ferritic SOFC interconnects. A long-term test on the $Mn_{1.5}Co_{1.5}O_4$ protection layer that was thermally grown on ferritic stainless steels indicated not only a significantly decreased contact resistance between a LSF cathode and the stainless steel interconnects, but also a much inhibited sub-scale growth on the stainless steels. The combination of the inhibited sub-scale growth, good thermal expansion matching between the spinels and the stainless steels, and the closed-pore microstructure may contribute to the observed excellent structural and thermomechanical stability of these spinel protection layers. Also, the spinel protection layers appeared to act effectively as a barrier to outward diffusion of chromium cations, preventing subsequent chromium migration into cathodes and contact materials.

ACKNOWLEDGEMENTS

The authors would like to thank Nat Saenz, Shelly Carlson, and Jim Coleman for their assistance in metallographic and SEM sample preparation and analysis. The work summarized in this paper was funded as part of the Solid-State Energy Conversion Alliance (SECA) Core Technology Program by the U.S. Department of Energy's National Energy Technology Laboratory (NETL). The authors would like to acknowledge helpful discussions with Wayne Surdoval, Lane Wilson, Don Collins, and Travis Schulz. PNNL is operated by Battelle Memorial Institute for the U.S. Department of Energy under Contract DE-AC06-76RLO 1830.

REFERENCES

1. W.J. Quaddakkers, J. Piron-Abellan, V. Shemet, and L. Singheiser, Mater. High Temp., 20, 115 (2003).
2. Z. Yang, K.S. Weil, D.M. Paxton, and J.W. Stevenson, J. Electrochem. Soc., 150, A1188 (2003).
3. W.Z. Zhu and S.C. Deevi, Mater. Sci. and Eng., A348, 227 (2003).
4. L. Antoni, Mater. Sci. Forum, 461-4, 1073 (2004).
5. J.W. Fergus, Mater. Sci. and Eng., A397, 271 (2005).
6. T. Malkow, U.V.D. Crone, A.M. Laptev, T. Koppitz, U. Breuer, and W.J. Quadakkers, in Solid Oxide Fuel Cells, U. Stimming, S.C. Singal, H. Tagawa, and W. Lehnert, Editors, PV 97-40, p. 1244, The Electrochemical Society, Pennington, NJ (1997).
7. C. Gindorf, L. Singheiser, and K. Hilpert, Steel Res., 72, 528 (2001).
8. K. Hilpert, D. Das, M. Miller, D.H. Peck, and R. Weiβ, J. Electroch. Soc., 143, 3642 (1996).
9. S.C. Paulson and V.I. Birss, J. Electrochem. Soc., 151, A1961 (2004).
10. Y. Matsuzaki and I. Yasuda, J. Electrochem. Soc., 148, A126 (2001).
11. S.P. Jiang, S. Zhang, and Y.D. Zhen, J. Electrochem. Soc., 153, A127 (2006).
12. W.J. Quadakkers, V. Shemet, and L. Singheiser, US Paten No. 2003059335 (2003).
13. J.P. Abeilan, V. Shemet, F. Tietz, L. Singheiser, and W.J. Quadakkers, in Proceedings of the 7th International Symposium on Solid Oxide Fuel Cells, S.C. Singhal and M. Dokiya, Editors, PV2001-16, p. 811, The Electrochemical Proceedings Series, Pennington, NJ (2001).
14. Z. Yang, J.S. Hardy, M.S. Walker, G. Xia, S.P. Simner, and J.W. Stevenson, J. Electrochem. Soc., 151, A1825 (2004).
15. S.P. Simner, M.D. Anderson, G. Xia, Z. Yang, and J.W. Stevenson, J. Electrochem. Soc., 152, A740 (2005).
16. Z. Yang, M.S. Walker, P. Singh, and J.W. Stevenson, J. Electrochem. Soc., 151, B669 (2004).
17. Z. Yang, G.-G. Xia, P. Singh, and J.W. Stevenson, Solid State Ionics, 176, 1495 (2005).
18. Y. Larring and T. Norby, J. Electrochem., Soc., 147, 3251 (2000).
19. Z. Yang, G. Xia, and J.W. Stevenson, Electrochem. & Solid State Lett., 8 A168 (2005).
20. Z. Yang, G.-G. Xia, S.P. Simner, and J.W. Stevenson, J. Electrochem. Soc., 152, A1896 (2005).
21. Z. Yang, G.-G. Xia, S.P. Simner, and J.W. Stevenson, in Ceramic Engineering and Science Proceedings, D. Zhu and W.M. Kriven, Editors, Vol.26, p.201, The Electrochemical Society Proceedings Series, Pennington, NJ (2005).
22. M. Zahid, F. Tietz, D. Sebold, and H.P. Buchkremer, in Proc. of 3rd Eur. SOFC Forum, M. Mogensen, Editor, p. 820, European Solid Oxide Fuel Cell Forum, Switzerland (2004).

23. X. Chen, P.Y. Hou, C.P. Jacobson, S.J. Visco, L.C. De Jonghe, Solid State Ionics, 176, 425 (2005).
24. L.A. Chick, L.R. Pederson, G.D. Maupin, J.L. Bates, L.E. Thomas, and G.L. Exarhos, Mater. Lett., 10, 6 (1990).
25. JCPDS-International Centre for Diffraction Data, #23-1237.
26. JCPDS-International Centre for Diffraction Data, #18-0408.
27. S. Naka, M. Inagaki, and T. Tanaka, J Mater. Sci., 7, 441 (1972).
28. J.L. Gautier, C. Cabezas, S. Barbato, Electrochim. Acta., 26, 1377 (1981).
29. A. Restovic, E. Rios, S. Barbato, J. Ortiz, and J.L. Gautier, J Electroanalyt. Chem., 522, 141 (2002).
30. E. Aukrust and A. Muan, J. Am. Ceram. Soc., 46, 511 (1963).
31. A. Holt and P. Kofstad, Solid State Ionics, 69, 137 (1994).
32. T. Sasamoto, N. Sumi, A. Shimaji, O. Yamamoto, and Y. Abe., J. Mater. Sci. Soc. Japan, 33, 32 (1996).
33. Z. Lu and J. Zhu, J. Am. Ceam. Soc., 88, 1050 (2005).

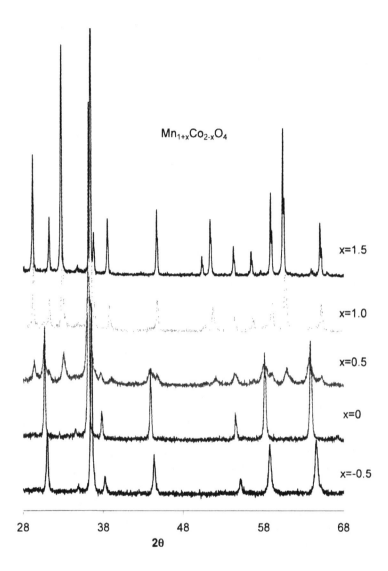

Figure 1. X-ray diffraction patterns of Mn$_{1+x}$Co$_{2-x}$O$_4$ spinel powders with x=-0.5, 0. 0.5, 1.0, 1.5. The power samples were prepared by a glycine-nitrate combustion synthesis approach that was followed by a calcinations at 800°C in air for 2 hours.

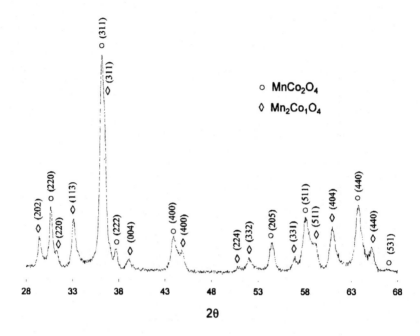

Figure 2. X-ray diffraction pattern of Mn$_{1.5}$Co$_{1.5}$O$_4$ was synthesized via a glycine-nitrate process. followed by calcination at 800°C for 4 hours in an air furnace.

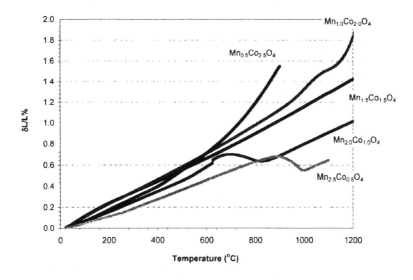

Figure 3. Thermal expansion of Mn$_{1+x}$Co$_{2-x}$O$_4$, x=-0.5, 0, 0.5, 1.0, 1.5.

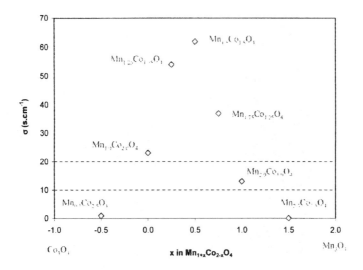

Figure 4. Electrical conductivity of Mn$_{1+x}$Co$_{2-x}$O$_4$, x = -0.5, 0, 0.25, 0.5, 0.75, 1.0, and 1.5, at 800°C in air. The measurements were carried out using a four-point resistance technique.

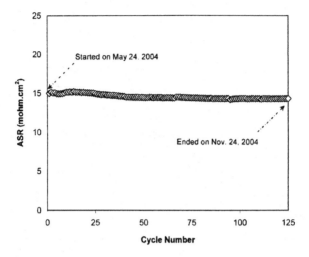

Figure 5. Contact ASR of La$_{0.8}$Sr$_{0.2}$FeO$_3$||La$_{0.8}$Sr$_{0.2}$Co$_{0.5}$Mn$_{0.5}$O$_3$||Crofer22 APU as a function of thermal cycle numbers. The Crofer22 APU was screen-printed with the Mn$_{1.5}$Co$_{1.5}$O$_4$ spinel paste and then heat-treated at 800°C in a reducing environment for 24 hours before being placed in the test stand and continuously tested for six months.

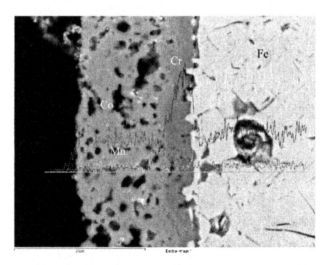

Figure 6. Microstructural and compositional analyses on the Mn$_{1.5}$O$_{1.5}$O$_4$ protection layer on Crofer22 APU that was subjected to a contact ASR measurement for a period of six months including thermal cycling.

FUEL CELL INTERCONNECTING COATINGS PRODUCED BY DIFFERENT THERMAL
SPRAY TECHNIQUES

E. Garcia and T. W. Coyle.
Centre for Advanced Coating Technologies, University of Toronto
184 College St.
Toronto, ON, M5S 3E4

ABSTRACT

Doped $LaCrO_3$ ceramic material is commonly used to produce interconnect coatings for solid oxide fuel cells (SOFCs). Three different thermal spray methods are used in this work to deposit $La_{0.9}Sr_{0.1}CrO_3$ interconnect coatings: high velocity oxy-fuel (HVOF) using a modified nozzle and two different atmospheric plasma spray (APS) torches. One of them is a commercial torch that uses Ar/H_2 as plasma forming gases and the other is a new torch design that uses gas mixtures based in CO_2. The spray parameters of each torch were set by studying the in-flight temperature and velocity of the particles as a function of the stand-off distance using a substitute powder (ZrO_2-TiO_2-Al_2O_3 composite) with similar physical properties to $La_{0.9}Sr_{0.1}CrO_3$. The process parameters that produced coatings with the lowest porosity were employed to deposit $La_{0.9}Sr_{0.1}CrO_3$ coatings on zirconium oxide substrates. Scanning electron microscopy (SEM) and X-ray diffraction analysis techniques were used to characterize the coatings produced by the three different torches. The microstructure features and crystalline phases present in the coatings are explained in terms of the process parameters and correlated with preliminary measurements of electrical resistivity of the as-sprayed coatings. In some cases, post-deposition heat treatments are studied in order to decrease the electrical resistivity.

INTRODUCTION

Solid oxide fuel cells (SOFCs) are an essential development in the ongoing push towards renewable energy sources. SOFCs are generally being developed for grid supplementing power generation or for power generation in remote locations.

There are two main designs of SOFC. In the planar design flat anode/electrolyte/cathode sandwiches are stacked on top of each other, with the cathode of one sandwich connect to the anode of the next by the interconnect material. The tubular cell design is based on a closed end tube made from the cathode material, with an electrolyte layer coating the outside of the cathode tube, and the anode coating the electrolyte layer (Fig. 1).

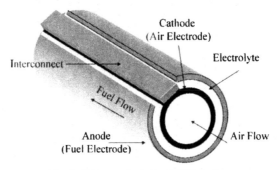

Fig. 1: Scheme of tubular SOFC design.

The outer anode layer of one tubular cell is connected to the inner cathode layer of the next tube by the interconnect material. An advantage of the tubular design is the slightly less severe demands placed on the interconnect materials, however fabricating the interconnect structures is a processing challenge. It must be a material with a thermal expansion coefficient that matches the other components of the cell and with an acceptable electrical conductivity at high temperature. The most common material choice is doped $LaCrO_3$[1,2]. A high density is required in the interconnects to prevent mixing of the air and fuel gases, however interconnects made of this material are difficult to process because of chromia evaporation at elevated temperatures which leads to poor densification[1].

One way of depositing this interconnect material is by thermal spray processes. They produce a fast deposition rate and it is easy to deposit patterned structures reducing in this way the manufacturing cost. However coatings produced by this processes may contain defects such as microcracks or pores that can reduce the desired gas tightness. In the same way, the high temperatures involved in this process may cause changes in the chemical composition of the coatings modifying the electrical properties of the original material. These problems must be minimized to ensure proper performance of the SOFC during service[3].

The objective of this work was to test three different thermal spray torches to produce $La_{0.9}Sr_{0.1}CrO_3$ interconnect coatings suitable for use in tubular SOFC. The techniques employed were high velocity oxy fuel (HVOF) that typically results in the highest density coatings of all common thermal spray techniques due to the high gas and particle velocities achieved and two atmospheric plasma spray (APS) torches. The APS process is characterized by the high temperature of the plasma plume. Two different APS torches were tested in this work in order to compare the effect of the plasma forming gases on the final properties of the coatings obtained.

First, the optimal spraying parameters to achieve the highest density of the coating were set using a substitute powder with similar physical properties to the expensive $La_{0.9}Sr_{0.1}CrO_3$ powder. After that, the interconnect material was sprayed and the coatings obtained were analysed to determine which techniques are suitable to produce interconnect coatings for tubular SOFC applications.

MATERIALS AND METHODS
Materials

The interconnect material tested in this work was $La_{0.9}Sr_{0.1}CrO_3$ powder (Praxair Specialty Ceramics, WA). This ceramic material has a high melting point (2500°C), electrical conductivity and chemical stability under oxidizing or reducing atmospheres that make it a good candidate to be used as interconnect. In addition, its thermal expansion is relatively similar to those of other SOFC components. The $La_{0.9}Sr_{0.1}CrO_3$ powder is produced by combustion spray pyrolysis followed by the spray agglomeration of the particles. In Fig. 2 is shown an example of the powder used in this work, formed by agglomerated particles with a submicron size. The complex process used to produce the powders and the rare alkali earth metals of its composition make $La_{0.9}Sr_{0.1}CrO_3$ powder expensive (~ $1000 U.S. / kg). The optimization of the process requires approximately a kilogram of powder, therefore a replacement powder that exhibited approximately the same melting temperature and particle size distribution as $La_{0.9}Sr_{0.1}CrO_3$ powder was used to reduce costs. The replacement powder was Metco 143 (Sulzer-Metco, Brampton, ON) a Zirconia, Titania, Yttria composite powder. The melting temperatures of both powders are approximately 2500 °C. The particle size of the Metco 143 powder is in the range of 5-75 μm comparable to the particle size of the $La_{0.9}Sr_{0.1}CrO_3$ powder that is in the range of 20 – 60μm.

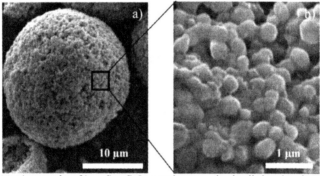

Figure 2: SEM micrographs of $La_{0.9}Sr_{0.1}CrO_3$ powder (a) and a detail showing the agglomerated particles (b).

The powders were sprayed over Yttria stabilized Zirconia (TZ-3YB, Tosoh Co, Japan) samples prepared by uniaxial pressing and sintering.

Thermal Spraying

Three different torches were employed. The Air Plasma Spray (APS) torches were a commercial SG-100 torch (Miller Thermal Inc., now Praxair Surface Technology, Indianapolis, IN) operated with Ar/H_2 gas mixtures to form the plasma and a new plasma torch developed at the Centre for Advanced Coating Technologies (CACT) based in CO_2 gas mixtures. For High Velocity Oxy Fuel (HVOF) spray deposition a DJ-2700 HVOF torch (Sulzer-Metco (USA) Inc. Westbury, NY) with propylene as a fuel was employed. As the melting point of the materials studied is close to the maximum temperature reached in HVOF spraying, the use of a protective diverging-converging shroud (Fig. 3) developed at the CACT[4] was tested in order to obtain the maximum temperature of the in-flight particles. The velocity and temperature distribution are homogenized within the jet and the entrainment of air is delayed. Thus, the average particle temperature is increased and the oxidation is reduced. These advantages have resulted in the deposit of dense alumina coatings[5].

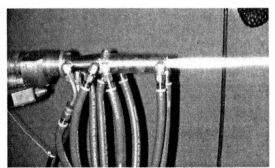

Figure 3: DJ-2700 HVOF torch with shroud attachment.

The set of parameter that produced the highest in-flight particle temperature and velocity distributions of the substitute Metco 143 powder was found with a DPV 2000 (Tecnar Lted. St-Bruno, QC, Canada) system. These parameters were used to spray the substitute powder on the zirconia substrates and evaluate the microstructure/porosity of the coatings obtained.

Ideal parameters for the Metco 143 powder were applied to spraying of the $La_{0.9}Sr_{0.1}CrO_3$ powder to confirm the results and to analyze the coatings.

Characterization of the Coatings

Coated substrates were cut with a diamond saw (Clemex, Brillian 221, Clemex Technologies Inc., Longueuil, QC, Canada). After mounting the specimens using a low viscosity resin, they were polished on a semi-automatic polisher (Imptech DPS 2000, Boksburg, South Africa) with 0.05μm alumina finish. The mounted samples were gold or graphite coated to perform SEM analysis.

The thickness, morphology, porosity and microstructure of the coatings were analyzed by scanning electron micoscopy (Hitachi S-2500, Hitachinaka-shi, Iberaky, Japan). The porosity of the coatings was determined by image analysis of secondary electron images using Clemex Vision^{TM} (Clemex Technologies Inc., Longueuil, QC, Canada) software. Energy Dispersive X-ray analysis (EDX) was employed to determine qualitatively the chemical composition of the coatings.

The crystalline phases were determined by X-ray diffraction (XRD) analyses (CN2651A1, Rigaku Co. The Woodlands, TX) using Cu Kα radiation. Scan parameters included a scan range $2\theta=20$-$50°$, a step size of $0.02°$ and a step time of 1.55 s.

Two point in-plane electrical resistance measurements (Keithley 25700 Multimeter/Data acquisition system, Cleveland, OH) were carried out on rectangular shaped coated substrates to evaluate the electrical resistivity of the as sprayed coatings. The resistivity (ρ) of the coating is given by Eq. 1.

$$\rho = \frac{R \cdot A}{l},$$ (1)

where R is the two point electrical resistance, A the cross sectional area of the coating and l the distance between the electrodes.

RESULTS AND DISCUSSION

Spray Parameters

In Table I are listed the spray parameters that produced the highest average values of temperature and velocity of the particles measured by the DPV 2000 system during spraying of the Metco 143 powder.

Table I: Optimal spray parameters for the different thermal spray techniques studied in this work for the spraying of Metco 143 powder. Average temperature and velocity of the particles measured by DPV 2000 system are also given.

Gases	HVOF (shrouded)		APS		CO_2 APS
	Propylene	O_2	Ar	H_2	CO_2 (mixture)
Flow (lpm)	77	252	50	2	47
Current(A)			628		250
Voltage(V)			56		136
Stand-off (mm)	280		50		75
T(°C)	2236 ± 234		2541 ± 232		2696 ± 168
V(m/s)	613 ± 175		248 ± 62		258 ± 67

The average particle velocity is much higher in the case of HVOF than in the case of the APS techniques. The highest average temperature for HVOF was obtained with the use of the shroud developed at CACT[4]. This temperature is ~200 °C lower than the melting temperature of the Metco 143 powder. More than 29% of the particles had a temperature higher than 2400 °C, even though the deposition efficiency was expected to be low, a coating could be formed.

The average particle temperature and velocity are higher in the case of the CO_2 APS than the conventional APS. A longer plasma flame is obtained with the new APS due to the high gas enthalpy. Thus the spraying standoff distance for the CO_2 APS must be longer than for the conventional torch as problems associated with overheating of the samples could appear.

Microstructure of the Coatings
Metco 143 Coatings

The polished cross-sections of the as sprayed coatings of Metco 143 powders on zirconia substrates using the parameters set in the previous section are shown in Fig. 4. It can be observed that the coatings obtained with the APS techniques after 16 passes are significantly thicker (~ 100 μm) than that obtained by HVOF after 12 passes (~ 15 μm). The average temperature of the particles could be the reason for this difference. With the APS techniques more sprayed particles are above the melting point of the Metco 143 powder than in the case of HVOF. As was expected, the deposition efficiency of the HVOF is the lowest.

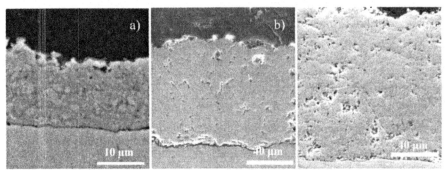

Figure 4: Secondary electron SEM micrographs of polished cross-sections of the coatings obtained by spraying Metco 143 powders using the optimal conditions with a) HVOF, b) CO_2 APS and c) conventional APS torches.

A set of 5 images of each coating was analyzed by Clemex software yielding porosity levels below 5% in all the coatings. As shown in Fig. 4, it can be observed that the pores are closed. The low porosity level and the morphology of the pores obtained with Metco 143 powders proved the suitability of these three techniques to spray the $La_{0.9}Sr_{0.1}CrO_3$ powder.

$La_{0.9}Sr_{0.1}CrO_3$ Coatings

Figure 5 shows the polished cross sections of the $La_{0.9}Sr_{0.1}CrO_3$ coatings sprayed using the optimal conditions listed in Table 1. The coatings obtained with the APS techniques after 16 passes are again significantly thicker (~ 110 μm) than that obtained by HVOF after 30 passes (~ 30 μm). Clemex image analysis showed that porosity values of all coatings were below 5%. The coating obtained with conventional APS torch shows larger amount of unmelted original particles. HVOF produced the denser coating but it is clear that the deposition efficiency is much lower than in the case of APS spraying techniques. As the average particle temperature obtained by HVOF is the lowest, a large amount of unmelted particles would be expected to reach the substrate. However, no evidence of their presence in the coating was revealed in the analysis of the cross-section. Apparently the unmelted particles reaching the substrate under HVOF conditions were not incorporated in the coating. This correlates with the low deposition efficiency observed for HVOF coatings. Coatings deposited by APS techniques show unmelted particles, which means that the spray parameters can be adjusted to obtain a denser coating. The amount of these unmelted particles is higher in the case of conventional APS torch and can be explained by the lower temperature recorded by the DPV 2000 during the spraying of the substitute Metco 143 powder.

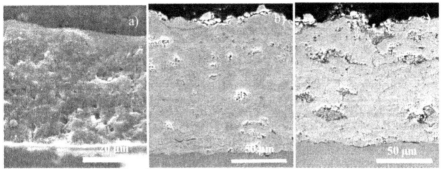

Figure 5: SEM micrographs of polished cross sections of the coatings obtained by spraying $La_{0.9}Sr_{0.1}CrO_3$ powder using the optimal conditions for Metco 143 powder with a) HVOF, b) CO_2 APS and c) conventional APS torches.

Qualitative EDX analysis of the coatings showed a reduction in the chromium content for the coatings produced by APS. This reduction was more evident in the case of the new APS where higher temperatures for the particles and the substrate itself were observed during spraying. The reduction of the chromium relative intensity is because of chromia evaporation at high temperature[1,3].

Some areas of the coatings obtained by APS exhibited vertical cracks running through the thickness of the coatings. Those cracks were more evident in the coating produced by the CO_2 APS. In Fig. 6 can be observed two cracks running through one coating produced by CO_2 APS. As the thermal expansion coefficients of the coating and the substrate are very similar the presence of those cracks could be due to the high thermal gradients present in the coatings during deposition. Therefore, the temperature of the coatings must be controlled to avoid the presence of cracks in order to obtain a gas tight coating and a good electrical conductivity in the coating.

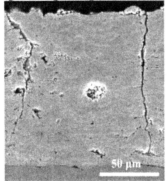

Figure 6: SEM micrograph of polished cross-section of a coating obtained by CO_2 APS showing vertical cracks.

X-ray Diffraction Analysis

Figure 7 shows the XRD patterns of the $La_{0.9}Sr_{0.1}CrO_3$ coatings obtained using the three different thermal spray torches.

Peaks related to $LaCrO_4$ are detected in the diffraction pattern of the coating sprayed with the HVOF torch, as is shown in Fig. 7. Previous works indicate that this phase is obtained when amorphous lanthanum chromate is heated to temperatures close to 400 °C[6,7]. The $LaCrO_4$ phase was characterized by a green color that is detected in the as sprayed samples. This verified that the HVOF technique was able to melt the $La_{0.9}Sr_{0.1}CrO_3$ powder and the temperature of the substrates during spraying was higher than 400 °C. One of the samples deposited by HVOF was heated at 850 °C[6,7] for 3 hours. The greenish color of the coating turned to brown, characteristic of the original perovskite phase. X-ray patterns recorded after the heat treatment showed a decrease in the $LaCrO_4$ phase. The absence of other phases from the decomposition of the original perovskite material can be explained by the short resident time in the flame due to the high velocity of the gases and the particles in the flame[8].

Diffraction patterns of the APS sprayed coatings (Fig 7b, 7c) are characterized by strong diffraction peaks related to the perovskite $La_{0.9}Sr_{0.1}CrO_3$ phase and the presence of the La_2O_3 crystalline phase. This phase is the result of the decomposition of the perovskite phase and the evaporation of chromia due to the high temperature reached by the powder during the spraying process[9]. The relative intensity of La_2O_3 peaks is higher in coatings obtained with the CO_2 APS torch than that obtained with the conventional one. This result can be explained by the higher temperature of the particles inside the plasma plume recorded by DPV 2000 and the higher temperature of the substrates during spraying in the case of the new APS torch. This result is in agreement with the qualitative result found by EDX analysis.

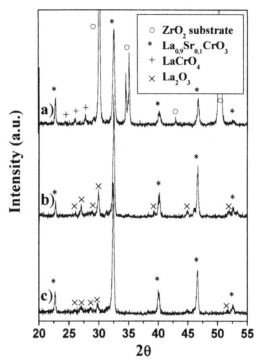

Figure 7: XRD patterns of coatings obtained by spraying $La_{0.9}Sr_{0.1}CrO_3$ powder using the optimal conditions for Metco 143 powder with a) HVOF, b) CO_2 APS and c) conventional APS torches.

Electrical Resistivity of the Coatings

Table II lists the values of the electrical resistivity obtained by Eq. 1. The electrical resistivity obtained for all coatings are higher than that found in the literature for a dense $La_{0.9}Sr_{0.1}CrO_3$ material[10]. The high resistivity of the as sprayed HVOF coating can be explained by the presence of the non-perovskite $LaCrO_4$ phase. Other factors must not be disregarded because after the heat treatment of the coating to transform that phase to the perovskite, the resistivity is still about four orders of magnitude higher than that of a dense sample. The electrical resistivity of the coatings obtained by APS techniques is almost two orders of magnitude lower than that obtained by HVOF. The electrical resistivity of the coating sprayed with the CO_2 APS torch is quite similar but slightly lower than that obtained with the conventional torch. This different can be attributed to the fact that less unmelted particles are found in the coating sprayed with the new APS torch. The electrical resistivity of both APS coatings is two orders of magnitude higher than that of the dense sample. The difference can be explained as a combined effect of microstructure defects (unmelted particles and vertical cracks) and the presence of the La_2O_3 phase in the crystalline composition of the coatings.

Table III: Electrical resistivity of the coatings obtained at Room Temperature by simple two point electrical resistance measurement of the coatings.

	ρ ($\Omega \cdot$m)
HVOF (as sprayed)	530
HVOF (heat treated)	52
APS	1
CO_2 APS	0.6
Dense[9]	0.008

CONCLUSIONS

The three thermal spray torches studied in this work produced dense coatings with closed porosity lower than 5 %

The use of a substitute powder is useful to find optimal spraying parameters of $La_{0.9}Sr_{0.1}CrO_3$ to get a dense coating and to reduce cost. A detailed study is necessary to minimize unmelted particles and undesirable crystalline phases.

The main crystalline phase found in the coatings produced by the three techniques is the perovskite $La_{0.9}Sr_{0.1}CrO_3$.

The interconnect coatings sprayed with HVOF are characterized by the presence of a strong peak associated with the zirconium oxide substrate and the non-perovskite $LaCrO_4$ phase. The substrate is detected due to the thin thickness of the coating. The heat treatment of the coatings showed that the non-perovskite $LaCrO_4$ phase can be transformed to the perovskite at temperatures close to 850 °C.

The presence of unmelted particles can be observed in the coatings produced by APS torches. These unmelted particles are more frequently found in the coatings produced by the conventional APS torch. Undesirable vertical cracks can be observed mainly in the coatings sprayed by the CO_2 APS torch. The temperature of the substrates must be controlled to minimize the thermal gradient during spraying avoiding the undesirable vertical cracks. The coatings produced by both APS torches are also characterized by a reduction of the chromium content of the coatings. This reduction is associated with the presence of the La_2O_3 phase detected in the coatings. A higher the relative intensity of the La_2O_3 phase was found in coatings obtained with the CO_2 APS torch. Spray parameters must be adjusted in order to reduce the number of unmelted particles and the La_2O_3 phase.

Coatings produced by both APS torches are characterized by an electrical resistivity two orders of magnitude higher than that of a dense material found in the literature but two orders of magnitude lower than that obtained with the HVOF technique. The differences between the electrical resistivity of the as sprayed coatings and that found in the literature for dense materials could be explained by the presence of vertical cracks and the La_2O_3 phase in the case of the APS coatings and the non-perovskite $LaCrO_4$ phase in the HVOF sprayed coatings. In this last case, the electrical resistivity of the as sprayed coating is reduced one order of magnitude after the heat treatment at 850°C due to the reduction of the $LaCrO_4$.

ACKNOWLEDGEMENTS

The authors thank OCE-MMO and Fuel Cell Technologies LTD. for financial support and Dr. L. Pershin and Mr. T. Lee for their help with the different thermal spray techniques.

Eugenio Garcia wishes to acknowledge to National Secretary of Education and Universities of Spain and European Social Fund for financial sponsorship.

REFERENCES

[1]S. M. Haile, "Fuel Cell Materials and Components", *Acta. Mater.* **51**, 5981-6000, (2003).

[2]W. Z. Zhu and S. C. Deevi, "Development of Interconnect Materials for Solid Oxide Fuel Cells", *Mat. Sci. Eng. A* **348**, 227-241, (2003).

[3]K. Okumura, Y. Aihara, S. Ito and S. Kawasaki, "Development of Thermal Spraying-Sintering Technology for Solid Oxide Fuel Cells", *J. Therm. Spray Technol.*, **93**, 354-359, (2000).

[4]A. Dolatabadi, J. Mostaghimi, and V. Pershin, "High Efficiency Nozzle for Thermal Spray of High Quality, Low Oxide Content Coatings", U.S. Patent No. US-2003-0178511-A1, (2003).

[5]A. Dolatabadi, J. Mostaghimi, and V. Pershin, A New Attachment for Controlling Gas Flow in HVOF Process. *J. Therm. Spray Technol.*, **14**, 91-99 (2005).

[6]P. Duran, J. Tartaj, F. Capel and C. Moure, "Formation, Sintering and Thermal Expansion Behavior of Sr- and Mg-doped $LaCrO_3$ as SOFC Interconnector prepared by the Ethylene Glycol Polymerized Complex Solution Synthesis Method", *J. Eur.Ceram. Soc.*, **24**, 2619-2629, (2004).

[7]K. Azegami, M. Yoshinaka, K. Hirota, and O. Yamaguchi, Formation and Sintering of $LaCrO_3$ Prepared by the Hydrazine Method, *Mater. Res. Bull.*, **33** , 341-348, (1998).

[8]R. Henne, G. Schiller, M Mueller, M. Lang, and R. Ruckdaschel, "SOFC Components Production- an Interesting Chalenge for Dc- and RF- Plasma Spraying", *Proceedings of the 15ᵗʰ International Thermal Spray Conference*, C. Coddet Ed. 25-29 May 1998, (Nice, France) ASM International 933-938, (1998).

[9]S. Simner, J. Hardy, J. Stevenson and T. Armstrong, Sintering Mechanisms in Strontium doped Lanthanum Chrmite, *J. Mater. Sci.*, **34**, 5721-5732, (1999).

[10]D. B. Meadowcroft, Some Properties of Strontium-doped Lanthanum Chromite, *Britt. J. Appl. Phys. (J. Phys. D)*, **2**, 1225-1233, (1969).

SURFACE MODIFICATION OF ALLOYS FOR IMPROVED OXIDATION RESISTANCE IN SOFC APPLICATIONS

David E. Alman and Paul D. Jablonski
U.S. Department of Energy
1450 Queen Ave. SW
Albany, OR, 97321

Steven C. Kung
SOFCo-EFS
1562 Beeson Street
Alliance, OH 44601

ABSTRACT

This research is aimed at improving the oxidation behavior of metallic alloys for SOFC application, by the incorporation of rare earths through surface treatments. This paper details the effect of such surface modification on the behavior of Crofer 22 APU, a ferritic steel designed specifically for SOFC application, and Type 430 stainless steel. Two pack cementation like treatments were used to incorporate Ce into the surface of the alloys. After 4000 hours of exposure at 800°C to air+3%H_2O, the weight gain of Crofer 22APU samples that were Ce surface modified were less than half that of an unmodified sample, revealing the effectiveness of the treatments on enhancing oxidation resistance. For Type-430, the treatment prevented scale spalling that occurred during oxidation of the unmodified alloy.

INTRODUCTION

Metallic alloys will be utilized for stack and/or balance-of-plant components in all Solid Oxide Fuel Cell (SOFC) systems. Cost efficiency requires the utilization of the lowest cost alloy to meet the performance needs. Cost considerations include factors associated with raw materials and fabrication. Based solely on raw materials, iron-based alloys (such as ferritic steels) are more cost effective than nickel-base alloys. Typically, ease of fabrication dictates alloys with relatively low aluminum and silicon contents regardless of whether they are iron-based or nickel-based. Therefore, oxidation resistant nickel- and iron-base alloys that are readily formable into sheet or tube and utilize chromium additions (to form chromia) for protection will be preferentially selected. For in-stack components, such as interconnects, this is desirable as Cr_2O_3 is an intrinsic semiconductor, and hence, electrically conductive under SOFC operating conditions. In addition, it is desirable to form a Cr-Mn spinel phase at the very surface of the oxide scale in order to reduce the vaporization of Cr containing species from the surface [1]. For balance-of-plant applications, Cr vaporization can result in stack poisoning, particularly from up stream components. Thus, techniques that can improve the oxidation resistance of Cr_2O_3 forming alloys may be cost enablers for SOFC systems development.

It is known that the addition of a small amount of reactive elements, such as the rare earths, Ce, La, Y, can significantly improve the high temperature oxidation resistance of both iron- and nickel- base alloys [2-7]. The incorporation of the reactive element can be made in the melt or through surface infusion treatment. Indeed, both Crofer 22 APU and Hitachi ZMG232, ferritic stainless steels specifically designed for SOFC interconnect application, contain small amounts of La additions. Surface modification allows for the concentration of the reactive element at the surface where the oxide will form, and thus, may have the most benefit. This research is aimed at evaluating the effectiveness of rare earth additions via surface treatments on the oxidation resistant chromia-forming alloys in SOFC environments. Previously we reported on the general effectiveness of rare earth surface treatments on a variety of Fe- and Ni-base (Crofer 22 APU, 430SS and Haynes 230) alloys [8]. This paper reports the results of ongoing research evaluating the effectiveness of the surface treatment with emphasis on the behavior of Crofer 22 APU.

EXPERIMENTAL PROCEDURES
        Crofer 22 APU sheet (1mm thick) was obtained from ThyessenKrupp VDM. The nominal composition of Crofer 22 APU is Fe-22Cr-0.5Mn-0.1Ti-0.1La (in weight percent). Oxidation coupons were machined from the sheet to dimensions of 25.4x12.5x1.0 mm^3. A 3.175 mm diameter hole was drilled into the upper portion of each sample to allow hanging on a quartz rack. Prior to treatment (and oxidation for untreated control coupons), the surfaces of the coupons were polished through a 600-grit finish and ultrasonically cleaned in alcohol.
        In an effort to further improve the oxidation resistance of the alloys, samples were treated to increase reactive element (i.e., Ce) content at the sample surface. A surface treatment was developed to modify the surface of the alloys and improve the oxidation resistance. There are many other methods for incorporating rare earths into the surface of an alloy, including the ones listed in the reference section [3,4, 9-11]. For comparison, some coupons were treated by a method described in a paper by Hou and Stringer [4]. This method consisted of spraying the surface of coupons preheated to 400°C with a cerium-nitrate salt slurry. The slurry consisted of the cerium-nitrate salt with HNO_3. The slurry was deposited on the surfaces of the coupons with a commercial air brush. After deposition, the coupons were oxidized at 400°C for 1 hour to decompose cerium-nitrate salt and infuse Ce and deposit CeO_2 on the surface.
        Prior to oxidation testing all the starting weights and physical dimensions of the coupons were measured and recorded. Both treated and untreated samples were tested. Further, one coupon was subjected to the thermal treatment portion of the infusion method without the Ce slurry. This coupon was subjected to the same thermal history and environment exposure as the Ce modified coupon; however, no Ce was incorporated into the surface. For oxidation testing, samples were hung on a quartz rack and placed inside a pre-heated furnace. Oxidation results are reported as weight gain normalized by the sample area and are referred to as Specific Mass Change. Oxidation tests were conducted in either moist or dry air for up to 4000 hours. The moist air was produced by bubbling dry air through a two-stage water column measuring ~2m total height prior to entering the furnace. This produced a gas consisting of air plus ~3% water vapor. The saturated water condition was verified qualitatively by the observation of water condensate on the exit side of the furnace gas stream. After a predetermined time interval, the coupons were removed from the furnace and the weight of each coupon was measured and recorded. The samples were then replaced into the furnace for the next cycle. X-Ray diffraction, scanning electron microscopy (SEM) in conjunction with wavelength dispersive x-ray spectroscopy (WDX) were used to characterize the oxide scale after exposure.

RESULTS AND DISCUSSION
        Figure 1 shows the oxidation behavior for the three different surface conditions for the Crofer 22 APU coupons. The filled circle is for the coupon that was surface modified with Ce (Crofer 22 APU+Ce). The open circle is for the coupon that was subjected to the Ce modification

thermal treatment but no Ce was incorporated into the surface (Crofer 22 APU+TC). The open triangle is for the coupon in the as-received condition (Crofer 22 APU). The Ce surface modification reduces the specific mass gain of the alloy during oxidation by about one-third (0.0008 g/cm^2 for the coupon with the Ce infusion compared to 0.0012 g/cm^2 for coupons without the Ce infusion). Further, the coupon that was subjected to the thermal portion of the infusion treatment but was not modified with Ce has essentially the same mass gain as the coupon tested in the as-received condition. This indicates that the thermal portion of the treatment does little to change the oxidation behavior. That is, exposing the surface of the alloy to the controlled atmosphere of the thermal treatment does not bias or "pre-oxidize" the surface to form a more protective oxide during subsequent oxidation. Improvement in oxidation resistance only occurred when Ce is incorporated into the surface.

Figure 2 shows the oxidation behavior of Crofer 22 APU at 800°C in moist air. Shown in this plot is data for the coupons treated with the Ce surface modification method described in a paper by Hou and Stringer [4]. In this Figure there are multiple samples (a total of 3) for each alloy and condition, identified by the same symbol with different shading. The multiple samples serve to give some idea of the variability, which is extremely low for all the samples and conditions; and to allow for samples to be preserved for scale analysis after 500, 2000 and 4000 hours. It is clear that both Ce treatments were effective in improving the oxidation resistance. The mass gains displayed by the treated Crofer 22 APU coupons were less than half that of the as-received coupon (0.0006 g/cm^2 for the treated coupons compared to 0.0013 g/cm^2 form the alloy in the untreated condition).

As expected from the mass gain results, a much thinner scale formed on the surfaces of the treated specimens relative to untreated coupon (Figures 3 through 6). X-ray diffraction scans from the surfaces of the coupons exposed for 4000 hours (Figure 7) revealed Cr_2O_3, TiO_2, a spinel and a mixed oxide formed as oxidation products. The intensity of the peak for the base Fe-Cr alloy is greater in the scans obtained from the two Ce treated coupons, due to the much thinner oxide scale that formed as a result of the surface modification. X-ray diffraction revealed CeO2 on the surface of the treated coupons, identified by the arrow "C" on the SEM cross section shown in Figure 6.

As expected, a Cr-Mn rich oxide formed at the gas-oxide interface (points 1 and 2 on Figure 6). This alloy was designed to form a Cr-Mn spinel at the gas-oxide interface layer to minimize Cr evaporation during oxidation [12-15]. In this spinel layer, the Mn concentration gradually decreases from a high value of about 20 atomic percent at the outer surface. Beneath the Cr-Mn rich spinel layer, the scale is composed predominately of Cr and is Cr_2O_3 (points 3 and 4 on Figure 6).

There is substantial internal oxidation that forms as small oxide particles beneath the external scale (point P in Figure 6) and along grain boundaries in the metal (as is clearly evident in the lower magnification images shown in Figures 3 to 5 and identified as point A in Figure 6). These are Al-rich oxide phases, and have been reported by other researchers [15-17]. "Metal-extrusions" are also observed in the scale (point 6 on Figure 6). The "metal extrusion" form as a consequence of the internal oxidation. Eventually, the discrete internal oxide areas, especially along grain boundaries, link up and surround un-oxidized metal regions. It is interesting to note, that the oxide immediately below the "metal-extrusions" (point 5 on Figure 6) is a Cr-Mn rich oxide, similar in composition to the phase at the gas-oxide surface.

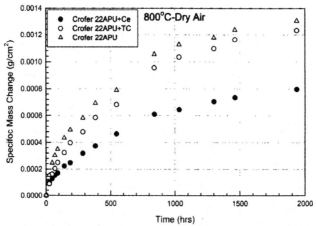

Figure 1. Oxidation behavior of Crofer 22 APU at 800°C in dry air is shown above. Crofer 22 APU+Ce = a coupon that was surface modified with Ce. Crofer 22 APU+TC = a coupon that was subjected to the thermal portion of the modification treatment, without any Ce. Crofer 22 APU = a coupon in the as received condition).

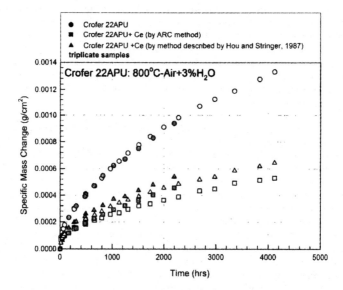

Figure 2. Effect of Ce surface modification on the oxidation behavior of Crofer 22 APU is shown above, (○) as-received Crofer 22 APU, (Δ) Crofer 22 APU+Ce by ARC method, (□) Crofer 22 APU+Ce by Hou and Stringer method.

Figure 3. Cross sections of the oxide scale that formed on Crofer 22 APU after 500, 2000 and 4000 hrs at 800°C in Air+3%H$_2$O are shown above.

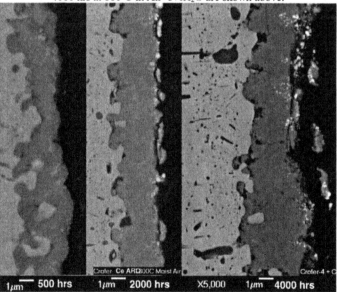

Figure 4. Cross sections of the oxide scale that formed on Crofer 22 APU+Ce (ARC method) after 500, 2000 and 4000 hrs at 800°C in Air+3%H$_2$O are shown above.

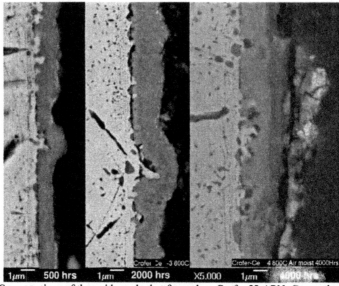

Figure 5. Cross sections of the oxide scale that formed on Crofer 22 APU+Ce (method described by Hou and Stringer, 1987 [4]) after 500, 2000 and 4000 hrs at 800°C in Air+3%H₂O are shown above.

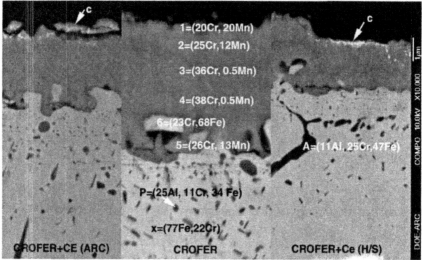

Figure 6. Comparison of the oxide scales that formed of Crofer 22 AP⏊ ⊔⎵⎓ ⎵ ⎵ APU +Ce by the Hou and Stringer method (H/S) and ARC method. Arrow "C" identifies CeO₂ particles. All compositions are in atomic percent and were determined by WDX analysis.

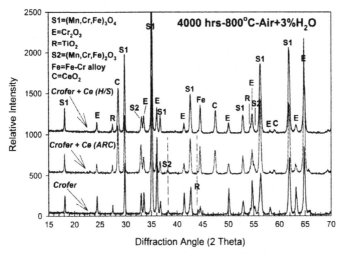

Figure 7. X-ray diffraction scans obtained from the surface of Crofer 22 APU and Crofer 22 APU+Ce by the ARC method and the Hou and Stringer method (H/S) after 4000 hrs exposure at 800°C to air+3%H₂O are shown above.

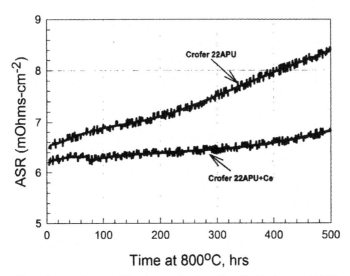

Figure 8. Effect of Ce surface modification on the Area Specific Resistance (ASR) at 800°C is shown above.

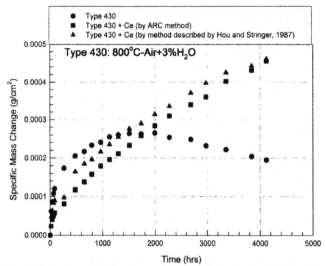

Figure 9. Effect of Ce surface modification on the oxidation behavior of Type 430 stainless steel is shown above.

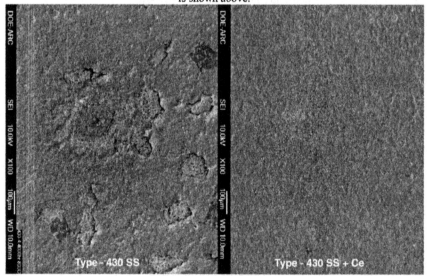

Figure 10. Surface views of the oxide that formed on Type 430 and Ce surface treated Type 430 after 4000 hours at 800°C in air+3%H₂O, is shown above. Notice that the oxide scale spalls from the surface of the Type-430, however, no spalling is observed in the Ce modified sample.

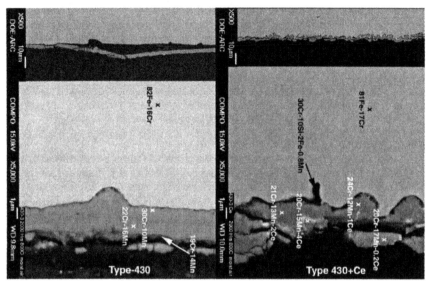

Figure 11. Comparison of the oxide scales that formed of Type 430 and Type 430 +Ce after 2000 hours at 800°C in air+3%H₂O.

Crofer 22 APU was designed as alloy for interconnect applications. Any treatment must not degrade the electrical performance of this alloy for this application. Area specific resistant (ASR) measurements were made on treated and untreated Crofer 22 APU. Tests were conducted at 800°C, and the results are shown in Figure 8. Crofer 22 APU+Ce treatment has a lower ASR compared to the untreated Crofer 22 APU. Clearly, long term testing is warranted; however, the surface treatment is effective in retarding oxide scale growth and does not increase the ASR, hence the surface treatment should enhance SOFC performance.

As reported previously, Ce surface treatments were effective in improving the oxidation resistance of a variety of other alloys. Figure 9 shows the results for Type 430 stainless steel (nominal composition: Fe-18Cr-1Mn). In this case the Ce surface treatment prevents the spalling observed in the base alloy (Figures 10). The oxide scale that formed in this alloy were similar in both surface conditions, untreated and Ce-treated (Figure 11), with a spinel forming at the gas metal interface. X-ray diffraction revealed both Cr_2O_3 and $(Cr,Mn)_3O_4$ as oxide products, with CeO_2 present on the surface of both Ce treated coupons. This treatment maybe enable lower cost alloys to be used in a variety of balance of plant applications that operate under extreme environments.

SUMMARY AND CONCLUSIONS

This research shows that under SOFC relevant conditions, i.e., 800°C moist air, an improvement in oxidation resistance and scale adhesion can be obtained by the incorporation of reactive elements (Ce) into the surface of the coupon. The improvement is attributed solely to the

Ce surface modification. A coupon that was subjected to the thermal portion of the infusion treatment but was not infused with Ce did not have a corresponding improvement in oxidation resistance. Further, two different pack-cementation processes were utilized and these resulted in essentially the same improvement in oxidation resistance. After 4000 hour: ....re at 800°C to air+3%H₂O, the weight gain of samples that were Ce surface modified .... half that of an unmodified sample. Further, initial electrical measurements indic.... Ce surface modification can reduce area specific resistance at the operating temperature. This is a consequence of the enhanced oxidation resistant and concomitant thinner oxide scales. Ce surface treatments were also found to mitigate oxide spalling that occurs in Type-430 SS.

ACKNOLEDGEMENTS

The assistance of P. Danielson with metallography, W.K. Collins and S. Matthes with SEM. R. Chinn and D. Smith with X-Ray Diffraction. all or the Albany Research Center, is greatly acknowledged. We also acknowledge G.G. Xia and Z.G. Yang of PNNL for assistance with ASR measurements.

REFERENCES

1.	H.U. Anderson and F. Tietz, in High Temperature Solid Oxide Fuel Cells, eds S.C. Singhal and K. Kendal. Elsevier, Oxford UK. 2004. p.173
2.	E. Lang (ed).. The role of active elements in the oxidation bel...
metal and alloys, Elsevier Applied Science, London, 1989.
3.	P.Y. Hou, and J. Stringer. Mat. Sci. and Eng. A202 1, (1995).
4.	P.Y. Hou and J. Stringer, J. Electrochem. Soc., 134 (7), 1836, (1989).
5.	K. Kofstad, High Temeparture Corrosion. Elsevier, London, 1988, p. 401.
6.	D.P. Whittle, J. Stringer. Philos Trans R Soc London Ser A, 295, 145 (1980).
7.	Pint BA, MRS Bull 19 (10), 26, (1994; 19 (10): 26.
8.	P.D. Jablonski , D.E. Alman and S.C. Kung, in Ceram. Engr. Sci. Proc.. Vol 26, ed N.Bansal, Amer. Ceram. Soc., Westerville OH, 2005. p. 193.
9.	F. Czerwinski and J.A. Szpunar, Thin Solid Films, 289, 213 (1996).
10.	R. Haugrud, "Corrosion Sci, 45. 211 (2003).
11.	C. Simon, M. Seiersten, and P. Caron, Mater. Sci. Forum., 251-254, 429 (1997).
12.	W.J. Quadakkers, V. Shemet L. Singheiser. U.S. Patent Application Publication No. US 2003/0059335 A1, March 27, 2003.
13.	W.J. Quadakkers, T. Malkow J. Piron-Abellan . U. Flesch, V. Shemet and L. Singheiser, in Proc. 4th Euro. SOFC Forum. Switzerland: Euro. Fuel Cell Forum, 2000, ed. E. McEvoy p. 827.
14.	J. Piron-Abellan, V. Shemet , F. Tietz, E. Wessel , WJ, Quadakkers, in SOFC VII. Electrochem. Soc, Pennington NJ, 2001. p. 793.
15.	J. Piron-Abellan, P. Huczkowski, S. Ertl, V. Shemet, L. Singheiser and W. Quadakkers, in Fuel Cell Seminar, 2004, San Antionio, TX, Nov 1-5, 2004.
16.	Z. Yang, G. Xia, P. Singh P and J.W., Stevenson, Solid State Ionics 176. 1495 (2005).
17.	G. Meier, 2004 SECA Annual Workshop and Core Technology Program Peer Review Workshop, May 11-13, 2004. Boston, MA, www.seca.doe.gov

Seals

COMPOSITE SEAL DEVELOPMENT AND EVALUATION

Matthew M. Seabaugh*, Kathy Sabolsky, Gene B. Arkenberg, and Jerry L. Jayjohn
NexTech Materials, Ltd.
404 Enterprise Drive
Lewis Center, OH, 43035

ABSTRACT

To achieve the high power densities in planar solid oxide fuel cell (SOFC) stacks, reliable seal technology is required. Current seal technology has been successful in laboratory stack testing, but in practice is hampered by reliability and lifetime issues, particularly with respect to thermal cycling. A composite approach, in which a crystalline ceramic phase is dispersed in a compliant viscous matrix, can potentially achieve good thermal, mechanical, and chemical stability through the stress relief and self-healing character of the viscous seal material and the interlocking nature of the crystalline phase. In this program, NexTech Materials has developed and tested a range of seal materials for application in SOFC stacks. This presentation will highlight our seal development effort to date.

INTRODUCTION

Many planar SOFC designs have been proposed using a number of sealing materials, with mixed degrees of success. The simplest methods have been single composition glass seals, which have been compositionally designed to melt at relatively low temperature and yet maintain sufficient viscosity at the operating temperature to provide a robust seal.[1,2] More complex approaches have attempted to use multiple glass phases in layers, or to add a minority fraction of ceramic particles to the mixture to fulfill the seal requirements.[3,4]

Glass-ceramic materials are a second sealing approach that has seen wide application in microelectronic packaging and glass-to-metal sealing applications.[5] Glass compositions with low melting temperatures and strong tendencies to devitrify are selected to bond glass and metal components with a seal that has a melting temperature much greater than the precursor glass. The *in-situ* solidification of the seal makes the fluid seal material rigid at the annealing temperature. Provided the solidified seal achieves hermeticity prior to crystallization, and withstands the stresses that develop during and after crystallization, this method can provide very good glass-ceramic seals.[6-9]

A final, pragmatic approach to seal design has been the selection of platy powders, such as mica and talc. These powders are tape cast to form a gasket that is then mechanically compressed during cell assembly and operation. In ongoing work, this approach has been used to seal test fuel cells.

State-of the-art sealing concepts for Solid Oxide Fuel Cells (SOFC) have been successful in providing oxidant and fuel separation in short term steady state tests of cells and small stacks. However, existing concepts have been shown to degrade rapidly when the stack components experience thermo-mechanical stresses due to thermal cycling. Successful commercialization and market entry of SOFC is partly dependent on the availability of a low cost and highly durable seal technology that can meet performance specification in terms of system transients

265

and service life.

These experiments focus on the development of composite sealing system for intermediate temperature solid oxide fuel cells (600-800°C) temperature range. These two-phase glass/ceramic mixtures structures are targeted to provide high durability, low cost and scalability for manufacturing. The proposed seals combine a crystalline ceramic phase that will provide a skeletal structure to the seal, and a glass matrix that will improve wetting at the seal interfaces and allow the seal to densify at lower operating temperatures. The composite structure is also anticipated to provide a degree of compliance to the stack at the operation temperature.

TECHNICAL APPROACH

A prospective composite seal must be chemically and mechanically robust. Additionally, it must demonstrate highly tailored and often contradictory properties. Seals must be chemically inert, with negligible interaction with the electrodes, electrolyte, and interconnect, but still form an intimate interface with the materials. The seal material must be compliant enough during manufacture and use to conform to irregularities in the component surfaces, but rigid enough during service to prevent creep and stack collapse. Finally, the seal must withstand tensile and compressive stresses that develop during thermal cycling of the fuel cell. Such resilience is particularly important for transportation applications.

Two approaches have been evaluated—the development of isotropic seal materials, in which equiaxed ceramic powders have been added to a glass matrix, and the development of textured seal materials, in which anisotropic ceramic particles have been oriented in a glass matrix. Examples of the resultant structures are shown in Figures 1 and 2. In both cases, it is expected that the majority crystalline phase will dominate the thermal expansion behavior and present a tortuous path for gas species with the possible added benefit of the textured seals in improved fracture toughness and creep resistance. The textured composites require close control of particle size distribution and content of the glass phase. While the seal pictured in Figure 2 has achieved significant alignment in the cast tape, it remains relatively porous (~50% theoretical density).

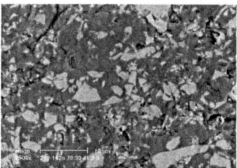

Figure 1. Example of Isotropic Composite Seal. The backscatter SEM image of equiaxed ceramic powder (light-gray particles) embedded in a non-crystalline glass matrix (dark-gray phase).

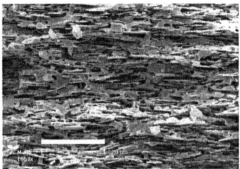

Figure 2. Example of Highly Textured Composite Seal. SEM
image shows fractured surface of seal material with highly
loaded anisotropic particles in non-crystalline glass matrix.

A range of particulate materials and glass compositions have been evaluated to determine
those best suited for SOFC sealing applications. This communication documents the
development of isotropic seal materials and their validation testing.

EXPERIMENTAL PROCEDURE

Seal Fabrication Approach

Modified aluminosilicate glass formulations have been selected for initial investigations
boron and phosphorous contents have been minimized to prevent volatilization of species during
operation. Alkaline-earth oxides were selected in order to increase the thermal expansion
coefficient of the composite materials into the target range of 8-12 ppm/°C, although Ba
containing glasses have been avoided due to reported interaction with metal interconnects during
operation under dual atmosphere conditions. The glasses were selected for intermediate
temperature solid oxide fuel cells with Tg in 700 – 800°C range.

Glasses were obtained in the form of glass frits (-60 mesh, ~2 m^2/g surface area) of the
desired composition. The received glass frits were subsequently ball milled for to reduce the
particle size to ~2 μm and to increase the surface area to ~ 4 m^2/g. Evaluation of glass powders
with finer particle sizes and higher surface areas were also completed. These powders exhibited
significant bloating of samples during sintering as gases were trapped in the structures by viscous
flow of the glass particles.

A range of equiaxed zirconia powders were evaluated as second phase additions to the
seal material, with surface area values ranging between 10 and 1.5 m^2/g. The impact of zirconia
characteristics on seal behavior will be discussed below.

The two powders were mixed by ball milling in ratios of 10:90 to 90:10. The sintering
behavior of the resultant mixtures were evaluated by sintering studies in pellet form. The
samples were produced by uniaxial pressing 1.25 cm diameter pellets. The sintered samples
were evaluated for density by Archemedes method and geometrically for shrinkage. As
promising compositions were identified, tapes were cast using a commercial tape casting system.

Seal Testing Approach

In order to evaluate these materials a testing system was designed to measure the seal leak rate by pressure decay method. This system is used to perform three measurements. a short-term screening test that measures the leak rate versus temperature, a longer term test that measures the performance of the seal versus time and a final test is a thermal cycle test that measures the integrity of the seal versus the number of cycles between 800-200°C. To perform the test under conditions similar to fuel cell conditions, each seal is tested at temperature using helium gas to simulate hydrogen and a mild compressive load (0.55 MPa). During the test. the interior of the seal area is pressurized to 13.8 kPa with helium. Upon reaching the testing pressure the inlet line is closed off from the gas inlet by a ball valve and the excess gas is sent to the exhaust (see schematic in Figure 3). Although the system pressure is higher than the backpressure that will be experienced by an actual fuel cell (2 kPa). the pressure allows direct and rapid measurement of the seal leak rate. The performance of each seal is determined by the leak rate (sccm) divided by the seal length (cm). The leak rate is calculated from the pressure drop across the closed system over time.

The current test apparatus design allows for three independent seals to be evaluated simultaneously. The helium pressure is controlled using metering valves to slowly increase the pressure on each seal. This used to mitigate any excess stress due to a large flux of gas into the seal from the regulator during pressurization. Both digital pressure gauges and pressure transducers measure the system pressure for each test stand. The primary function of the digital pressure gauges is to monitor the pressure during the pressurization of each experiment. The pressure transducers are used to measure the seal performance during the each individual experiment. This information is sent to a digital chart recorder that continuously records these values along with temperature of each seal. Stainless steel tubing is used throughout the system to deliver the helium to the test apparatus.

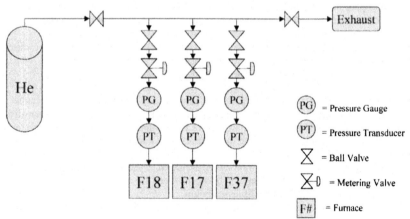

Figure 3. Seal Testing System Diagram

The seal samples are produced as tape cast layers and laminated to form seals 500μm thick. These laminates are punched into discs of following dimensions: OD = 2.3cm and ID = 1.3 cm. A glass ink is applied to the top and bottom surfaces consisting of the same glass powder as used in preparing the seal tapes, to aid in wetting and adhesion of the seal. The screen printed glass layer is 10-20 μm thick.

The sample is tested by placing the seal between a doped zirconia disc and the metal manifold shown in Figure 4. This manifold consists of an Inconel 600 tube to which a Crofer 22 APU disc has been welded. Prior to each test, the manifold is polished to remove oxide scale, and then heated to 700°C in an air atmosphere for one hour to achieve a continuous oxide coating. The manifold is placed such that the holes in the seal sample and the tube align.

Once the seal is aligned with the tube the test apparatus is then lowered onto the sample (Figure 5). The entire system is then aligned using a level to ensure that the Crofer 22 APU disk is in contact with the sample over the entire seal area (Figure 6). The seal is cured by heating the sample 850°C where it holds for three hours to ensure proper sealing. Once the seal has been given time to cure it is then cooled at 5°C/minute to the desired test temperature, and short term testing performed. For seals being tested to evaluate stability, the seals are held at 800°C.

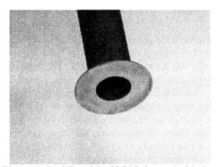

Figure 4. Seal Test Manifold Prior to Oxidation

Figure 5. Aligned Sample in Furnace Chamber

Figure 6. Assembled Sample in Furnace Chamber

EXPERIMENTAL RESULTS AND DISCUSSION

Sintering Studies

Figure 6 shows the results of the sintering study for the composite mixtures of the selected seal material (#41) and zirconia powder at 850°C. The surface area of the seal material and zirconia powder was 4 m^2/g and 1.5 m^2/g, respectively. The two powders were mixed in varying ratios by ball milling mixtures of the glass and zirconia powder. The study showed that the composites with seal material content less than 50% saw little to no improvement in density over the pure zirconia samples. The composites with seal material content greater 70% achieved >95% density with shrinkage over 14%. The pure glass material sample achieved 100% density at 850°C with ~13% shrinkage. Based on this assessment, high glass content was selected for these two components. Alternative formulations (particularly those using higher surface area glasses use higher concentrations of the crystalline component).

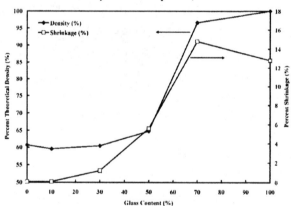

Figure 6. Impact of Glass Content on Composite Densification and Shrinkage

Seal Performance Testing

Figures 7 and 8 provide composite seal performance data obtained using the 70% glass formulations identified in Figure 6. In Figure 7, a screening test of short term performance, the calculated leak rate for two composites utilizing similar compositions to that described in Figure 6 but with varying glass formulations is compared. Both seals perform well below the targeted leak rate of 0.094 sccm/cm at all temperatures and below reported values for seals tested under similar conditions. [10,11]

The seal test target leak rate was calculated based upon the metric of 1% fuel loss. A fuel flow rate of 376 sccm was calculated based upon a fuel cell with the following specifications: area of 81 cm^2, current density of 0.5 A/cm^2, and 75% fuel utilization. The amount of fuel loss was then determined to be 3.76 sccm. The seal length (40 cm) is the perimeter of a 10 cm square seal. Finally the target seal leak rate of 0.094 sccm/cm is obtained by dividing the fuel loss by the seal length.

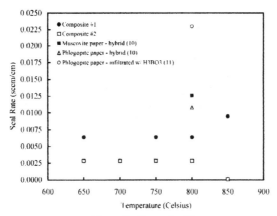

Figure 7. Screening Tests

Figure 8 shows pressure decay curves for a promising seal material, demonstrating stable seal performance over 50 hours with multiple pressurization cycles. The pressure decay data has been analyzed to determine the leak rate over several hours, before the pressure was again raised to the starting value on a periodic basis. Based on the pressure decay data, it is clear that the seals retain their hermeticity over the 50 hour test. Ongoing evaluations are being performed to determine the applicability of the reported seals during fuel cell operation.

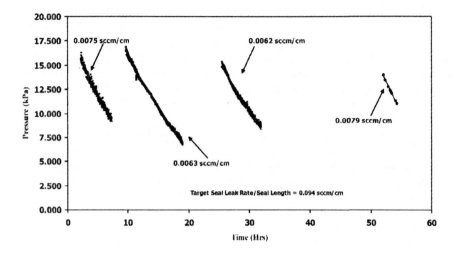

CONCLUSIONS

Using a combination of glass formulations and crystalline materials, a range of seal materials have been developed. Using *in-situ* measurements of seal performance, materials offering good hermeticity have been identified. Continuing evaluations of performance under fuel cell operation are underway.

REFERENCES

[1] A. Bieberle, et al., "Glass Seals," *Oxygen Ion and Mixed Conductors and Their Technological Applications*, Tuller, et al., ed. Kluwer, Boston MA (2000)
[2] I.D. Bloom and K.L. Ley, U.S. Patent No. 5,453,331
[3] Y. Harufuji. Japanese Patent JP04-47,672
[4] L.A. Xue, US Patent No. 6,271,158
[5] I.W. Donald, J. Mat. Sci, **28** 2841-2886, (1993)
[6] I. Mukaisawa, Japanese Patent JP06-060891
[7] K. Eichler, et al., J. European Ceram. Soc. **19**, 1101-1104, (1999)
[8] P.H. Larsen, et al., J. Non-Cryst. Sol. **244**, 16-24 (1999)
[9] P.H. Larsen, et al., J. Mat. Sci., **33**, 2499-2507 (1998)
[10] Y.S. Chou et al., J. of Power Sources 112 (2002) 130-136
[11] Y.S. Chou et al., J. of Power Sources 135 (2004) 72-78

ACKNOWLEDGEMENTS
The authors would like to acknowledge the support of the U.S. Department of Energy, Contract Number DE-FG02-02ER83528

INVESTIGATION OF SOFC-GASKETS CONTAINING COMPRESSIVE MICA LAYERS UNDER DUAL ATMOSPHERE CONDITIONS

F. Wiener, M. Bram, H.-P. Buchkremer, D. Sebold

Forschungszentrum Jülich GmbH, Institute for Materials and Processes in Energy Systems, IWV, D-52425 Jülich, Germany

ABSTRACT

Mica gaskets and also composite gaskets containing compressive mica interlayers are under consideration as sealing material in solid oxide fuel cells (SOFC).
The results of leak tests on commercially available mica papers under stationary and thermal cycling conditions are presented. The elastic recovery and spring constant at 1073 K was also determined.
To study potential interactions between the interconnect steel Crofer22APU and the mineral phases vermiculite (exfoliated) $(K,Mg,Fe)_3(Si,Al)_4O_{10}(OH)_2)$ and talc $(Mg_3Si_4O_{10}(OH)_2)$, corrosion experiments were conducted in simulated SOFC conditions. In these experiments, the opposite walls of the gaskets were simultaneously exposed to air and wet H_2.
A substantial increase in the thickness of the oxide layers formed by the interconnect steel is observed in contact with mica. The $Cr_2O_3/(Cr,Mn)_3O_4$ duplex layer normally formed on the Crofer is replaced by a thicker (factor of 5-10) layer of a complex microstructure that is assumed to contain Cr_2O_3, $(Cr,Mn,Mg,Fe)_3O_4$ and Fe_2O_3 phases. The modified microstructure is found in the entire air manifold, with an increased thickness up to a distance of 300 μm from the mica.

1. INTRODUCTION

SOFCs are used for energy conversion since they generate electrical power by the electrochemical reaction of an oxidant and a fuel gas. Currently two basic designs for SOFC applications exist: tubular and planar. Stacks of planar SOFCs are believed to offer the potential for higher cost efficiency, and higher power density per volume or mass when compared to tubular designs. However, a high temperature sealing concept ensuring stability and low leakage during long-term operation with thermal cycles remains a critical challenge. An increase in leak rates reduces the efficiency of the system. Furthermore, larger leaks lead to gas-phase oxidation of the fuel, forming so-called "hot spots". These can potentially damage stack components.

Sealants in SOFC stacks need to fulfill a variety of requirements [1]. Long-term stable separation of oxidant and fuel gases has to be achieved. The seal has to maintain integrity through thermal cycling operation either by compensating or by tolerating mechanical and thermal mismatches of the stack components. The gasket materials need to withstand oxidizing and humid reducing atmospheres without degradation at an operating temperature of typically 1073 K for several thousand hours. Also the seals should not degrade the materials they are in contact with nor form volatile compounds with the potential for electrode poisoning. Electrical insulation is required for designs with metallic interconnects to avoid short circuiting.

Currently glass-ceramics are the sealing materials most commonly used. They offer gas tightness, and the potential for adaptation to the coefficient of thermal expansion (CTE) of other stack components by controlling the crystallized phase content. Typically the electrical resistance and chemical inertia of glass-ceramics are sufficient for this application. However, due to the inherent brittleness and the rigid interfaces, crack growth in the sealant or the sealant/ interconnect interface may occur during thermal cycling operation [2-4].

To overcome the drawbacks of rigid bonding of sealant materials with the interconnect or electrolyte, work began on developing a compressive sealing concept suitable for SOFC temperatures and the limited acceptable sealing force on SOFC stacks [5-8]. Another potential advantage of compressive seals over rigid designs is better electrode contacting in stack operation, resulting in reduced ASR (area specific resistance) over the cell.

Compressive seals based on mica show a high potential for SOFC applications [1]. The goal of this paper is the determination of the elastic behavior and leakage characteristics of commercially available mica papers at the temperature range from room temperature up to SOFC temperatures (1073-1223 K) to help in materials selection for gasket design.

The mica papers used as a compressive layer consist of minerals that contain alkaline and earth alkaline elements. It is claimed that some of these elements interact with Cr_2O_3 –forming ferritic interconnect steels by accelerating chromium evaporation through volatile phases. [10]. Accelerated rates of chromium evaporation pose two critical dangers. Cr impurities in the LSM or LSCF cathodes show a detrimental effect on the catalytic activity of the cathode material [11]. This cathode poisoning potentially affects system aging rates by orders of magnitude. Furthermore, a depletion of the Cr content below a threshold value of approximately 16 at.% causes catastrophic breakaway oxidation in Crofer [13].

Therefore, the influence of the the mica on oxide scale growth of the interconnect material Crofer 22 APU is investigated. The sample was selected based on the outcome of the leakage investigations. Potential consequences of these interactions on the continuous operation of SOFC stacks are discussed.

2. EXPERIMENTAL

In this work, three commercial mica papers were investigated: Thermiculite 815, Thermiculite XJ 766 (both Flexitallic, UK) and Statotherm HT (EagleBurgmann Germany). To enhance its stability, Thermiculite 815 consisits of exfoliated vermiculite and has a 1.4401 steel inlay. Thermiculite XJ 766 is a mixture of vermiculite and talc. Statotherm HT consists of pure mica platelets stabilized by an organic binder.

To study the elastic behavior of these mica papers, square sheets (20 mm x 20 mm) were prepared and precompressed with 80 kN (200 MPa) at room temperature. Precompaction was found to be necessary in former investigations due to the strong irreversible plastic deformation of mica paper during the first load cycle [7]. The target thickness of the papers after precompaction was 300 μm for Thermiculite XJ 766 and 800 μm for Statotherm HT and Thermiculite 815. The thickness of the samples was controlled by using spacers. Compression tests were done in an INSTRON 1362 high-temperature testing machine. Starting with a preload of 20 N, the samples were compressed at 50 N/min to 900 N (2.25 MPa) and subsequently unloaded at 200 N/min. The loading/unloading cycles were repeated four times. The elastic recovery was determined between the maximum applied load and an arbitrarily defined low load of 50 N.

The leak rates of flat, compressive seals were determined in a test bed at high temperatures (800-1300 K) as described in detail in [6]. A scheme of the apparatus is shown in figure 1. Two punches made of Crofer 22 APU 1st /1.4770 (Thyssen VDM, Germany) were used. The surface of the punches was treated with emery paper of grade P800 before each test. The three mica papers already mentioned before were tested. The samples had a square shape with 50 mm outer length. The width of the sealing face was 4 mm, resulting in a sealing length of 168 mm, and a surface of 736 mm². The samples were laser cut from sheets and precompressed again to 300 μm (Thermiculite XJ 766) and 800 μm (Thermiculite 815, Statotherm HT). The leak rates were constantly monitored with 20 kPa relative gas pressure of Ar. The leak rates of all mica papers were measured under stationary conditions at 1073 K with a sealing load of 0.68 MPa.

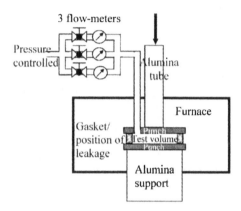

Figure 1. Schematic of test bed for high temperature leak rate determination.

In an additional test, the potential of seals for an application in a 20 kW system, which is under construction in Juelich, was evaluated. In this system seals are exposed to different operating temperatures ranging from 700 to 1200 K. Therefore, the temperature was stepwise increased with 3.3 K/min to 473 K, 673 K, 873 K, 1073 K and 1223 K. The dwell time at each temperature was 240 min. Afterwards the temperature was decreased to room temperature using the same steps. Considering the sealing conditions of components in this 20 kW system we deviated from standard sealing load of 0.68 MPa and employed 0.8 MPa instead.

It is assumed that compressive seals are more suitable to tolerate thermal cycling from SOFC conditions to room temperature than rigid seals (glass-ceramic or metal braces). For confirmation, a square gasket in the abovementioned configuration was subjected to 34 thermal cycles from room temperature to 1073 K with ramps of 10 K/min.

All specimens investigated fulfilled the requirements regarding electrical insulation. The last crucial requirement for SOFC applications is chemical inertia with respect to the components in contact with sealing surfaces. Based on the outcome of the leakage investigation, this study was restricted to the mica with the lowest leak rates. Therefore, the corrosion experiment was conducted with Thermiculite XJ 766 mica paper placed between sheets of the ferritic alloy Crofer 22 APU /1.4760 (Thyssen VDM, Germany). The sample had the same geometry described for the former test. Thermiculite XJ 766 consists of a matrix of exfoliated vermiculite $((K,Mg,Fe)_3(Si,Al)_4O_{10}(OH)_2)$ filled with talc $(Mg_3Si_4O_{10}(OH)_2)$. The procedure for simultaneous corrosion testing in reducing and oxidizing atmospheres is described in detail in [12]. The testing was conducted for 400 hours at 1073 K with air in one manifold and with hydrogen with 3 vol.% H_2O in the other.

After the corrosion test, cross sections of the gaskets were prepared for SEM/EDX (Zeiss, Gemini 1530) to analyze the morphology and chemical composition (qualitative) of the samples. Another test was conducted to determine weight loss and to prepare samples for follow-up analyses (not discussed in the paper). Therefore air was passed over a sample of 1.142 g Thermiculite XJ 766 in a tube furnace at 1073 K for 400 hours at a rate of 5 $l_n \cdot min^{-1}$.

3. RESULTS AND DISCUSSION

The characteristic deformation behavior of compressive mica seals is exemplary shown in figure 2 and summarized in table 1.

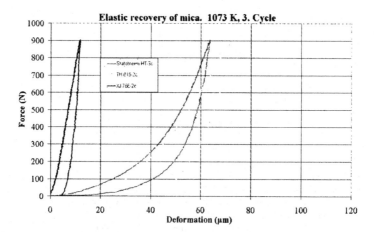

Figure 2. Exemplary results of load-deflection experiments. The 3. load cycle at 1073 K is shown.

Table I: Elastic spring-back (in μm) and spring constant (in N/μm) of mica papers at 1073 K during 4 load cycles with 0.9 kN and spring constant. The starting thickness was 300 μm (Thermiculite XJ 766) and 800 μm (Thermiculite 815, Statotherm HT).

	Thermiculite 815 (μm), (N/μm)		Thermiculite XJ766 (μm), (N/μm)		Statotherm HT (μm), (N/μm)	
1. Cycle	39,0				32,3	
2. Cylce	35,1	52	11,4	167	33,2	82
3. Cycle	31,0	59	10,3	243	34,9	79
4. Cycle (after 2 h holding)	19,1	83	6,9	587	33,9	81

If the samples were loaded without a dwell time at maximum load, the elastic recovery stays almost constant. The reduced elastic recovery of Thermiculite XJ 766 compared to the other samples is mainly caused by the lower thickness of the sample. The introduction of a dwell time at maximum load leads to a clear reduction of elasticity for both Thermiculite materials, while it does not influence the behavior of Statotherm HT. In case of Thermiculite 815, the loss of elasticity is caused by the metallic inlay, which starts to creep at SOFC operating temperature. For Thermiculite XJ 766, a decomposition of the talc filler material is assumed to be a reason for the ongoing deformation. The decomposition reaction is discussed later on.

The spring constants are determined from the ration of deformation over the applied load in the area of constant slope (500-900 N). Low spring constants are preferred in compressive sealing concepts. Considering this, Thermiculite 815 initially offers superior elastic behavior, but is influenced by mechanical load cycles. The properties of Statotherm HT indicate enhanced long term stability of the recovery, but this advantage is diminished by the limited stability hampering its handling. Therefore, Thermiculite 815 was choosen as elastic component in the Juelich´s composite seal [7].

The outcome of the leakage measurements (raw data) is shown in figure 3. Generally, at a given cross section of the leakage paths an increase in the measured leak rates is observed at lower temperatures using the experimental set-up shown in figure 1. Cooling from SOFC operating temperature (1073 K) to room temperature is coupled with an increase of the gas volume (approximately factor 4) and gas viscosity (approximately factor 3) leading to a clear increase of the mass flow. Therefore, for the following discussion, leak rates normalized to 1073 K with respect to gas volume and gas viscosity are considered.

The leak rate of Statotherm HT shows two ranges. During initial heating up to 673 K the gas flow remains constant. On heating to 873 K the leakage constantly increases till reaching a stable level approximately twice as high as before at 673 K. This constant value is observed during the following heating, and during cool down. It is assumed that the increase in gas flow is a consequence of organic binder burn-out. Statotherm HT seems equally suited as a gasket material at the entire temperature range up to 1223 K.

Thermiculite 815 shows a steady increase of normalized leak rates during heat up and at the ramps. At the 873 K ramp a maximum is reached which is maintained through further heating and cooling to 873 K. At lower temperatures the normalized leak rates are again decreased. Therefore, Thermiculite 815 shows better sealing performance at lower temperatures than at SOFC temperatures, but does not approach the values of Statotherm HT under any conditions.

With Thermiculite XJ 766 a steady decrease of the leakage up to a 1073 K is observed. This observation is supposedly the result of orientation of the sheet–like mica particles and talc filler material. No changes in the normalized leak rate are observed during further heating/holding to 1223 K or during cooling. At the entire temperature range the material highlights leak rates of an order of magnitude lower than Statotherm HT. This is attributed to the highly compacted microstructure of the material shown in figure 4.

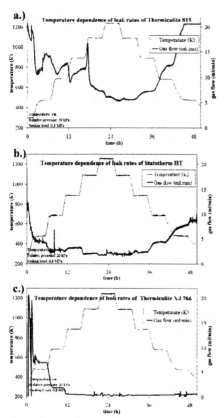

Figure 3. Raw data of gas flow through mica gasket with 168 mm sealing face at 20 kPa relative pressure and sealing load 0.8 MPa a) Statotherm HT; b) Thermiculite 815; c) Thermiculite XJ 766.

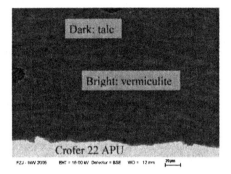

Figure 4. Micrograph of Thermiculite XJ 766 showing mineral phases vermiculite and talc.

Characteristic values for elastic recovery and leak rates of 3 mica papers at 1073 K are normalized over the gaskets inner circumference and summarized in table II. These values are in good agreement with related investigations published recently [1]. In practice the observed leak rates on the order of 10^{-2} sscm cm^{-1} correspond to a loss of less than 0.1 vol.%. fuel gas. This leakage is homogenously distributed over the entire gasket length as can be shown at room temperature with leak detection spray. Therefore the risk of "hot spots" in stack operation can be ruled out with compressive mica gaskets. In the future, it has to be demonstrated that the sealing concept can be transferred from simple test geometries to SOFC stacks.

Table II. Leak rates of mica papers (Statotherm HT, Thermiculite XJ766 and Thermiculite 815). The gas pressure difference is 20 kPa, compressive stress 0.68 MPa (+ promising, O barely acceptable, - not suitable).

Gasket	Sealing behavior	Gas flow/sealing lenght, (sscm cm^{-1})	Spring-back	Potential applications
Thermiculite 815	-	$4,5*10^{-1}$	+	Elastic layer in composite gasket
Statotherm HT	O	$1,3*10^{-1}$	+	electrical insulation in composite gaskets
Thermiculite XJ766	+	$1,5*10^{-2}$	O	Auxiliary power units APU (Air manifold)

With respect to leak rates, Thermiculite XJ 766 shows the most promising behavior. Therefore it was the material selected for more time-intensive follow-up investigations. One key requirement for SOFC operation in mobile applications is thermal cyclability. The response of Thermiculite XJ 766 to 34 cycles is shown in figure 5.

In the first two cycles the leakage increases by 40%. The following 32 cycles show a gradual increase of the leak rate totaling 15%. The dotted lines indicate the trend of gas flow in the cycling operation. Again, the measurement is influenced by the increase of gas volume and gas viscosity at higher temperatures. This results in different magnitudes.

Chemical interactions between the gasket material and the interconnect steel are most clearly manifested in compositional and morphological changes of the oxide scales formed on the metal. The Crofer 22 APU interconnect steel was developed to form a duplex-oxide layer [13]. On top of the protective Cr_2O_3 layer, a stable $(Cr,Mn)_3O_4$-spinel is assumed to inhibit chromium evaporation from the chromia scale. Overall thickness of both layers after 400 hours is generally 2-3 μm on the air side and 1-2 μm on the fuel side. For future reference, the typical surface morphology of this steel (1.4760) after 250 hours at 1073 K in air is shown in figure 6.

The changes of morphology and phase content of the oxide scales on the interconnect steel in contact with mica paper under air and wet hydrogen are compared in figure 7. Figure 7f indicates the location of the investigated sample areas. Table III summarizes the observed phase content.

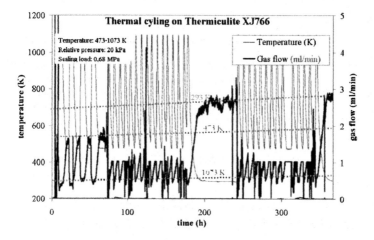

Figure 5. Gas flow (raw data) through Thermiculite XJ 766 gasket during thermal cycling with 10 K/min. The dashed lines show trends at 293 K, 473 K and 1073 K.

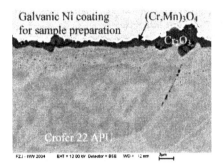

Figure 6. Duplex oxide morphology of Crofer22 APU /1.4760 after 250 hours at 1073 K.

On the air side in direct contact with the mica an area of increased oxide thickness (20-25 μm) extending up to a distance of 250-300 μm from the gasket is formed (Fig. 7a). The dense oxide displays an obvious layer sequence: A Cr(Cr,Mn,Fe)$_3$O$_4$ of thickness < 5 μm with an unusually low concentration of Mn is formed directly on the Crofer 22 APU interconnect material. Then a strip of iron-rich chromium oxide follows. The top layer is a 15-20 μm thick, homogeneous region of Cr oxide, with significant amounts of Mg, Fe and Mn. In this oxide Mn clearly shows enrichment towards the air manifold, whereas for Fe and Mg no obvious gradient exists. However, there is a sharp peak of Mg at the interface with air determined with linescans. The EDX-linescans were continued inside the Crofer 22 APU steel up to a depth of 195 μm to investigate whether the formation of the abnormal oxide scale is accompanied by Cr depletion (Table IV).

The obvious decrease of Cr (18.2 wt.%) towards the interface enhances the risk of catastrophic oxidation as evident from the formation of unstable Fe-containing oxides. The area shown in Fig. 7a is the only position in the sample where a clear Cr depletion was found. Fig. 7b and 7c shows the oxide scale taken at a distance of 500 and 3800 μm, respectively. Neither of these structures is as developed as in the reference (Fig. 6). At a distance of 500 μm the scale separates into two phases (Fig. 7b) which can be distinguished by phase contrast. The brighter phase, almost pure Cr$_2$O$_3$, is embedded in a matrix of (Cr,Mn.Mg) oxide, which is most likely a solid-solution spinel.

The oxide scale at a distance of several mm from the gasket fails to develop a layered duplex morphology. Instead, a single phase with somewhat increased thickness (3-5 μm) of (Cr,Mn) oxide is observed (Fig. 7c). The Mn concentration of this phase is homogeneous, but at a significantly lower level than the spinel phase of the reference sample. The phase diagram for the system Cr$_2$O$_3$-Mn$_2$O$_3$ suggests Cr$_2$O$_3$ containing Mn in solid solution [14].

On the fuel side, the oxide thickness increases to 8-10 μm in direct contact with the mica (Fig. 7d). Under anode conditions, the microstructure remains a layered duplex. Mg is detected in the spinel, but not in chromia. This area of increased oxide growth extends 400 μm starting from the edge of the mica. At distances of more than 400 μm, the oxide scale exhibits the usual morphology and composition as known from the reference samples under anode conditions.

The sample treated in a tube furnace for 400 hours at 1073 K showed an overall mass loss of 7.3 wt.%. As usual with Thermiculite XJ 766 embrittlement is observed.

Summarizing the experimental observations, the presence of talc-containing Thermiculite XJ 766 influences oxidation behavior strongly on the air side, and moderately on the anode manifold. The effect is most profound in the direct contact area interconnect/mica. This area shows an oxide thickness of 20-25 μm, mainly consisting of $(Cr,Mn,Mg,Fe)_3O_4$. Chromium depletion of the Crofer 22 APU steel is evident. At distances of more than 500 μm from the mica no significant increase of oxide scale thickness is observed, but the material fails to develop a protective duplex layer.

On the fuel side, the usual duplex layer is developed. At areas in direct contact with mica, Mg is detected in the top layer, the spinel. The thickness of the scale increases to 8-10 μm, while no iron in the oxide nor chromium depletion of the steel have been determined. On both sides, Mg can be detected in the oxides whose morphology is unusual.

Based on the outcome of the corrosion test further experimentation (XRD, DTA/TG) on as-received and heat treated Thermiculite XJ 766 was triggered to determine the cause and mechanism of the accelerated oxide growth. It was determined that talc decomposed to enstatite, quartz and water at temperatures higher than 950 K [15]. This decomposition appears to be the path for mobilisation of volatile $Mg(OH)_2$. The Mg-ions from this metastable phase are readily incorporated into the oxide scale of the interconnect steel.

In conclusion, the obvious deterioration of interconnect oxidation behavior is not tolerable. The formation of iron-containing oxides even after 400 hours at 1073 K under dual atmosphere conditions points to the possibility of catastrophic breakaway oxidation in operation. On the air side the absence of a duplex layer poses the danger of accelerated chromium evaporation and cathode poisoning. In the reference sample the outermost $(Cr,Mn)_3O_4$ immobilizes chromium, while the investigated sample appears to contain less stable solid solutions of Mn, Mg and Fe in chromia. As a consequence for the development of a compressive gasket a full encapsulation of the mica is desired to reduce the risk of undesired volatile constituents.

Air manifold　　　　　　　　　　　**Fuel manifold**

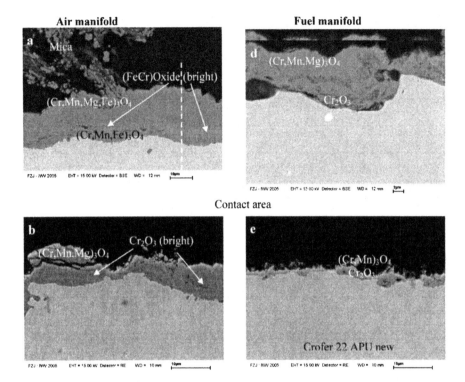

Contact area

Distance from mica: 500μm

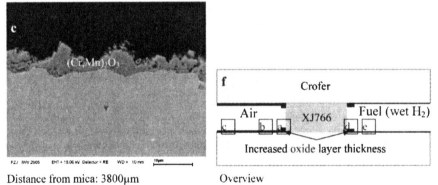

Distance from mica: 3800μm　　　　　Overview

Figure 7. Overview of findings in gasket A (XJ766) after 400 h at 1073 K. The left-hand side shows micrographs taken on the air manifold, and the right-hand side micrographs on the fuel manifold.

Table III. Phase content of the oxide scales at different sample areas after heat treatment at 1073K/400 h. Arranged according to increasing distance from interconnect.

Manifold, figure	Distance from mica (μm)	Assumed phase content	Total thickness of oxide scale (μm)
Air, 7a	0	$(Cr,Fe,Mn)_3O_4$, (Fe,Cr)-oxide, $(Cr,Mn,Fe,Mg,)_3O_4$	20-25
Air, 7b	500	Two phases, mixed, $(Cr,Mn,Mg,)_3O_4$, Cr_2O_3	3-5
Air, 7c	3800	Single phase (Cr,Mn) oxide	3-5
Fuel, 7d	0	Duplex Cr_2O_3, $(Cr,Mn,Mg)_3O_4$	8-10
Fuel, 7e	500	Duplex Cr_2O_3, $(Cr,Mn)_3O_4$	1-2

Table IV: Chromium concentration in Crofer 22 APU interconnect und⸱ ⸱⸱⸱ oxide scales in the air manifold (Fig. 7a), determined by EDX.

Distance from oxide (μm)	Cr (wt.%)	Distance from oxide (μm)	Cr (wt.%)
6	18.2	111	24.6
27	19.5	132	24.5
48	21.9	153	23.8
69	23.1	174	23.6
90	23.8	195	23.8

4. CONCLUSIONS

Based on the result of leakage investigations mica paper is discussed as a promising sealing material for SOFC stacks. To achieve low leakage rates, the use of talc as a filler phase is prerequisite. At 1073 K this composite shows smaller elastic springback (~2.3%) than pure mica gaskets (~2.4% and ~4.2%)

A test set-up where mica is exposed to a dual atmosphere simulating SOFC conditions reveals that the gasket undergoes interactions with the interconnect at SOFC operating temperatures of 1073 K. This results in a modification of ferritic steel corrosion. On the oxidant manifold Mg is incorporated into oxide scales, modifying an ordered duplex layer to a disordered microstructure. In the triple point of mica/steel/air the modified composition is accompanied by rapid oxide growth of iron-containing oxides, and chromium depletion in the steel.

ACKNOWLEDGEMENTS

The authors would like to thank V. Haanappel for conducting the experiments in the simulated SOFC environment, and M. Kappertz for the metallographic preparation. We would like to thank J. Moench and R.W. Steinbrech for determination of elastic behavior as well as W.J. Quadakkers and P. Huczkowski for reference micrographs and helpful discussions.

REFERENCES

[1]J.W. Fergus, "Sealants for solid oxide fuel cells," J. Power Sources **147**, 46-57 (2005).
[2]K.L. Ley, M. Krumpelt, R. Kumar, J.H. Meiser, and I. Bloom, "Glass-ceramic sealants for solid oxide fuel cells: Part I. Physical properties," J. Mater. Res. **11** [6] , 1489-1493 (1996).
[3]S. Ohara, K. Mukai, T. Fukui, and Y. Sakaki, "A new sealant material for solid oxide fuel cells using glass-ceramic," J. Ceram. Soc. Jap. **109** [3] 186-190 (2001).
[4]N.Lahl, L, Singheiser, K. Hilpert, K. Singh, and D. Baradur, "Aluminosilicate glass ceramics as sealant in SOFC stacks," Electrochemical Society Proceedings Vol. 99-19 1057-1066 (1999).
[5]Y.-S Chou, J.W. Stevenson, and L.A. Chick, "Novel compressive mica seals with metallic interlayers for solid oxide fuel cell applications," J. Am. Ceram. Soc. **86** [6] 1003-1007 (2003).
[6]M.Bram, S. Reckers, P. Drinovac, J. Moench, R.W. Steinbrech, H.P. Buchkremer, and D. Stoever, "Characterization and evaluation of compression loaded sealing concepts for SOFC stacks," Proc. Electrochem. Soc. 2003-07 (SOFC VIII) 888-897 (2003).
[7]M. Bram, S. Reckers, P. Drinovac, J. Moench, R.W. Steinbrech, H.P. Buchkremer, and D. Stoever, "Deformation behavior and leakage tests of alternate sealing materials for SOFC stacks." J. Power Sources **138** 111-119 (2004).
[8]Y.-S. Chou, and J.W. Stevenson, "Long-term thermal cycling of phlogopite mica-based compressive seals for solid oxide fuel cells," J. Power Sources **140** 340-345 (2005).
[10]R.Weiß, D.Peck, M.Miller, and K. Hilpert, "Volatility of chromium from interconnect material," Proceedings of the 17th Risø International Symposium on Materials Science, Roskilde, Denmark, 1996, 479-484 (1996).
[11]Y. Matzsuzaki, and I. Yasuda, J. Electrochem. Soc. **148** [2] A126-A131 (2001).
[12]V.A.C. Haanappel, I.C. Vinke, and H. Wesermeyer, "A novel method to evaluate the suitability of sealing materials for SOFC stacks," Proceedings of the 6th European SOFC Forum, Lucerne, Switzerland, 2004, 784-791 (2004).
[13]P. Huczkowski, V. Shemet, J. Piron-Abellan, L. Singheiser, W. J. Quadakkers, and N. Christiansen, "Oxidation limited life times of chromia forming ferritic steels, Materials and Corrosion," Materials and Corrosion **55** 825-830 (2004).
[14]D. H. Speidel, and A. Muan, "The System Manganese Oxide -Cr_2O_3 in Air." *J. Am. Ceram. Soc.* **46** [12] 577-578 (1963).
[15]F. Wiener, M. Bram, H.-P. Buchkremer, D. Sebold, Chemical interaction between Crofer 22 APU and mica-based gaskets under simulated SOFC conditions, J. Mat. Sci. in press.

PERFORMANCE OF SELF-HEALING SEALS FOR SOLID OXIDE FUEL CELLS (SOFC)

Raj N. Singh and Shailendra S. Parihar
Department of Chemical and Materials Engineering
University of Cincinnati
Cincinnati, OH-45221-0012

ABSTRACT

A self healing glass-sealing concept is developed and used for making metal-glass-ceramic seals for potential application in solid oxide fuel cells (SOFC) in order to enhance reliability and life of cell. In this study, a glass displaying self-healing behavior is used to fabricate seals and performance of these seals under long-term exposure at higher temperatures coupled with thermal cycling are characterized by leak tests. Self healing ability of these glass seals is also demonstrated by leak rate measurements. These results demonstrated excellent performance and self-healing ability of the seals for times exceeding 2500 hours and 275 thermal cycles between 25^0C and 800^0C.

INTRODUCTION

Solid Oxide Fuel Cell (SOFC) is a solid state energy conversion device that produces electricity by electrochemically combining fuel and oxidant across an ionic conducting oxide membrane at higher temperatures.[1] Planar configuration of SOFC is superior to other configurations in terms of efficiency and power density, but requires hermetic seals to prevent fuel-oxidant mixing and provide electrical insulation to stacks.[2,3] Seals required for a planar SOFC can be classified as metal-metal, ceramic-ceramic, and metal-ceramic seals. Among these seals, metal-ceramic seals are particularly challenging because of their severe functional requirements as well as difficulty in selection of material and associated processing optimization. Glasses are most widely used to make metal-ceramic seals because these can be modified to have very close match of thermal expansion with other fuel cell components, and glass seals show good hermeticity along with good thermal and environmental stability.[4, 5, 9, 10] Although glasses are good option to seal SOFC components, they suffer from their inherent brittleness. Due to the brittleness of the glass seal, cracks can be developed in the seal during thermal cycling or shock which can cause leakage from the seal leading to degradation in cell performance. Fortunately, there are ways to minimize this problem by using some innovative approaches like layered composite seals[6] and self-healing glass seals.[7, 8]

In this study, a silicate glass, which shows self-healing behavior at the fuel cell operating conditions, is used to seal a ferritic stainless steel (Crofer 22 APU) against 8 mol% yttria stabilized zirconia (YSZ). An approach for selecting glass and ability of seal made with this glass to self-heal is demonstrated by leak testing. Performance of seal tested under different atmospheric conditions and thermal cycling between room temperature and 800^0C is also demonstrated with the particular emphasis on hermeticity and ability of the seal to self-heal.

EXPERIMENTAL

Main criterion for glass selection was its coefficient of thermal expansion in relation to YSZ and metal. A particular glass composition was chosen to ensure very close expansion match between glass and YSZ. The sealing glass is a SiO_2-based glass that has SiO_2, Al_2O_3, and other minor constituents.

Crofer 22 APU was chosen as a metal as it does not have very high difference in thermal expansion with YSZ and it can be more realistic material choice for SOFC due to its reasonable cost, higher resistance to oxidation, and less problem with chromia evaporation, which can cause cathode poisoning. Metal used in this study was 0.3 mm thick sheet.

Glass tapes were made by tape casting. Then these tapes were laminated together to desired final thickness. Glass laminates were sintered to make dense rectangular bars of 1-inch length for thermal expansion measurement and healing study. For thermal expansion measurement of metal, metal sheet was rolled and pressed and 1 inch long sample was cut from it. YSZ membranes were made using tapes fabricated from 8-mol% yttria stabilized zirconia (YSZ) powder. After binder removal YSZ tapes were sintered at 1485^0C for 3 hours to make dense YSZ membrane. Dense samples of YSZ were used for making seals and expansion measurements.

Thermal expansion measurements on all the samples were done in an inert atmosphere at a heating rate of $3^0C/min$ using a differential dilatometer. For self healing study, a crack was intentionally made on the surface of the glass sample, which was placed inside a furnace equipped with a CCD camera. Video imaging of the crack was monitored and recorded while it was heated at a rate of $5^0C/min$ to the desired temperature.

YSZ was sealed to Crofer metal using this glass. This was done by placing the glass between YSZ and Crofer and heating this assembly to 700^0C-900^0C. After seal was made it was first tested at room temperature to check its hermeticity.

The seals were tested at room temperature and higher temperatures by pressurizing the seal from one side and monitoring the pressure of that side as a function of time using a pressure transducer. A hermetic seal shows a very little or no change in the pressure as a function of time. The seals were also tested after thermal cycling at different temperatures to assess the effect of thermal cycling on hermeticity. Seal was attached to steel housing with which a steel tube was connected. This assembly of seal and steel housing was placed inside a tube furnace with one closed end. Other end of the tube was sealed against a metal flange with openings for vacuum pump, thermocouple, gas entry into the furnace and gas entry into the steel housing assembly. This set up gave the ability to control the temperature and pressure of the furnace as well as the pressure inside the housing. By controlling the pressure in the furnace and housing one can control the pressure difference (ΔP) experienced by the seal. Initial seal testing was done at room temperature for 20 hours at a $\Delta P \sim 150$ Torr. Higher temperature leak tests were done for 1 hour at ΔP of 10-40 Torr. A simplified diagram of seal testing set up is shown in the Fig.1.

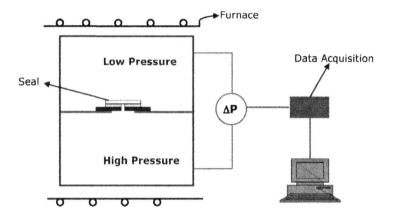

Figure 1: Seal testing set up

RESULTS AND DISCUSSION

Thermal expansion characteristics of Crofer, glass, and YSZ are shown in Fig.2. Crofer expands more than the glass and YSZ and hence it is under tensile stress after seal is cooled down from seal making temperature to room temperature. On the other hand, the glass is under compression, which is desirable. Glass becomes sufficiently fluid at higher temperature, which enables it to flow and conform to Crofer and YSZ surfaces at seal making temperature. In order to flow, it is required that the glass does not crystallize too much at seal making temperature. Glass remains amorphous after 500-hour exposure at 800^0C, which was confirmed by X-ray study.

Self-healing study of the glass alone using a video imaging approach shows that glass starts to heal over a temperature range of 550^0C-600^0C. It takes nearly 5 minutes for crack to completely disappear, which indicates that the rate of healing is sufficiently rapid. Still photographs from the recorded video of the crack on the glass sample before and after healing are shown in Fig.3. It shows that the crack is completely healed.

Room temperature seal testing

Seals of (Crofer- glass - YSZ) were successfully made and tested at room temperature under a ΔP ~150 Torr. These seals were completely leak tight even after 24 hours of testing. Data on leak rate for one of the seals is shown in Fig. 4.

Expansion

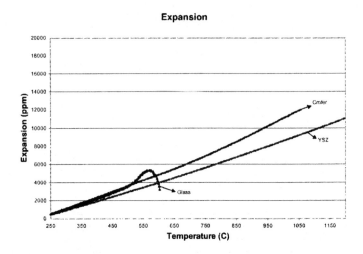

Figure 2: Thermal expansion behavior of Crofer, glass, and YSZ.

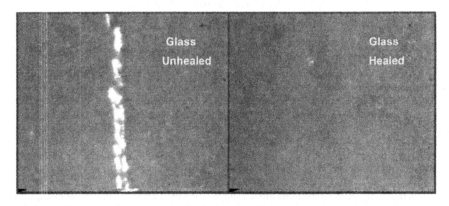

Figure 3: Still photograph of crack on glass sample before and after healing.

High temperature seal testing in Argon atmosphere

A seal designated as Seal 27 was tested for leak at higher temperatures during thermal cycling between room temperature and 800^0C. Data on leak rate of this seal at different temperatures during first thermal cycle is shown in Fig.5. Seal was completely leak tight at different temperatures. Atmosphere on both sides of the seal was argon gas.

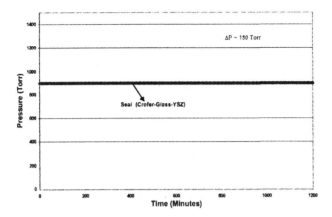

Figure 4: Leak check data of a Crofer-glass-YSZ seal.

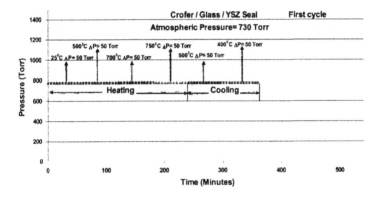

Figure 5: High temperatures leak check data of the seal during first thermal cycle.

After continuous cycling between room temperature and 800^0C for 2 more cycles Seal# 27 started leaking at 800^0C during the third cycle. To heal the seal it was kept at 800^0C for 30 minutes in a seal test under no pressure difference across it. After the healing, this seal became leak tight again, as shown in Fig.6. The cause of leak could be cracks or movement of the glass at high temperatures.

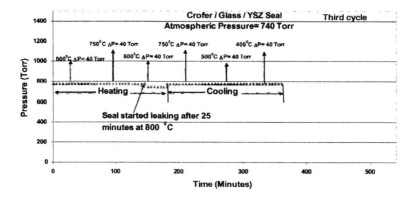

Figure 6: High temperatures leak data of the seal during third thermal cycle.

Subsequent to the healing treatment, this seal remained leak tight for 195 thermal cycles between room temperature and 800°C. Total exposure time for the seal at 800°C was 1185 hours. Healing of the seal and its further hermetic behavior demonstrated that this seal can be repaired even if it started to leak, just by keeping it at higher temperature for some time. After 195 thermal cycles there was no degradation in seal hermeticity in argon atmosphere, hence seal environment was changed to (argon+ 4% hydrogen) gas mixture, so as to test the performance of the seal under reducing atmosphere.

High temperature seal test under (argon + 4%hydrogen) gas mixture

In the (argon+ 4% hydrogen) atmosphere seal was tested for 22 thermal cycles between room temperature and 800°C. Total exposure of the seal at 800°C under this condition was 400 hours. Seal did not show any sign of degradation in leak tightness in this condition. Data of leak test of some of the cycles under (argon+ 4% hydrogen) atmosphere is shown in Fig.7.

High temperature seal test under (argon + 6% H_2O + 4 % hydrogen) gas mixture

After seal was tested under (argon+ 4% hydrogen) atmosphere for 22 thermal cycles and it remained totally leak tight, testing atmosphere was again changed to make it more severe. (Argon+ 4% hydrogen) gas mixture was bubbled through a water bubbler to humidify the gas mixture with 6-mol% water vapor. Seal was tested under this atmosphere for 15 thermal cycles and total exposure at 800°C under this condition was 300 hours. In this atmosphere also seal did not show any sign of degradation and it remained totally leak tight. Seal leak data under humid (argon+ 4% hydrogen) is shown in the Fig.8.

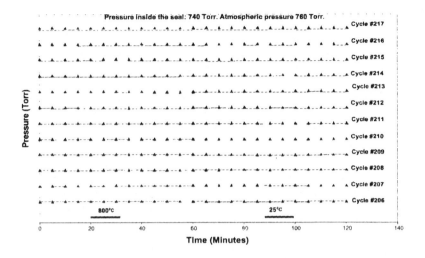

Figure 7: High temperature leak test data of the seal in (Argon+4% hydrogen) atmosphere.

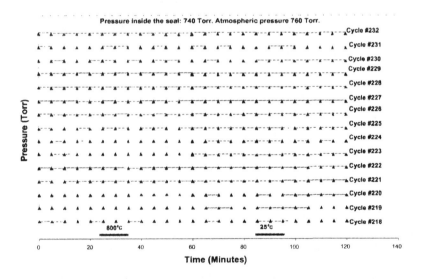

Figure 8: High temperatures leak data of the seal in (Ar+ 4%H$_2$+ 6% H$_2$O) atmosphere.

High temperature seal test under dual atmosphere

In the next step, seal was tested under dual atmosphere. One side of the seal was kept in air atmosphere while other side was kept in humid (argon+ 4% hydrogen). Seal was tested in this condition with continued thermal cycling for 39 cycles between room temperature and 800^0C and total exposure time at 800^0C was 760 hours. Seal remained totally leak tight for 37 cycles but it started leaking in the 38th cycle. In 38th cycle (which was cycle# 270 for the seal under all type of testing conditions) seal was leaking both at room temperature as well as at 800^0C. Again seal was healed by heating it to 825^0C for 2 hours in the seal test. In the very next cycle seal again became leak tight. Data on leak rate under dual atmosphere is shown in Fig.9. Self healing of the seal even after as much as 270 thermal cycles and total exposure time of more than 2500 hours clearly indicates that these seals are capable of maintaining hermeticity for a very long time. Testing of the seal is still going on and it has completed 275 cycles and 2700 hours of exposure at 800^0C. The posttest examinations of the seal is planned to understand the cause of leaks and mode of healing in relation to changes in the properties of the glass.

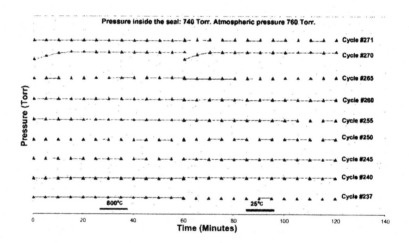

Figure 9: High temperatures leak data of the seal tested in dual atmosphere.

CONCLUSIONS

Hermetic seals were made using a glass that shows the self-healing characteristics at elevated temperatures. Self-healing behaviors of the glass as well as of seals made using this glass were clearly demonstrated. Stability of these glass seals under different testing conditions was shown by leak tight responses even after as many as 275 thermal cycles between room temperature and 800° C. Along with the stability, the glass seals retained their ability to self heal even after very long exposure times of more than 2500 hours at higher temperatures.

ACKNOWLEDGEMENTS

This project was partly supported by University of Cincinnati and US Department of Energy-SECA program through grant number DE-FC26-04NT 42227. The guidance of Dr. Travis Shultz and Mani Mannvanan, Program Managers- Fuel Cells (DOE-NETL) are very much appreciated. Useful discussions with Drs. Don Collins, Wayne Surdoval of DOE-NETL, Prabhakar Singh, Matt Chou and Jeff Stevenson of PNNL, N. Minh of GE-Energy, and Pinakin Patel of FuelCell Energy are also appreciated.

REFERENCES

1. N.Q. Minh, "Ceramic Fuel Cells," *Journal of the American Ceramic Society*, **76** [3] 563-88 (1993).
2. ZhenguoYang, Guanguang Xia, Kerry D. Meinhardt, K. Scott Weil, and J. W. Stevenson, "Chemical Stability of Glass Seal Interface in Intermediate Temperature Solid Oxide Fuel Cell," *J.Mat.Eng.Perform.* 13 327-334 (2004).
3. S.B. Sohn, S.Y. Choi, G.H. Kim, H.S. Song, G.D. Kim, Stable Sealing Glass for Planar Solid Oxide Fuel Cell, *J. Non Cryst Solids*, 297 103–112(2002).
4. R. N. Singh, "High Temperature Seals for Solid Oxide Fuel Cells," *Proceedings of 29th International Conference on Advanced Ceramics and Composites*, Cocoa Beach, 25(3), 299-307 (2004).
5. R. N. Singh, "High Temperature Seals for Solid Oxide Fuel Cells," *Proceedings of ASM International Conference*, Columbus, Oct 18-20, 2004.
6. R. N. Singh and S. S. Parihar, "Layered Composite Seals for Solid Oxide Fuel Cells," *Proceedings of 29th International Conference on Advanced Ceramics and Composites*, Cocoa Beach, (2005).
7. S. S. Parihar and R. N. Singh, "Self Healing Glass Seals for Solid Oxide Fuel Cell," *Proceedings of the 107th Annual Meeting of The American Ceramic Society*, (2005).
8. R. N. Singh and S. S. Parihar, "High Temperature Seals for Solid Oxide Fuel Cells (SOFC),"*SECA Core Technology Workshop*, Tampa, FL, January 27, 2005.
9. Y.S. Chou, J.W. Stevenson, and L.A. Chick, "Ultra Low Leak Rate of Hybrid Compressive Mica Seals for SOFC," *J. Power Source*, 112, 130 (2002).
10. S. Tanaguchi, M. Kadowaki, T. Yasuo, Y. Akiyamu, Y. Miyaki and K. Nishio, " Improvement of Thermal Cycle Characteristics of a Planar-Type SOFC by Using Ceramic Fiber as a Sealing Material," *J. Power Sources*, 90, 163 (2000).

PROPERTIES OF GLASS-CERAMIC FOR SOLID OXIDE FUEL CELLS

S. T. Reis[1], R. K. Brow[1], T. Zhang[1], and P. Jasinski[1,2]

[1] Department of Materials Science & Engineering, University of Missouri-Rolla, Rolla, MO
[2] Department of Biomedical Engineering, Gdansk University of Technology, 80-952 Gdansk, Poland

ABSTRACT

The thermal stability of 'invert' glass-ceramics, based on crystalline pyro- and orthosilicate phases, developed for hermetic seals for solid oxide fuel cells (SOFC) are described. The effects of long-term (up to 60 days) high temperature (up to 800°C) heat treatments on the properties of the glass-ceramics in oxidizing and reducing environments were evaluated by dilatometric analyses, x-ray diffraction, weight loss measurements, and impedance spectroscopy. The glass-ceramics have thermal expansion coefficients in the range $9.5\text{-}11.5\text{x}10^{-6}/°C$ and can be sealed to SOFC materials, including Y_2O_3-stabilized ZrO_2 (YSZ) and Cr-steel interconnect alloys, at or below 900°C. The thermal expansion characteristics of some glass-ceramics remain essentially unchanged after >40 days at 800°C, and the glass conductivity at 750°C remains constant (in the range $0.3\text{-}1.0\text{x}10^{-7}$ S/cm) in forming gas.

INTRODUCTION

Solid oxide fuel cells (SOFCs) convert chemical energy to electricity via an electrochemical reaction and are projected to become important alternative energy sources because of their high efficiency and low emissions[1-2]. Planar SOFC designs require simple manufacturing processes, have relatively short current paths and produce higher power densities and efficiencies than tubular designs[2]. In order for a planar SOFC to properly operate, a suitable sealant is required to prevent the fuel gas and air from mixing. The sealant must possess thermomechanical characteristics that are compatible with other SOFC components (i.e., the electrolyte and interconnects), must resist deleterious, high temperature interfacial reactions with those components, must be an electrical insulator, and must remain thermochemically stable in fuel cell operating environments, which include a range of environments (p_{O2} and p_{H2O}), temperatures on the order of 800°C, for times up to 50,000 hours[1-4]. In many ways, the seal performance will control the structural integrity and mechanical stability of the SOFC stack, and could also determine the overall stack performance.[3]

There have been many reports on the development of a variety of compositional systems to form suitable glass and glass-ceramics seals for SOFCs, including silicates, aluminosilicates, borosilicates, and aluminophosphates[4-14]. Most of these sealing materials have shortcomings. Some fail to remain thermomechanically stable under SOFC operational conditions, and others undergo deleterious interfacial reactions with other SOFC components. One such reaction occurs between BaO-containing sealants and the Cr-oxide scale that forms on interconnect alloys, resulting in the formation of a $BaCrO_4$ interfacial phase that can adversely affect the mechanical integrity of the seal[3,15].

This work describes the thermal and electrical properties of 'invert' glasses and glass-ceramics developed for hermetic seals for solid oxide fuel cells. Invert glasses are those with relatively low concentrations of glass-forming oxides (<45mol% SiO_2) and so possess structures with continuous networks of modifying polyhedra[16]. These compositions fall outside of compositional ranges reported for other SOFC sealing systems, which generally have greater silica contents and often possess significant concentrations of BaO[10,13].

EXPERIMENTAL PROCEDURES

Glasses were prepared from mixtures of reagent grade alkaline earth carbonates and silica, with concentrations of other oxides, including ZnO, to modify melt and glass properties. The batches were melted in platinum crucibles in air for fours hours, typically at 1550°C. A typical melt size was approximately 50 grams, although melts as large as 1 kg have been made. Melts were quenched on steel plates and glasses were annealed for six hours near the appropriate glass transition temperature. The compositional ranges for the glasses under investigation are (in mol%) (0-30)CaO, (0-30)SrO, (0-30)ZnO, (1-7)B_2O_3, (2-4)Al_2O_3, (0-2)TiO_2, and (35-45)SiO_2.

Glass powders (sieved to 45-75 μm) were used for differential thermal analyses (DTA), using a Perkin-Elmer DTA-7. The powders were heated in air at 10°C/min to determine the glass transition temperature, T_g, and crystallization temperature, T_c. Glass powders were crystallized by heat treating at temperatures around T_c for one hour in an argon atmosphere and the crystalline phases were identified by x-ray diffraction (XRD), using a Scintag XDS200X. The thermal expansion characteristics of glass and crystallized samples were determined by dilatometric analyses, using an Orton model 1600D. Crystallized samples were prepared by first sintering, then crystallizing glass powders in graphite moulds (25 x 10 mm) under argon for the indicated time and temperature. Dilatometric data was collected by heating samples in air at 3°C/min, to determine the coefficient for thermal expansion (CTE) and dilatometric softening points (T_d).

Sealing tests were performed by reacting glass tapes with SOFC component materials, including Y_2O_3 (8 mol%)-stabilized zirconia (YSZ) substrates and the ferritic stainless steel Crofer 22APU. A dispersed solution was prepared from glass powders (diameter ~ 10μm), mixed with ethanol, toluene, and fish oil in a ratio of (wt%) 2: 1: 1: 0.03. After dispersing for 24 hours, the solution was mixed with PVB binder to form a slurry for tape-casting. Tapes about 100μm thick were sandwiched between clean Crofer 22APU and YSZ substrates, this assembly was then heated in an alumina muffle furnace in air to 450°C for 2h, to remove organics, then to the desired sealing temperature under flowing argon. In general, the glass melts, spreads and bonds to both substrates, then crystallizes to form the desired 'glass-ceramic' phases, at 800-900°C in 1-2 hours. Some reaction couples were further subjected to long-term heating at 800°C in air. Reaction couples were prepared for microscopic evaluation. Samples were cross-sectioned with a diamond saw, then polished to a sub-micron finish and evaluated by analytical scanning electron microscopy (Hitachi 4700 SEM with Phoenix EDAX system).

Chemical stability of crystallized glass samples in reducing environments was evaluated by measuring the weight loss from bulk samples (10mm x 10mm x 10mm) held in wet, flowing forming gas (10%H_2, 90% N_2) at 800°C for up to 28 days. The samples were polished to a 600 grit finish with SiC paper, cleaned with acetone and supported by a Ni/Cr wire in the furnace. The measurements were made for each crystallized glass sample and the average weight loss, normalized to the glass surface area, was determined.

Glass pastes were also deposited and sintered on YSZ substrates in order to measure the glass electrical conductivity. The samples were initially sealed at 850°C for 1h in argon, and then gold electrodes with platinum wire-bonded conductors were applied. Both AC and DC techniques were employed for sample conductivity measurements. Impedance spectra were measured with SI 1296 dielectric interface and an SI1260 impedance analyser in the frequency range from 100kHz to 0.1Hz, and a Keithley 6517A electrometer was used for DC resistance measurements. Measurements were made at 750°C in flowing forming gas (10% hydrogen and 90% nitrogen).

RESULTS AND DISCUSSION
Thermal Properties
    Figure 1 compares the dilatometric results for glasses G#27 and G#50, after crystallization during typical sealing runs, with the CTE results for YSZ and Crofer 22APU. The 'as sealed' G#27 exhibits an inflection in the expansion curve near 700°C, due to softening of a residual glass phase. The dilatometric and DTA results for these compositions, and two others, are summarized in Table I. (CTE results for 'as melted' glasses and 'as sealed' glass ceramics' are shown; the glass-ceramics were formed by heating the glasses to 850-900°C for one hour.)

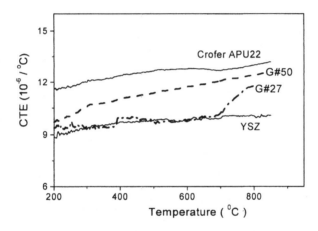

Figure 1:   Dilatometric results for crystallized glass #27 (850°C for 1h), crystallized glass #50 (900°C for 1h), YSZ and Crofer 22APU. Note the inflection in the expansion curve near 700°C for crystallized glass#27.

Table I: Thermal properties of SOFC sealing glasses.

Glass ID	T_g(°C) (DTA)	T_d(°C) (dilatom.)	T_c(°C) DTA	CTE/Glass (100-600°C) (ppm/°C)	CTE/Crystal. (100-700°C) (ppm/°C)
27	700	730	904	9.5	10.0
35	730	756	907	11.0	11.3
36	745	750	956	10.9	12.5
50	754	761	1020	10.9	11.7

Figure 2 (top) shows the XRD data collected from G#27 after an initial sealing cycle (850°C/1 hour). Two crystalline phases were identified from this sample: $CaSrAl_2SiO_7$ and $Ca_2ZnSi_2O_7$. Other compositions yield crystalline orthosilicates, including $CaSrSiO_4$, Sr_2SiO_4 and Zn_2SiO_4. Glasses with greater silica contents form more polymerized crystalline phases, including metasilicates.[4,10]

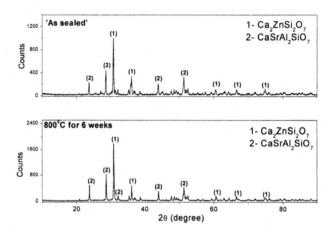

Figure 2: X-ray diffraction patterns for glass#27 after crystallization at 850°C for one hour (top) and after 42 days at 800°C (bottom).

Thermal and Chemical Stability

Figure 3 shows the average CTE (between 200 and 700°C) for three crystallized compositions as a function of time at 800°C. Also shown for comparison are the 'as received' CTE values for YSZ and Crofer 22APU. Glass #27 has a stable CTE at 800°C. X-ray diffraction indicates that this material exhibits no discernible change in the crystalline phase distributions as a result of the heat-treatment (figure 2, bottom). The CTE of glass #35, on the other hand, drops from $11.3 \times 10^{-6}/°C$ to $9.3 \times 10^{-6}/°C$ after 42 days at 800°C (Figure 3). This decrease in CTE is accompanied by an increase in the relative fraction of $CaSiO_3$ in this glass-ceramic, as shown by the XRD data in Figure 4. ($CaSiO_3$ has a relatively low CTE[17], $\sim 6.5 \times 10^{-6}/°C$). Glass#50 retains a relatively high CTE ($11.4 \times 10^{-6}/°C$) after 42 days at 800°C (Figure 3); the XRD collected from these samples (not shown) reveals no discernible changes in the distributions of the crystalline phases (Sr_2AlSiO_7 and $CaSrSiO_4$) formed during crystallization.

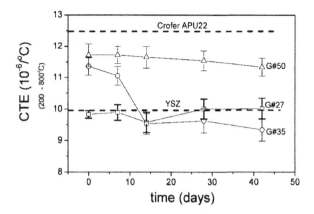

Figure 3: CTE **data** for crystallized glasses #27, #35, and #50 held at 800°C for up to 42 days. The 'as received' CTE values for YSZ and Crofer 22APU are shown for comparison.

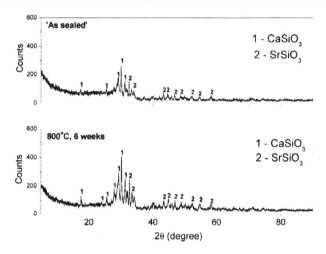

Figure 4: X-ray diffraction patterns for glass#35 after crystallization at 850°C for one hour (top) and after 42 days at 800°C (bottom).

Figure 5 shows cumulative weight losses from solid glass-ceramic samples, with B_2O_3-contents between 2 and 7 mole%, held in wet, flowing forming gas at 800°C for up to 28 days. In general, glasses with greater concentrations of B_2O_3 exhibit greater volatilization rates under these conditions. In addition, materials with a greater fraction of residual glass tend to exhibit greater volatilization rates (not shown). These volatilization experiments provide clues about the long-term stability of a proposed sealing system and so place constraints on compositional design. Based on this constraint, the most promising sealing glasses have low B_2O_3-contents.

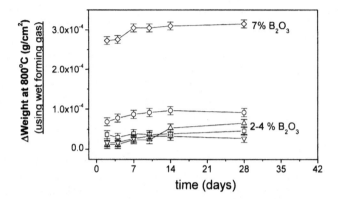

Figure 5: Sample weight losses for up to 28 days at 800°C in wet forming gas.

Interfacial Characterization

Figure 6 shows a scanning electron micrograph of the interface between crystallized G#27 and Crofer22 APU, after 10 days at 800°C. The bonding interface appears to be uniform, with no obvious heterogeneities. Microscopic characterization of the glass/YSZ interface (not shown) indicates no obvious heterogeneities. Despite the CTE-mismatch between G#27 and the Crofer 22APU substrates, sandwich seals with these glass-ceramics between YSZ and Crofer 22APU have survived for up to 28 days in air at 800°C without separating. More severe thermal cycling experiments are presently underway using G#27 and G#50. Both compositions are BaO-free and so do not form the $BaCrO_4$ interfacial reaction product that has been reported to develop in other SOFC sealing systems; e.g., reference 3.

Conductivity Measurements

Figure 7 shows the electrical conductivity of three different glass-ceramics, collected at 750°C in forming gas. For each sample, conductivity initially increases and then stabilizes after about one day. The conductivities of glasses #27 and #50 stabilize in the range ~0.3-1.0×10^{-7} S/cm, whereas the conductivity of glass #36 increases (>1.0×10^{-7} S/cm) with time. It is unclear what controls the time-dependence of conductivity in these systems, although it is worth noting that the two systems with 'stable' conductivities have 'stable' crystalline phase assemblages (as discussed above), whereas glass #36 slowly crystallizes over the course of the conductivity experiment.

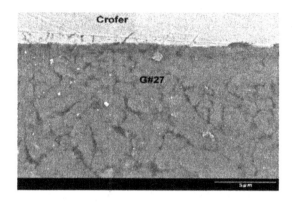

Figure 6: Scanning electron micrograph of the bonding interface between Crofer 22APU and Glass #27 after ten days at 800°C.

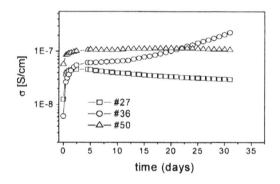

Figure 7: Electrical conductivity of sealing glass-ceramics in forming gas at 750°C.

CONCLUSIONS

The thermal properties and thermo-chemical stability of invert silicate glasses that crystallize to form stable pyro- and orthosilicate phases make them attractive candidates for SOFC sealing applications. Promising BaO-free compositions crystallize to form sealing materials with desirable thermomechanical properties, including thermal expansion characteristics. that are compatible with YSZ substrates and interconnect materials like Crofer 22APU, and that remain stable under SOFC operational conditions. The glass-ceramics bond to YSZ and Cr-steel alloys without forming the deleterious interfacial reaction products that have been noted for other sealing systems. Optimized glass-ceramics remain electrically insulating under operational conditions.

ACKNOWLEDGEMENTS

The authors thank Xiao-Dong Zhou (now at PNNL) and Harlan Anderson (UMR) for providing materials used in the sealing tests and for their useful advice during the course of this research. This work was supported by the US Department of Energy, Solid State Energy Conversion Alliance Core Technology Program, contract DE-FC26-04NT42221.

REFERENCES

[1]S. Singhal "Ceramic fuel cells for stationary and mobile applications", *Am. Ceram. Soc. Bulletin*, **82 (11),** 9601-9610 (2003).

[2]N. Q. Minh, "Ceramic fuel cells", *J. Am. Ceram. Soc.*, **76,** 563-588 (1993).

[3]Z. Yang, J. W. Stevenson and K. D. Meinhardt, " Chemical interactions of barium-calcium-aluminosilicate-based sealing glasses with oxidation resistant alloys ", *Solid State Ionics*, **160,** 213-225 (2003).

[4]S. B. Sohn, S. Y. Choi, G. H. Kim, H. S. Song, G. D. Kim, *J. Non-Cryst. Solids*, " Stable sealing glass for planar solid oxide fuel cell", **297,** 103-112 (2002).

[5]T. Taniguchi, M. Kadowaki, T. Yasuo, Y. Akiyama, Y. Miyake, K. Nishio, " Improvement of thermal cycle characteristics of a planar-type solid oxide fuel cell by using ceramic fiber as sealing material "*J. Power Source*, **90,** 163-169 (2000).

[6]K. Eichler, G. Solow, P. Otschik, W. Schaffrath, "BAS(BaO.Al$_2$O$_3$.SiO$_2$)-glasses for high temperature application", *J. Eur. Ceram. Soc.*, **19** 1101-1104 (1999).

[7]P. H. Larsen, P. F. James, "Chemical stability of MgO/CaO/Cr$_2$O$_3$-Al$_2$O$_3$-B$_2$O$_3$-phosphate glasses in solid oxide fuel cell environment", *J. Mater. Sci.*, **33** 2499-2507 (1998).

[8]S. P. S. Badwal, "Stability of solid oxide fuel cell components", *Solid State Ionics*,. **143** 39-46(2001).

[9]S. V. Phillips, A. K. Data, L. Lakin, in: *Proceedings of the 2nd International Symposium Solid Oxide Fuel Cells*, Athens, Greece, July 2-5, 1991, p.737.

[10]K.D. Meinhardt, et al., "Glass-ceramic material and method of making," US patent 6,430,966, Aug. 13, 2002; K.D. Meinhardt, et al., "Glass-ceramic joint and method of joining," US Patent 6,532,769, March 18, 2003.

[11]P. H. Larsen, F. W. Poulsen, R. W. Berg, "The influence of SiO$_2$ addition to 2MgO-Al$_2$O$_3$-3.3P$_2$O$_5$ glass ", *J. Non-Cryst. Solids,* **244** 16-24 (1999).

[12]S. B. Sohn and S. Y. Choi, "Suitable Glass-Ceramic Sealant for Planar Solid-Oxide Fuel Cells", *J. Am. Ceram. Soc.*, **87,** 254-260 (2004).

[13]P. Geasee et al., "Investigation of glasses from the system BaO-CaO-Al$_2$O$_3$-SiO$_2$ used as sealants for the SOFC," *Proc. Int. Congr. Glass*, Edinburgh, Scotland, Vol. 2 (2001) 905.

[14]J. W. Fergus, "Sealants for solid oxide fuel cells", *Journal of Power Sources*, **147** 46-57 (2005).

[15]Z. Yang, K. D. Meinhardt, and J. Stevenson, "Chemical Compatibility of Barium-Calcium-Aluminosilicate-Based Sealing Glasses with the Ferritic Stainless Interconnect in SOFC", *J.Electrochem.Soc.*,**150**(8), A1095-A1101(2003).

[16]H.J.L Trapp and J.M.Stevels, "Conventional and invert glasses containing titania. Part I", *Phys .Chem. Glasses*, **1** 107-118 (1960).

[17]W. Höland and G. Beall, "Glass Ceramic Technology" American Ceramic Society, Westerville, OH (2000).

MECHANICAL BEHAVIOR OF SOLID OXIDE FUEL CELL (SOFC) SEAL GLASS-BORON NITRIDE NANOTUBES COMPOSITE

Sung R. Choi*, Narottam P. Bansal, Janet B. Hurst, and Anita Garg
NASA Glenn Research Center, Cleveland, Ohio 44135

ABSTRACT
    Barium calcium aluminosilicate glass composites reinforced with about 4 weight percent of BN nanaotubes (BNNTs) were fabricated by hot pressing. Flexure strength and fracture toughness of the glass-BNNT composites were determined at ambient temperature. The strength of the composite was higher by as much as 90 % and fracture toughness by as much as 35 % than those of the unreinforced G18 glass. Pullout of the BNNTs was observed from fracture surfaces of the composite. Elastic modulus, density, and microhardness of the composite were lower than those of the unreinforced glass.

INTRODUCTION
    Carbon nanotubes (CNT) show many properties which are superior to those of graphite. Boron nitride nanotubes (BNNT) have similar structure as CNT and exhibit many similar properties. Pure BNNTs were first synthesized using a plasma arc discharge method yielding mainly double walled BNNTs or a CNT substitution reaction giving multiwalled BNNTs. BNNTs are potential candidate materials for nanosized electronic and photonic devices with a large variety of electronic properties. BNNTs show thermal stability in air at 700 °C: Some thin nanotubes with cylindrical structure may be stable up to 900 °C [1]. BNNTs are, therefore, suitable as reinforcement for composite materials for applications at elevated temperatures in oxidizing environment. By contrast, CNTs oxidize in air at ~400 °C and burns completely at 700 °C, limiting their high temperature applications.
    A number of studies are currently available on the metal, ceramic and polymer matrix composites reinforced with CNTs as seen from the recent review articles [2-6]. However, to the best of our knowledge, no studies have been done to date on the subject of BNNTs reinforced composites. Processing and some basic properties of solid oxide fuel cell (SOFC) seal glass reinforced with 4 wt % BNNTs have been first reported by the current authors in a recent study [7]. This paper, as an extension to the previous study, presents ambient-temperature mechanical behavior of the seal glass-BNNT composite. The properties include flexure strength, fracture toughness, elastic modulus, density and microhardness. Some microstructural features examined using XRD, SEM and TEM are also presented.

EXPERIMENTAL PROCEDURES
Materials
    Synthesis of BNNTs and processing of BNNTs-reinforced glass composite were reported earlier [7] and are briefly described here. Amorphous boron powder was mixed with several weight

---

* Now with Naval Air Systems Command, Patuxent River, MD 20670; sung.choi1@navy.mil

percent of fine iron catalyst particles in a hydrocarbon solvent and ball-milled in a polythene bottle using ceramic grinding media. The milled material was applied to various high temperature substrates such as alumina, silicon carbide, platinum and molybdenum and reacted in a flowing atmosphere of nitrogen containing small amount of ammonia. Nanotubes of significant length and abundance were formed on heat treatments at temperatures from 1100 to 1400 °C for 20 minutes to 2 hrs. Batch sizes of typically 2 g were produced but the process should be easily scalable to larger amounts.

The starting materials for composite processing were a barium calcium aluminosilicate (BCAS) [8] glass G-18 powder of composition (mol %) $35BaO-15CaO-5Al_2O_3-10B_2O_3-35SiO_2$ or $56.4BaO-8.8CaO-5.4Al_2O_3-7.3B_2O_3-22.1SiO_2$ (wt %) with an average particle size of 14.2 μm. This glass has a glass transition temperature of 619 °C, dilatometric softening point of 682 °C, and coefficient of thermal expansion 10.5×10^{-6} /°C (from room temperature to 500°C) and 11.8×10^{-6} /°C (20-800°C). Appropriate quantity of BCAS glass and 4 wt % of as-synthesized BNNTs were slurry mixed in acetone and ball milled for ~24 h using zirconia milling media. Acetone was evaporated and the powder dried in an electric oven. The resulting mixed powder was loaded into a graphite die and hot pressed in vacuum at 630 °C under 10 MPa for 15 min into 50 mm x 25 mm plates using a mini hot press. The applied pressure was released before onset of cooling. Glass powder was also hot pressed at 630 °C. Grafoil was used as spacers between the specimen and the punches.

The hot-pressed plates were machined into flexure bar test specimens with nominal depth, width and length of 2.0 mm x 3.0 mm x 25 mm, respectively. Machining direction was longitudinal along the 25mm-length direction. The sharp edges of test specimens were chamfered to reduce spurious premature failure emanating from those sharp edges.

Microstructures of the polished cross-sections were observed in a Hitachi S4700 field emission scanning electron microscope (FESEM) equipped with a super thin window EDAX Genesis System energy dispersive spectrometer (EDS). In addition, TEM and X-ray diffraction analyses were also carried out.

Mechanical Testing

Strength testing for the G18 glass-BNNT composite was conducted in flexure at ambient temperature in air. A four-point flexure fixture with 10 mm-inner and 20 mm-outer spans was used in conjunction with an electromechanical test frame (Model 8562, Instron, Canton, MA). A stress rate of 50 MPa/s was applied in load control using the test frame. A total of 10 test specimens were tested. One of two 3 mm-wide sides of each test specimen was subjected to maximum tension or compression. Although the specimen configuration was not the same, testing, in general, was followed in accordance with ASTM test standards C 1161 [9].

Fracture toughness using flexure test specimens was determined at ambient temperature in air using single edge v-notched beam (SEVNB) method [10]. The same method was used previously to determine fracture toughness of unreinforced G18 glass and alumina or zirconia-reinforced G18 glass composites [11]. This method utilizes a razor blade with diamond paste, grain size of 9 μm, to introduce a final sharp notch with a root radius ranging 10-20 μm by tapering a saw notch. The sharp v-notched specimens with a notch depth of about 1.0 mm along the 3 mm-wide side of test specimens were fractured in a four-point flexure fixture with 10 mm-inner and 20 mm-outer spans using the

 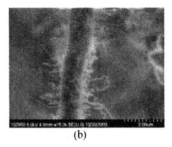

(a)   (b)

Figure 1. Typical field emission scanning electron microscope (FESEM) micrographs of as-synthesized boron nitride nanotubes (BNNTs): (a) overall; (b) a single boron nitride nanotube.

Instron test frame (Model 8562) at an actuator speed of 0.5 mm/min. Three specimens were tested. Fracture toughness was calculated based on the formula by Srawley and Gross [12].

Other testing to determine elastic modulus, density and Vickers microhardness were also performed. Elastic modulus of the glass-BNNT composite was determined at ambient temperature by the impulse excitation of vibration method, ASTM C 1259 [13] using the flexure test specimens. Density was measured with a bulk mass/volume method using the same flexure specimens that were used in elastic modulus experiment. A total of fourteen specimens were used for elastic modulus as well as in density measurements. Both elastic modulus and density measurements were made prior to flexure strength or fracture toughness testing. Microhardness was evaluated at ambient temperature with a Vickers microhardness indenter with an indent load of 9.8 N using three flexure specimens with five indents for each specimen, in accordance with ASTM C 1327 [14].

RESULTS AND DISCUSSION
Microstructures

Figure 1 shows field emission SEM micrographs of as-synthesized BNNTs. BNNTs were observed to have diameter ranging from tens to hundreds of nanometers and significant lengths (tens of micrometers). A typical grown nanotube is also shown in the figure.

Microstructure of the G18 glass-BNNT composite is presented in Figure 2, where the microstructure of the unreinforced G18 glass is also included for comparison. The composite showed well dispersed black regions and strings and a few phase particles (white regions) in the glass matrix. Energy dispersive spectroscopy (EDS) analysis indicated that the G18 glass consisted of Si, Ba, Ca, Al and O, as expected, where else the black regions contained B and N. The white particles contained Zr and O probably due to ZrO_2 particles from an impurity in the glass composition and/or from wear debris of zirconia grinding media.

X-ray diffraction pattern taken from the surface of the hot pressed G18 glass-BNNT composite is presented in Figure 3 showing that it consists of two large halos indicating amorphous nature of the composite material, the same pattern for the unreinforced G18 glass. In spite of the availability of large interface between the BNNTs and the glass matrix, which would be conducive

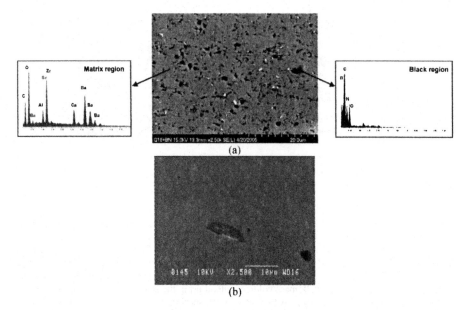

(a)

(b)

Figure 2. Typical field emission scanning electron microscope (FESEM) micrograph of G18 glass composite reinforced with 4 wt% boron nitride nanotubes, including energy dispersive spectroscopy (EDS) analysis. The microstructure of the unreinforced G18 glass is included for comparison in (b).

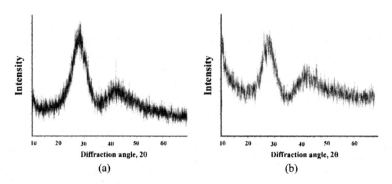

Figure 3. X-ray diffraction patterns for (a) G18 glass reinforced with 4 wt % boron nitride nanotubes (BNNTs) and (b) G18 glass.

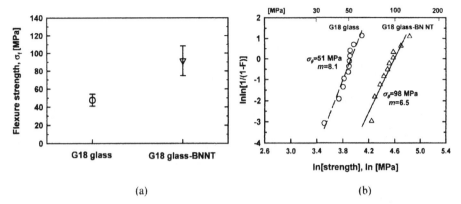

Figure 4. Flexure strength (a) and Weibull strength distribution (b) of G18 glass reinforced with 4 wt % boron nitride nanotubes (BNNTs). The strength data on G18 glass are included for comparison [7]. Error bars indicate ±1.0 standard deviation. σ_θ: characteristic strength; m: Weibull modulus.

to nucleation and crystallization of the glass, it did not crystallize during hot pressing as the temperature was lower than its glass transition temperature. TEM of the BNNT composite showed that the lattice structure of nanotubes was multiwalled nanotubes, most of which were intact by the compaction during the elevated-temperature processing. The interface between the BNNTs and the glass was clean with no porosity or any unusual feature.

Flexure strength

Figure 4 shows the results of flexure strength testing for the G18 glass-BNNT composite. The figure also includes the strength of G18 glass [11] for comparison. The increase in strength with 4 wt % nanotube reinforcement (σ_f = 92 ± 17 MPa) was evident and amounted to 90% improvement compared to the unreinforced G18 glass strength (= 48 ± 7 MPa). This significant strength improvement of the glass with BNNT reinforcement is notable, compared to a moderate strength increase of 40-60 % for the G18 glass reinforced with 5 mol% alumina platelets or zirconia particulates [11]. Despite the insufficient number of test specimens used, Weibull modulus (m) ranged from m = 7-8 for both G18 glass-BNNT composite and unreinforced G18 glass, indicative of the same degree of flaw population presented for the two materials, as seen in Figure 4.

In general, failure origins were from either surface or volume flaws mostly associated with pores ranging in size approximately from 50 to 100 μm. Somewhat significant variation in density with a coefficient of variation of 3.5% was observed from specimen to specimen for the G18-BNNT composite and initially thought to be a major factor to control the magnitude of fracture strength. However, this appeared to be untrue, as shown in Figure 5, where strength of each individual

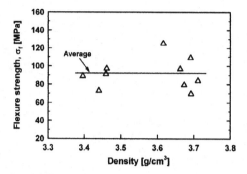

Figure 5. Flexure strength as a function of density for each individual test specimen of G18 glass reinforced with 4 wt % boron nitride nanotubes (BNNTs).

specimen was plotted as a function of its individual density. Strength did not show any trend that exhibits strength increase with increasing density, typical of many brittle materials [15].
This again indicates that only dominant flaws presented in the material might have been associated in controlling strength of the BNNT composite.

Fracture Toughness
    The results of fracture toughness testing by the single edge v-notched beam (SEVNB) method are shown in Figure 6. Fracture toughness shows a similar trend as strength. However, the increase in fracture toughness was less significant than that in strength. Facture toughness of the G18 glass-BNNT composite (K_{Ic} = 0.69 ± 0.09 MPa√m) increased only by 35% as compared with that (= 0.51 ± 0.03 MPa√m) of G18 glass. This increase in fracture toughness for the composite is comparable to those for the G18 glass composites having similar content of alumina or zirconia [11].
    Limited availability of the BNNT composite in this work did not allow to evaluate R-curve or crack growth resistance of the material. However, it should be mentioned that the G18 glass reinforced with alumina platelets or zirconia particulates have exhibited a rising R-curve with its degree being increased with increasing reinforcement content [11]. Bridging and/or crack defection [16] was considered to be as a major strengthening or toughening mechanism operative for those composites [11]. The similar mechanism can also be considered to be attributed to increase in strength and fracture toughness of the G18-BNNT composite, as can be seen from typical fracture surfaces showing pullouts of BNNTs and their troughs in Figure 7.

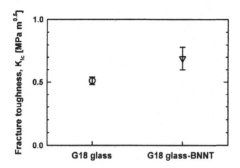

Figure 6. Fracture toughness of G18 glass reinforced with 4 wt % boron nitride nanotubes (BNNTs). Fracture toughness of G18 glass is included for comparison [11]. Error bars indicate ±1.0 standard deviation.

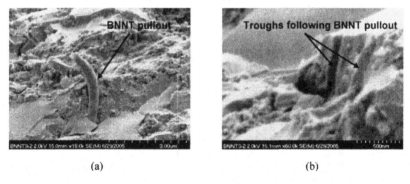

(a)   (b)

Figure 7. Fracture surfaces showing pullout of a boron nitride nanotube (BNNT) (a) and troughs following BNNTs pullout (b) [7].

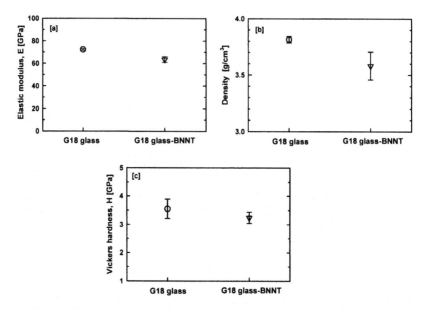

Figure 8. Elastic modulus (a), density (b), and Vickers microhardness (c) of G18 glass reinforced with 4 wt. % boron nitride nanotubes (BNNT). Data on unreinforced G18 glass [11] are included for comparison. Error bars represent ±1.0 standard deviation.

Elastic Modulus, Density, and Microhardness

The results of density, elastic modulus, and microhardness measurements are presented in Figure 8. The bulk density of the BNNT composite, 3.582 ± 0.124 g/cm^3 was lower than that (= 3.814 ± 0.028 g/cm^3) of the G18 glass [11]. Also note a significant scatter in density for the BN nanotube composite as compared with the G18 glass: 3% (BNNT composite) *versus* 0.7% (G18 glass) in coefficient of variation. This suggests some careful considerations in material processing which was not observed or an issue for other glass composites reinforced with alumina platelets or zirconia particulates [11]. The value of elastic modulus was E = 64 ± 3 GPa, lower than that (= 72 ± 1 GPa) of the G18 glass. Although density and elastic modulus decreased with addition of light BNNTs, the rule of mixture was not applicable to either density or elastic modulus of the BNNT composite (with a consideration of BN density as 2.2 g/cm^3 [17]). Elastic modulus of the BNNT composite is a little lower than that ($\approx$70 GPa) of most silicate glasses.

Vickers microhardness, H, was found to be H = 3.2 ± 0.2 GPa, a little lower than 3.6 ± 0.3 GPa of the G18 glass. Addition of BNNTs to G18 glass did have an appreciable effect on elastic modulus, density, and microhardness.

FUTURE WORK

The work presented in this paper has been done primarily for the purpose of determination of the ambient-temperature mechanical properties of G18 glass reinforced with 4 wt% boron nitride nanotubes. However, SOFCs operate in a typical temperature range of 700-1000 °C, so that thermal and mechanical properties and environmental durability of the developed glass composites needs to be evaluated at elevated temperatures, as done previously for YSZ/alumina composites [18]. Needed work includes constitutive relation (viscosity)/sealability [19], coefficient of thermal expansion, intermediate-temperature strength, thermal fatigue, life limiting factors, and stability of material including crystallization [8], etc. Also, statistical reproducibility of the material in terms of mechanical behavior is also needed. Some of these factors will be investigated in the near future and others are long-term efforts.

CONCLUSIONS

1. A barium calcium aluminosilicate (BCAS) glass G-18 composite reinforced with 4 wt. % boron nitride nanotubes was fabricated by hot pressing.
2. Reinforcement with boron nitride nanotubes improved both the flexure strength and fracture toughness of the G18 glass. The strength of the reinforced composite was higher by as much as 90 % and fracture toughness by as much as 35 % than those of the G18 glass.
3. Elastic modulus, density, and Vickers microhardness of the G18 glass reinforced with 4 wt% boron nitride nanotubes decreased, as compared to the G18 glass. The rule of mixture was not applicable to either density or elastic modulus of the nanotubes-reinforced composite.

Acknowledgements

The authors are grateful to Ralph Pawlik for mechanical testing and John Setlock for composite processing. This work was supported by Low Emission Alternative Power (LEAP) Program, NASA Glenn Research Center, Cleveland, OH.

REFERENCES

1. Y. Chen, J. Zou, S.J. Campbell and G. Le Caer, "Boron Nitride Nanotubes: Pronounced Resistance to Oxidation," *Appl. Phys. Lett.*, **84**[13], 2430-2432 (2004).
2. W.A. Curtin and B. W. Sheldon, "CNT-Reinforced Ceramics and Metals," *Materials Today*, 7 [11] 44-49 (2004).
3. P. J. F. Harris, "Carbon Nanotube Composites," *Internatl. Mater. Rev.*, **49**[1], 31-45 (2004).
4. H. D. Wagner and R. A. Vaia, "Nanocomposites: Issues at the Interface," *Materials Today*, 7 [11] 38-42 (2004).
5. E. T. Thostenson, Z. Ren, and T. W. Chou, "Advances in the Science and Technology of Carbon Nanotubes and Their Composites: A Review," *Comp. Sci. Technol.*, **61**[13], 1899-1912 (2001).

6. E. T. Thostenson, C. Li, and T. W. Chou, "Nanocomposites in Context: A Review," *Comp. Sci. Technol.*, **65**[3-4], 491-516 (2005).
7. N. P. Bansal, J. B. Hurst, and S. R. Choi, "Boron Nitride Nanotubes-Reinforced Glass Composites," *J. Am. Ceram. Soc.*, **89**[1] 388-390 (2006).
8. N. P. Bansal and E. A. Gamble, "Crystallization Kinetics of a Solid Oxide Fuel Cell Glass by Differential Thermal Analysis," *J. Power Sources,* **147**[1-2] 107-115 (2005).
9. ASTM C 1161, Test Method for Flexural Strength of Advanced Ceramics at Ambient Temperature," *Annual Book of ASTM Standards*, Vol. 15.01, American Society for Testing & Materials, West Conshohocken, PA (2005).
10. J. Kübler, (a) "Fracture Toughness of Ceramics Using the SEVNB Method: Preliminary Results," *Ceram. Eng. Sci. Proc.*, **18**[4] 155-162 (1997); (b) Fracture Toughness of Ceramics Using the SEVNB Method; Round Robin," VAMAS Report No. 37, EMPA, Swiss Federal Laboratories for Materials Testing & Research, Dübendorf, Switzerland (1999).
11. S. R. Choi and N. P. Bansal, "Mechanical Properties of SOFC Seal Glass Composites," *Ceram. Eng. Sci. Proc.*, **26**[4] 275-283 (2005).
12. J. E. Srawley and B. Gross, "Side-Cracked Plates Subjected to Combined Direct and Bending Forces," pp. 559-579 in *Cracks and Fracture*, ASTM STP 601, American Society for Testing and Materials, Philadelphia (1976).
13. ASTM C 1259, "Test Method for Dynamic Young's Modulus, Shear Modulus, and Poisson's Ratio for Advanced Ceramics by Impulse Excitation of Vibration," *Annual Book of ASTM Standards*, Vol. 15.01, American Society for Testing and Materials, West Conshohocken, PA (2005).
14. ASTM C 1327, "Test Method for Vickers Indentation Hardness of Advanced Ceramics," *Annual Book of ASTM Standards*, Vol. 15.01, American Society for Testing and Materials, West Conshohocken, PA (2005).
15. S. R. Choi, W. A. Sanders, J. A. Salem, and V. Tikare, "Young's Modulus, Strength and Fracture Toughness as a Function of Density of In Situ Toughened Silicon Nitride with 4 wt % Scandia," *J. Mater. Sci. Lett.*, **14** 276-278 (1995).
16. K. T. Faber and A. G. Evans, "Crack Deflection Processes," *Acta. Metall.*, **31**[4] 565-576 (1983).
17. D. W. Richerson, *Modern Ceramic Engineering*, pp.132-133, Marcel Dekker, Inc., New York (1992).
18. S. R. Choi and N. P. Bansal, "Flexure Strength, Fracture Toughness, and Slow Crack Growth of YSZ/Alumina Composites at High Temperatures," *J. Am. Ceram. Soc.*, **88**[6] 1474-1480 (2005).
19. B. M. Steinetz, N. P. Bansal, F. W. Dynys, J, Lang, C. C. Daniels, J. L. Palko, and S. R. Choi, "Solid Oxide Fuel Cell Seal Development at NASA Glenn Research Center," presented at the 2004 Fuel Cell Seminar, San Antonio, TX, November 1-5, 2004; Paper No. 148

MECHANICAL BEHAVIOUR OF GLASSY COMPOSITE SEALS FOR IT-SOFC APPLICATION.

K.A. Nielsen, M. Solvang, S.B.L. Nielsen, D. Beeaff
Risoe National Laboratory, DK-4000 Roskilde, Denmark

ABSTRACT
Glass-based sealants have been developed with emphasis on filler material and surface treatment of the sealing components in order to optimise their mechanical and functional behaviour during the initial sealing process as well as during thermal cycling of the SOFC-stack after exposure to operating conditions. The bonding strength and microstructure of the interfaces between composite seals and interconnect materials were investigated as a function of surface treatment of the sealing surfaces, glass matrix composition, sealing pressure and temperature. The initial sealing performance and resistance to thermal cycling were then investigated on selected combinations of materials after ageing.

Strongest bonding between sodium aluminosilicate glass composite and steel surfaces was obtained for sealing at 850°C. For the strongest interface, having shear strength of 2.35 MPa, rupture occurred in the glass matrix, meaning that the glass-steel interfaces are, in this case, even stronger. Application of transition metal oxide coatings on etched surfaces of Crofer 22APU steel showed a significant improvement in the development of a seamless transition zone between metal and glass, whereas the same coatings on a sanded surface showed no influence on the bonding strength, which on the other hand were all recorded at a fairly high level, only 15-20% less than the 2.35 MPa seen for the glass. Ageing and thermal cycling of sealed samples did not deteriorate the recorded strength.

INTRODUCTION
Glass is a versatile material and has been widely used for sealing in the ongoing development of intermediate temperature solid oxide fuel cell technology where the physical and chemical properties of a glassy seal can be tailored to the other components, i.e. metallic interconnect and ceramic fuel cell, even though the coefficients of thermal expansion (CTE) usually must be in the range 11-13 ppm/K and higher than the CTE of most base glasses. Although high-expansion glasses are known, other demands, such as deformability at the sealing temperature and long-term stability of the seal, often limit the number of applicable glass systems.

Literature reports on glass and glass-ceramic compositions within the groups of alkali silicates, alkali aluminosilicates and alkaline earth aluminoborosilicate[1-10], illustrating two equally viable sealing strategies. One route, which focuses on composite seal materials from a compliant glass and a dispersed filler material with high CTE, e.g. MgO, has shown promise in terms of exact matching the CTE between the sealant and the seal surfaces, and at the same time attain suiTable values of Tg and sufficient wetting of the surfaces. Further, it shows a high stability and a slow crystallisation behaviour[5]. Similar performance was recently reported also for metallic filler materials in alkali borosilicate glass[11]. The other route, using glass-ceramic materials with fast crystallisation behaviour and high final CTE[7-10] results in a solid bond, which ideally stays unaffected during the life-time of the stack. The need for long-term compliancy through a slow crystallisation behaviour in the glass matrix mainly depends on the magnitude of thermally induced stress between the components, for instance as consequence of stack design and accuracy of the thermo-mechanical match between sealant and the neighbour stack components.

Partial dissolution of the filler material had been minimised in earlier seal composites of sodium aluminosilicate (NAS) glass with MgO-filler by using only larger MgO-grains, d=90-200μm[4], because magnesium is expected to increase surface tension of the glass[12] and, dependent on concentration, increase viscosity[13]. However, as a consequence of such partial dissolution an improved overall wetting and bonding behaviour was observed[6] even though corrosive reactions at the glass-to-steel interfaces complicated the analysis of bonding mechanisms.

This paper focuses on the optimisation of the bonding between interconnect ferritic steel materials and composite seals based on a NAS-glass (Tg = 515°C) with MgO-filler. Exploratory studies on the mechanical and chemical interactions between the glassy seal materials and ferritic interconnect steels, Crofer 22APU from Krupp Thyssen which develops a dual-layer protective oxide scale and Fe22Cr-type test melts from Sandvik, which develop a single layer of protective chromia scale, generally showed promising results for a number of combinations, and particularly good bonding was observed for transition metal coatings on the steel surface[14, 15]. Therefore, further exploration of the interface chemistry and optimisation of the bonding strength as a function of sealing parameters, e.g. temperature, sealing pressure and ageing, is expected to provide a foundation for predictions of sealing performance. By minimising the thickness of the sealing in stack design, the risk of rupture initiation at the seal due to thermal mismatch between the seal itself and the neighbour components may be minimised as well. The sealing also has to transfer stresses created in the stack as a consequence of thermal mismatch created on a larger scale, e.g. between cell and interconnect components and testing the adhesion in a shear stress mode, as shown schematically in Figure 1, seems to produce a relevant measure of the mechanical sealing performance.

EXPERIMENTAL

The NAS glass was melted from analytical grade chemicals SiO_2, Al_2O_3 and Na_2CO_3 in an alumina crucible at 1500°C for 2 hours, crushed and re-melted in order to ensure homogeneity, reaching a final composition: Na_2O: 17.8mole%, Al_2O_3: 9.4mole%, SiO_2: 72.8mole%. The glass was then milled into fine powder, d_{50}=5.5μm, which was combined with 30 vol% filler material of MgO powder (d<39μm) and suspended in an organic vehicle before being shaped as thin films having dried thicknesses of ~260μm by tape casting. The tapes were sandwiched between either interconnect materials or fuel cell support materials for seal testing.

Two types of ferritic steel was used, Krupp Thyssen Crofer 22APU (2nd batch) and a test melt from Sandvik (Fe: ~77%, Cr: 22.3%, Mn: 0.1% + other elements), which were machined to dimensions and, except for the 'as-received' and sanded surfaces, etched by HNO_3/HF-solution in an ultrasonic bath. All samples were then rinsed in ethanol and dried. Sanded metal surfaces were prepared by using SiC-paper (grit 1000). Pre-oxidation of metal surfaces was performed by heating in air to 850°C for 10 minutes. Heating and cooling rates were 3 °C/min.

Slurries for spray-coating the metal surfaces were prepared by ball milling metal oxides (Co-, Ni-, Mn, Cu-oxide, respectively) suspended in an organic vehicle until an average particle size in the range 0.8-5μm were obtained, after which they were applied to etched samples, having dried layer thicknesses in the range 5-7μm.

Interconnect samples and thin film samples of seal composites were stacked in sequences and run simultaneously to expedite the experimental work. The sandwiches were subjected to heat treatments at two different temperature levels (850°C, 950°C) to evaluate the sealing performance, durability and reactivity between interconnects and glass composites within the sealing temperatures of interest. For exploratory sealing experiments, a pressure of 120 kPa was applied to the samples and heating was performed at constant rate, 2.5°C per minute, until the soak temperature, which was then held for 160-170 hours before cooling to room temperature at nominally 2.5°C per minute. For measuring the bonding strength, which was done at room temperature, a standard procedure[16] was used for design of specimens and mechanical testing. Each sample comprised three identical metal strips which were joined symmetrically by two equally large glass tape interfaces (2.5 by 2.5 cm) on both side of the central strip, protruding 65 mm to one side of such double joints, whereas the two other strips protruded to the opposite side, cf. Figure 1. The samples were produced in series of five and heated at constant rate, 3°C per minute, to the sealing temperature, 850°C or 950°C, held for 4 hours and in some cases, cf. Table 1, further aged at 850°C for 168 hours followed by duplicate thermal cycling between RT and 850°C. Two levels of sealing load were applied, 100 kPa or 800 kPa. Tensile testing (Instron 8532) was done with a cross head speed of 83μm per second, which typically results in a load rate in the order of 15-20 N/s down to 0.5-4 N/s in cases where the steel started to yield before rupture. Loads were measured by a (±) 5kN load cell (Instron UK802).

 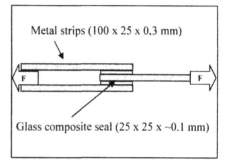

Metal strips (100 x 25 x 0.3 mm)

Glass composite seal (25 x 25 x ~0.1 mm)

Figure 1: Sample fabrication geometry and specimen set-up for sealing experiments.

Particle sizes were measured on a Beckman Coulter LS particle size analyser. Coefficients of thermal expansion (CTE) were measured in a Netzsch DIL 402C ramping at 2°C per minute. Microscopy and element analysis was done on polished cross sections of samples in a JEOL JSM 5310 LV microscope equipped with energy dispersive X-ray detector.

RESULTS AND DISCUSSION

During the initial screening, no significant difference in the metal reactivity was observed between as-received, sanded and slightly oxidised surfaces, cf. Table 1, which is probably because the surfaces will get uniformly pre-oxidised during the heat-up sequence, before the glass starts to flow and exclude the air from the metal surface, but etched surfaces of the Crofer steel appeared much more reactive. During heat up, the steel and the glass matrix starts to react with the result that the sodium concentration in the glass near the steel-to-glass interface is significantly reduced and the chromium concentration simultaneously is increased. The reactivity has been qualitatively evaluated by measuring the width of such reaction zones, as observed on polished cross sections, and numbers for the width are given in Table 1. By coating the steel surface with a thin layer of transition metal oxide powder the reaction between the glass matrix and the steel surfaces were reduced significantly, cf. Table 1. The metal oxides reacted with the chromium from the steel and formed an intermediate layer of mixed oxides, which partly prevented the reaction between chromium and sodium in the glass. The affinity between the metal oxide and chromium decreased in the order: Mn_3O_4, Co_3O_4, NiO, if evaluated by the width of the reaction zone. In contrast, the bonding between the glass composite and the coated steel decreased in the order: Co_3O_4, NiO, Mn_3O_4, and particularly good bonding was observed when the Crofer steel was coated with nickel or cobalt oxide, as seen on Figure 2, where strong and gradually changing interfaces are formed. Although the addition of MgO-filler material significantly reduced the porosity of the glass matrix[6], relatively large pores still exists as may be seen from Figure 2.

Table 1: Qualitative assessment of the bonding behavior between NAS-glass composite with 30vol% MgO filler and interconnect steel (Crofer 22APU and Sandvik test melt) with different surface coatings and oxide layers. Five levels were assigned to adhesion: Excellent (4x), good (3x), fair (2x), poor (x) and none (0) and listed together with the average width of the chromium-sodium-diffusion zone ($d_{(zone)}$).

		850°C / 170h		950°C / 170h	
Metal	**Coating**	**Adhesion**	**$d_{(zone)}$**	**Adhesion**	**$d_{(zone)}$**
Sandvik, as recieved		x	10μ	3x	30μ
Sandvik, sanded		3x	15μ	2x	20μ
Crofer, as recieved		2x	10μ	x	25μ
Crofer, sanded		2x	15μ	2x	20μ
Crofer, pre-oxidised		2x	12μ	x	27μ
Crofer, etched		x	10μ	0	80μ
Crofer, etched	5μm NiO			3x	25μ
Crofer, etched	5μm Co_3O_4			4x	15μ
Crofer, etched	5μm Mn_3O_4			x	6μ

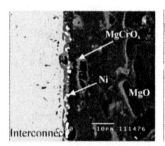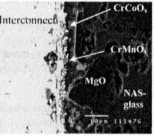

Figure 2: Interfaces developed after 170 hours at 950°C between thin film NAS glass composite (30 vol% MgO-filler) and Crofer 22APU interconnect steel coated with 5μm NiO (left), 5μm Co_3O_4 (centre) or 5μm Mn_3O_4 (right).

The micrographs in Figure 2 were formed by back-scattered electron images and illustrate the distribution of various elements. The interconnect steel at the left side in the micrographs, in all cases exhibits internal oxidation of the aluminium and titanium content of the steel, which show up as bands of nodular precipitates in the metal matrix. At the interface between seal and steel, chromia and chromium-manganese spinel phases are observed, cf. Figure 2, right. When coated with nickel oxide, metallic nickel is formed and deposited either as nodular particles or as a continuous band on top of the chromia-rich oxide scale, cf. Figure 2, left. When coated with cobalt oxide, cobalt forms mixed oxide phases with chromium or in few cases even form small isolated particles of metallic cobalt, similarly to what is observed for the nickel. Coatings of manganese oxide react only slightly with the glass and seem to stay fairly intact as a granular layer separating the glass and steel surfaces. During sample preparation, this layer is often removed (not shown in the figure), which leaves an open flaw between steel and glass.

In the glass seal composites, light grey and whitish deposits on the MgO-filler particles were dominated by chromium oxides with a lesser amount (<15 at%) of MgO, often with traces of oxides from the glass (Si, Al) or iron oxides from the metal. These deposits were always concentrated on the side of the MgO-grains that faced the metal-glass interface and looked as if they were formed by condensation of diffusing substances. The glassy transition zone between the innermost row of MgO-grains and the steel surface, cf. Figure 2, was gradually enriched in chromium when moving from the areas of the pristine glass towards the metal surface and at the same time gradually depleted in sodium, often to a degree where sodium was absent close to the metal surface. The formation of sodium chromate (Na_2CrO_4), a relatively low-melting compound ($M_p=792°C$)[17], was expected to destabilise the protective chromia scale on the interconnect, but sodium and chromium oxides seemed never to co-exist in the glass-steel interface region. Sodium chromate is stable in air to only 974°C and readily decomposes at 950°C if the oxygen partial pressure is reduced from ambient conditions by less than 5%[18]. In contrast to expectations that sodium depletion will cause poor wetting and adhesion due to the increasing viscosity and tendency towards crystallisation in the glass, the glass at the interface often looked vitreous and with good wetability. Dissolution of magnesium from the filler into sodium-depleted (Na)-Al-Si-O glass is reported[19] to dramatically change the temperature sensitivity of the glass viscosity dependence and may contribute to the improved wetability. The microstructure of the seals was evaluated after testing and showed extensive primary crystallization, which was assumed to be the Na-Al-Si-oxides, Albite and Nepheline according to the results from concurrent monitoring of principal crystallisation in the composite by XRD.

Based on qualitatively promising behaviour, a selection of the interfaces listed in Table 1 were tested further, cf. Table 2, with the aim of obtaining quantitative measures for the bonding strength at room temperature as a function of steel surface coating, sealing temperature, orthogonal sealing load and moderate ageing plus thermal cycling. An example of the traces recorded of load versus cross head travel distance has been shown in Figure 3. Weaker interfaces disintegrated before reaching the load at which the steel strip itself started to yield. Stronger interfaces passed this transition point and experienced a significantly lower load rate before rupture. None of the interfaces were so strong that the metal strip itself was elongated more than ~4% or ruptured. In general, the results for load at rupture showed varying degrees of deviation from the average value within each group of 5 samples, cf. Table 2, which may lead to the conclusion that the microstructure of either the steel-scale interface or the scale-glass interface containing the strength limiting defects, cf. Table 2, is not as homogeneous as may be wished for.

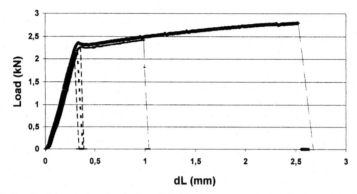

Figure 3: Results from testing 5 samples of Crofer 22APU, sanded surface, exemplifying the traces recorded in load-til-rupture testing.

Table 2: Results of bonding strength between NAS-glass composite sealing, coated and un-coated steel surfaces, evaluated by measuring the shear strength on 5 identical samples within each group after different heat treatments as shown in the table, including 4 hours at a sealing temperature plus in some cases further ageing, indicated by an 'A', for 168 hours at 850°C followed by two thermal cycles between 850°C and RT. The final thickness of the glass seal (height) is shown, and abbreviated signatures for steel (S), protective scale on the steel (Sc) and glass matrix (G) indicates the position of the weakest interface, e.g. between steel and protective scale (S-Sc) where the rupture occurred.

Steel	Surface/Coating	Sealing pressure (kPa)	Sealing temp. (°C)	Seal height (µm)	Shear strength (MPa)	Weak interface
Crofer 22APU	Sanded	800	850	210	1.87 ±13%	Sc – G
Crofer 22APU	Sanded	800	850 + A	100	1.88 ± 3%	Sc – G
Crofer 22APU	Sanded + Co₃O₄	800	850	200	2.03 ±15%	S – Sc
Crofer 22APU	Sanded + Co₃O₄	800	850 + A	110	2.09 ± 2%	S – Sc
Crofer 22APU	Sanded + Co₃O₄	800	950	120	1.81 ±10%	S-Sc (Sc-G)
Crofer 22APU	Sanded + Co₃O₄	100	950	140	1.30 ±20%	Sc – G
Crofer 22APU	Sanded + Co₃O₄	800	950 + A	115	1.93 ±10%	G – G
Crofer 22APU	Sanded + NiO	800	850	120	0.92 ±28%	Sc – G
Crofer 22APU	Sanded + NiO	800	850 + A	110	1.86 ± 3%	G – G
Crofer 22APU	Sanded + NiO	800	950	120	1.57 ± 10%	Sc – G
Crofer 22APU	Sanded + Mn₃O₄	800	950 + A	80	1.85 ± 1%	S – Sc
Crofer 22APU	Sanded + CuO	800	850	75	2.06 ± 4%	S – Sc
Sandvik, test	As received	800	950	65	1.21 ±23%	S – Sc
Sandvik, test	As received	100	950	90	1.12 ±24%	Sc – G
Sandvik, test	As received	800	850 + A	75	2.36 ± 2%	G – G
Sandvik, test	As received	800	950 + A	70	1.85 ±25%	S – Sc

The shear strength results of the Sandvik samples seem to exhibit a higher standard deviation than the Crofer samples. The higher deviation may be a consequence of the lower seal height also observed for this group of samples because the extruded glass may cause unknown edge effects and spread over a slightly larger area in an uncontrolled manner. Extruded glass was only removed from the outer edges and was not removed from in between the outer set of metal strips, cf. Figure 1. The Sandvik group of samples did not exhibit a generally higher level of shear strength than the Crofer set of samples and based on the limited number of tests it may be concluded, at least tentatively that residual stresses in the individual samples did not have a major influence on the strength level and creation of edge flaws.

Sealing at different orthogonal sealing pressure did have an effect on the bonding strength and the thickness of the seal, although the influence is different for the two different conditions investigated. For the Crofer with Co_3O_4 coating, the thickness of the seal was almost unchanged when increasing the sealing pressure, but the bonding strength increased by nearly 40%. For the Sandvik samples, the seal thickness decreased on increasing sealing pressure by almost 30% but the strength remained virtually unchanged. This may reflect the different impact from the metal on the formation of the microstructure in the seal, as it indicates that the extensive crystallisation seen in the glass occurs faster when exposed to the Crofer alloy than when exposed to the Sandvik alloy. Crofer 22APU releases greater amounts of both chromium and manganese, both into the oxide scale on the metal surface and into the glass adjacent to the interfaces. A possible detrimental action of the manganese may be a higher susceptibility to reduction in the glass leading to a destabilisation of the glass network. Differences in the crystallisation behaviour as a function of the surface or surface coating is also seen through the seal thickness for the Crofer samples coated with manganese or copper oxides that are at the same level as seen generally for the Sandvik alloy and it all indicates a complexity of the conditions required for bonding the metal to glass.

Sealing Crofer 22APU (sanded) with and without metal oxide coatings at 850°C gave no significant difference in bonding strength, around 2 MPa, except for the NiO coating, which showed relatively poor strength after sealing for 4 hours but reached a strength around 2 MPa after ageing at the same temperature for 168 hours. Fracture for this particular set of samples was observed to occur within the glass seal as opposed to between the glass and the metal-oxide scales as seen for the same interface sealed for longer time at 750°C and samples sealed at 950°C. The glass matrix appears much weaker in the samples with NiO sealed for 4 hours than in all other samples, but the reason for this discrepancy is not clear at the moment. It could be hypothesised that reducing the nickel oxide and creating a dense microstructure may require more than four hours at 850°C. Generally, the weakest interface was found between the steel and the scale for the coated samples, but between the scale and the glass for the uncoated Crofer samples. The difference in qualitative bonding behaviour seen for etched and coated samples, cf. Table 1, has not been reproduced by the sanded plus coated samples, cf. Table 2, where probably the better performance of the sanded surface dominates the overall bonding behaviour.

The difference in measured strength, both for Crofer with NiO or Co_3O_4 coatings and Sandvik, when sealed at different temperatures, 850°C or 950°C, indicates that a stronger microstructure is obtained by sealing at the lower temperature. For the Sandvik samples, the weak interface, also shifts from being between the steel and the scale, cf. Figure 4, or between the scale and the glass after sealing at 950°C to being inside the glass matrix after sealing at 850°C, which agrees with the expectation that minimising the interaction between the glass and the steel, as observed for the lowest temperature, cf. Table 1, will create a thinner, hence stronger, transition zone.

Ageing for 168 hours and subsequent thermal cycling does not seem to shift the bonding strength to lower values, which means that the initial assumption that stresses created through thermal mismatch between the seal composite at a thickness of ~100µm and the neighbour components is negligible for the applications in question.

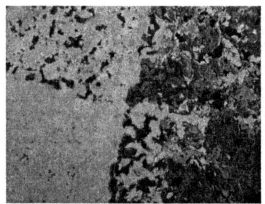

Figure 4: Ruptured surface of Sandvik steel after sealing at 950°C and subsequent ageing at 850°C. In the left part of the micrograph, the metal surface is visible whereas the right part features glass and the lower side of the corrosion scale from the adjacent metal strip, on which parts the metal surface is visible.

CONCLUSION

The strongest bond between NAS-glass seal composite and steel surfaces was obtained by sealing at the lower sealing temperature, 850°C, and the strongest interface, having shear strength of 2.35 MPa, was found on the Sandvik steel. For this sample, rupture occurred within the glass matrix and the interfaces are therefore stronger. Ageing and thermal cycling of sealed samples did not deteriorate the measured strength. Application of transition metal oxide coatings on etched surfaces of Crofer steel showed a significant improvement in the development of a seamless transition zone between metal and glass, whereas the same coatings on a sanded surface showed no influence on the bonding strength, which was determined to be 15-20% less than the above mentioned strength for the glass itself. Crystallisation occurs in the glass during sealing as a consequence of exposure to the steel surfaces, and is faster when exposed to Crofer steel than to Sandvik steel, so loss of compliance may be a problem if a proper coating has not been applied to the steel surfaces prior to sealing. The shear strength results obtained for the various interfaces primarily applies to the durability on cooling down from the service temperature as a self-healing mechanism, hence a degree of compliancy, of the glass composite material has been observed earlier in stack testing[6].

Further refinement of the sealing parameters and the composition of metal oxide coatings on the steel surfaces should yield better reproducibility, i.e. lower scatter of the strength measurements, and lead to more uniform microstructures in the transition zone between steel and glass.

ACKNOWLEDGEMENTS

The work was part of the DK-SOFC project funded by the Danish Power Suppliers (ELKRAFT system) and the Danish Energy Authority under contract no: 103594 (FU3403/4), who are acknowledged together with Haldor Topsøe A/S, and the staff at Risø National Laboratory, for their inspiring collaboration.

LITERATURE
(1)   S.B. Adler et al., "Reference Electrode Placement and Seals in Electro-chemical Oxygen Generators", *Solid State Ionics*, **134** (2000), p. 35.

(2) J.G. Larsen, P.H. Larsen and C. Bagger, "High Temperature Sealing Materials". US Patent Ser. no. 60/112039.

(3) P.H. Larsen, "Sealing Materials for Solid Oxide Fuel Cells" Ph.D. thesis, Materials Research Department, Risø National Laboratory, Roskilde, Denmark. (1999).

(4) K.A. Nielsen, M. Solvang, F.W. Poulsen, P.H. Larsen, *Ceramic Engineering and Science Proceedings* **25**, 309-314, (2004).

(5) W. Höland and G. Beall, "Glass-Ceramic Technology", The American Ceramic Society, Westerville, OH, (2002).

(6) M. Solvang, K.A. Nielsen, A.R. Dinesen, P.H. Larsen, "Optimisation of Glass Composite Sealant for Intermediate Temperature Solid Oxide Fuel Cells", p. 1914 in ECS, Solid Oxide Fuel cells IX, ed. S.C Singhal & J. Mizusaki (2005)

(7) S.M.Gross et al. "Glass-ceramic composite as a new sealing material for SOFCs" p. 1924 in ECS, Solid Oxide Fuel cells IX, ed. S.C Singhal & J. Mizusaki (2005).

(8) Z. Yang, J.W. Stevenson, K.D. Meinhardt, "Chemical interactions of barium–calcium–aluminosilicate-based sealing glasses with oxidation resistant alloys" *Solid State Ionics*, **160**, 213-225 (2003).

(9) Z. Yang, K.D. Meinhardt, J.W. Stevenson, " Chemical compatibility of barium-calcium-aluminosilicate-based sealing glasses with the ferritic stainless steel interconnect in SOFCs" *J. Electrochem. Soc.* **150**, A1095-A1101 (2003).

(10) Z. Yang, K. Scott Weil, D.M. Paxton, J.W. Stevenson, "Selection and evaluation of heat-resistant alloys for SOFC interconnect applications", *J. Electrochem. Soc.* **150**, A1188-A1201 (2003).

(11) C.C. Beatty, "Compliant Glass-Silver seals for SOFC application" p. 1949 in ECS, Solid Oxide Fuel cells IX, ed. S.C Singhal & J. Mizusaki (2005)

(12) Donald, I.W., "Preparation, properties and chemistry of glass and glass-ceramic-to-metal seals and coatings. *J. Mat. Sci.* **28**, p. 2841-86 (1993)

(13) Volf, M. B., Chemical approach to glass, Elsevier (1984)

(14) K.A. Nielsen, M. Solvang, S.B.L.Nielsen, A.R. Dinesen, D. Beeaff, P.H. Larsen "Glass composite seals for SOFC application", *J. of the European Ceramic Society* (in press)

(15) Eppler, R.A. "Glazes and glass coatings" The American Ceramic Society, Westerville, Ohio (2000)

(16) Standard test method for strength properties of double lap shear adhesive joints by tension loading, ASTM D 3528 – 96, PO Box C700, PA 19428-2959, USA.

(17) D.R. Lide, Handbook of Chemistry and Physics, 82nd ed. CRC press, (2002)

(18) FactSage, ver. 5.3.1, www.factsage.com

(19) B. Mysen, Structure and properties of silicate melts, Elsevier Science Publishers B.V. Amsterdam (1988)

MECHANICAL PROPERTY CHARACTERIZATIONS AND PERFORMANCE MODELING OF SOFC SEALS*

Brian J. Koeppel, John S. Vetrano, Ba Nghiep Nguyen, Xin Sun, and Moe A. Khaleel
Pacific Northwest National Laboratory
P.O. Box 999/MS K5-22
Richland, WA 99352

ABSTRACT

The objective of this work was to provide a modeling tool for the design of reliable seals for SOFC stacks. The work consisted of experimental testing to determine thermal-mechanical properties of a glass-ceramic sealing material and numerical modeling of stack sealing systems. The material tests captured relevant temperature-dependent property data for Pacific Northwest National Laboratory's (PNNL) G18 sealant material as required by the analytical models. A viscoelastic continuum damage model for this glass-ceramic sealant was developed and implemented in the MSC MARC finite element code and used for a detailed analysis of a planar SOFC stack under thermal cycling conditions. Realistic thermal loads for the stack were obtained using PNNL's multiphysics solver SOFC-MP. The accumulated seal damage and component stresses were evaluated for multiple thermal loading cycles. The seals nearest the stack mount location were most susceptible to damage which began during the first operational cycle and accumulated during shutdown. Viscoelastic seal compliance was also found to beneficially reduce the stresses in the anode.

INTRODUCTION

Hermetic sealing of solid oxide fuel cells (SOFCs) remains a strong technical challenge. The air and fuel streams in the SOFC must be segregated by reliable seals. Rigid glass-ceramic seals have excellent hermeticity, but are susceptible to damage and cracking caused by operational thermal stresses as well as cyclic thermal loading. Experimental observations have indicated that even a few thermal cycles can cause strength reduction[1] and cracking of the seals, which leads to reduced performance and loss of cell integrity. To ensure seal designs are robust, it is necessary to determine the most critical areas of the seal structure and how they can be improved to meet stack durability goals. In addition to failure of the seal material, the degradation of the interfaces between the joined dissimilar materials must also be addressed. A sealing assembly may have multiple interfacial layers including an oxide scale layer, coatings applied for corrosion protection, and reaction zones with formation of phases, depletion of elements, or void generation[2]. The reaction zones and porosity typically create brittle, weakened regions that affect the structural integrity of a stack. Therefore, in order to characterize the performance of the sealing system, the behavior of the seal material and the interfacial layers must be captured.

The objective of this work is to develop constitutive models from experimental material data that capture the relevant mechanical response of the glass-ceramic materials to evaluate seal performance in SOFC stacks. To this end, experimental material characterization and testing were performed for a glass-ceramic variant used for SOFC sealant purposes. The material selected for this study was the barium-calcium-aluminosilicate (BCAS)-based glass designated "G18" and developed at Pacific Northwest National Laboratory (PNNL)[3]. This material composition was developed to provide a closely matched coefficient of thermal expansion (CTE) with yttria-stabilized zirconia (YSZ) electrolytes and ferritic stainless steels to minimize thermally-induced stresses. Testing was conducted based on ASTM standards[4-8] to determine relevant temperature-dependent physical and mechanical data needed by the modeling such as thermal expansion, strength, fracture toughness, and relaxation behavior. Next, a viscoelastic damage model[9] that extends the continuum damage mechanics model previously developed for the same sealant material was generated to capture the time-dependent behavior of the material due to the relaxation phenomenon at high temperatures[10]. The viscoelastic damage response was obtained by modifying the basic Maxwell's model[11] to allow for damage that affects the elastic modulus. The model was implemented in MSC MARC and was applied to the structural analysis of an SOFC stack under thermal cycling conditions. The prediction of potential areas of seal damage constitutes a significant step to remedy the stack design.

EXPERIMENTAL TESTING

Glass-Ceramic Sealant Material

To form G18 specimen materials, the glass powder and an organic solvent were tape cast, dried, laminated, and then fired to remove the organics. The glass was then fired at 850°C for one hour and heat treated at 750°C for four hours to form the blanks from which specimens were machined. During heat treatment the glass partially devitrifies to form various ceramic crystalline phases within a residual glassy matrix. Additional heat treatment causes further devitrification, but a significant fraction of residual glass (up to 30%) remains for up to 1000 hr at 750°C[1].

Monotonic Physical and Mechanical Property Measurements

The elastic properties of the G18 materials with 4 hr and 1000 hr heat treatments were evaluated at room and elevated temperatures. Room temperature elastic measurements were performed using a pulse-echo ultrasonic test (ASTM E494[4]). Elevated temperature measurements using the dynamic resonance method (ASTM C1198[5]) were performed up to 600°C. Above this temperature the viscous response of the glassy phase produced invalid results, so elastic modulus estimates for temperatures above 600°C were made from four-point bend flexure tests. The elastic property results are summarized in Figure 1.

The monotonic strengths of the G18 material for 4 hr and 1000 hr heat treatments were evaluated at room and elevated temperatures using four-point bend flexure tests (ASTM C1211[6]). The strength as a function of temperature is shown in Figure 2. The strength of the glass-ceramic was good up to 600°C but then dropped dramatically above 700°C. The performance of the glass-ceramic seal is not wholly characterized by the material strength behavior but also the interfacial strength. The glass-ceramic creates a stable bond with the electrolyte material, but reactions and lower strengths with the metallic interconnects have been observed. The interface strengths of seal assemblies with a Crofer 22 APU interconnect were also evaluated under tension and shear loading (modified ASTM D2095[7], ASTM F734-95[8]). The interface strengths were significantly smaller than the flexural strengths as shown in Figure 2.

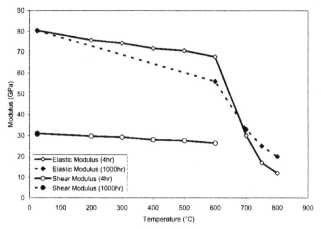

Figure 1. Temperature-dependent elastic and shear modulus for G18 glass-ceramic.

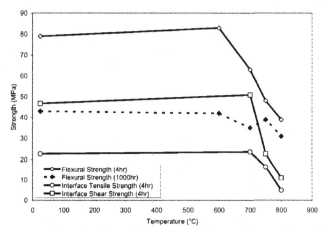

Figure 2. Temperature-dependent flexural strength, interface tensile strength, and interface shear strength for G18 glass-ceramic.

Stress Relaxation Measurements

The partially crystallized sealant has a residual glass content that will become viscous at operating temperatures. To adequately capture the seal stresses during transient loading, the relaxation of the seal must be included to prevent exaggerated stresses predicted by an elastic constitutive model. Constant strain compressive tests were performed to evaluate the time dependent stress relaxation and estimate the viscosities[9] as shown in Figure 3.

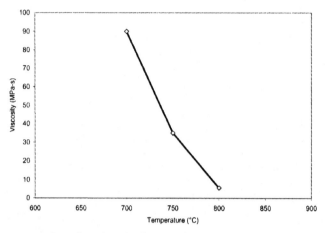

Figure 3. Temperature-dependent viscosity for G18 glass-ceramic with 4 hr heat treatment.

NUMERICAL MODELING
Viscoelastic Continuum Damage Model

A continuum damage mechanics (CDM) model previously developed to capture the inelastic response of the glass-ceramic seals under monotonic loading at stack operating temperatures[10] is briefly summarized here. The heterogeneous G18 material was replaced by an equivalent homogeneous material with effective properties. The phenomenological damage model relies on thermodynamics of continuous media and uses a scalar damage variable governing the reduction of the elastic modulus from the initial "undamaged" state to a saturation value corresponding to the last point on the monotonic stress/strain curve. The elastic deformation energy was taken as the thermodynamic potential, and the constitutive relations, thermodynamic force associated with the damage variable, and specific entropy were obtained by deriving the thermodynamic potential with respect to the strain, damage variable, and temperature. The Clausius-Duhem inequality defines the dissipation criterion. A damage criterion was defined which depends on a damage threshold function determined using the relation for modulus reduction and the experimental stress-strain curves. The damage evolution law was obtained using the damage criterion and the consistency conditions. The damage evolution law together with the dissipation criterion imposes the requirement that damage evolution obeys the second law of thermodynamics. This damage model was then extended to include the viscoelastic behavior at elevated temperatures due to the residual glass content[9]. The improvement led to a modified Maxwell model where the viscous strain rate depends on the stress and material viscosity while the damage variable affects only the elastic modulus.

PNNL experiments have shown that the stress/strain responses of G18 glass at temperatures up to 700°C were linear until failure, whereas their responses were nonlinear for temperatures higher than 700°C[10]. Therefore at high temperatures the damage variable evolves according to the damage evolution law to a critical (or saturation) state at which failure occurs and the material can no longer carry loads, but at lower temperatures the ratio of the maximum normal stress to the experimental failure stress was used to predict failure. After either failure criterion is met, crack propagation is then coarsely captured using a failure model and a vanishing element technique[10].

The interfaces between the glass seal and other constituent layers were modeled by very thin layers with zero Poisson's ratio and elastic properties averaged from those of the glass-ceramic and the other constituent. A special form of the three-dimensional finite element of MARC was used to represent an interface. The stiffness tensor of these elements was diagonal since the coupling between the strain components was suppressed by prescribing zero Poisson's ratio. In this way, the interface serves only to transfer the tractions across it. Delamination was predicted using the maximum opening stress criterion or the critical stress criterion accounting for interfacial shears.

Stack Stress Analysis Model

A stack finite element model was created using the default SOFC geometry from the current parametric version of the Mentat-FC[12] graphical user interface (GUI) being jointly developed by Pacific Northwest National Laboratory and MSC Software for SOFC analysis. The model was a planar stack containing an anode-supported cell mounted in a frame structure with integral inlet and outlet manifolds for the air and fuel streams (Figure 4a). For each cell in the stack, there were two glass seals. The first attached the outer edge of the electrolyte to the frame, and the second attached the interconnect structure (Figure 4a). Each seal was represented by three layers of continuum elements. The largest middle layer represented the bulk glass-ceramic material while the thinner layers on each side represented the interfaces which could be assigned separate material properties. Both single cells and a stack containing three cells to provide insight into cell-to-cell variations (Figure 4b) were evaluated.

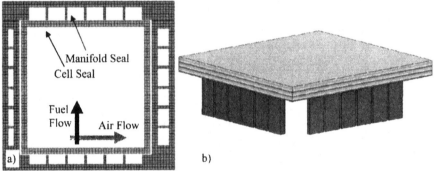

Figure 4. a) Sealing footprint and b) three dimensional view of the multi-cell SOFC stack.

For the electrochemistry solution, it was assumed that the thermal history of the stack was characteristic of laboratory conditions, where the stack is placed within a furnace and temperature changes occur slowly to minimize gradients. For the model, it was assumed that the stack exchanged heat within the furnace via external radiation and free convection. The free convection was modeled with empirical relations implemented through MARC's "FILM" user subroutine. The furnace temperature history was defined and used as the sink temperature for convective and radiation heat transfer. The fluid properties of air were interpolated for the local film temperature which is the average of the surface and ambient temperatures at each element face. The convection coefficient for each face was computed from the average Nusselt number, which was obtained using correlations for free flow around immersed flat plates as a function of the Rayleigh number. The external radiation was also included in the subroutine assuming diffuse gray radiation was representative of the furnace environment. By linearizing the radiation equation, an effective convection term was computed and added to the term that was computed for the buoyant flow at each face. The subroutine returned the total convective coefficient to the MARC solver for computation of the heat flux. This computational procedure was adequate for a representation of external heat exchange with the furnace environment, and can be extended for forced convection or addition of stack insulation.

The furnace thermal history begins with a uniform temperature of $1065°K$ ($792°C$) as the initial condition of the stack. This is the estimate of the stress-free temperature of the stack for the glass seals which have devitrified to form a significant crystal volume. The temperature was then reduced over 1 hr to the furnace operation temperature assumed to be $1023°K$ ($750°C$). The electrochemical behavior was then activated and ramped up over a period of 60 s. It was then allowed to remain in operation for 1.5 hr to reach steady-state before being similarly ramped down. The furnace temperature was reduced to room temperature $298°K$ ($25°C$) over 2 hr and allowed to sit for an additional 3 hr. For the next cycle, the furnace was heated back to $1023°K$ ($750°C$) over a period of 2 hr. This provided a reasonable transient temperature input to smoothly allow the stack to reach its steady-state temperature distributions at operating and room temperatures. This temperature history was then repeated for multiple cycles.

The gas flows within the cell were the second major method for exchange of heat. In the model, the fuel and air streams were controlled by inlet temperature and velocity specifications. The inlet air and fuel flow temperatures were equated to the furnace temperature history assuming that the gases had been sufficiently heated in the manifold prior to entering the stack. The velocities were held constant throughout the entire simulation. The model computed the heat exchanged by convection between the stack surfaces and the gas flows through the use of rigid contact surfaces, and the fluid velocity then determined the heat removal rate through the regions of gas flow. The electrochemical analysis was handled by the in-house multiphysics solver SOFC-MP[12-13]. This analysis returned a volumetric heat flux that was uniformly distributed through the anode, electrolyte, and cathode layers. Default electrochemistry parameters and a Butler-Volmer formulation were used.

The solution procedure used an uncoupled sequential analysis. First, the transient thermal-flow-electrochemistry solution was obtained for a single complete cycle. Second, the temperature distribution results were used as loads in the quasi-static structural analysis to evaluate the thermal stresses. Appropriate portions of the thermal cycle were used to generate an arbitrary number of structural analysis cycles assuming that mechanical deformations did not significantly influence the thermal behavior. The sequentially coupled approach was computationally efficient because the thermal and electrochemical problems were solved for only a single cycle rather than for each iteration. Post-processing was performed to evaluate the thermal, electrochemical, and structural responses.

Stack Thermal Analysis Results

The thermal distribution changed considerably during its transient cycle. During heat-up the hot air blown through the stack heated the inlet side cell (Figure 5a) until it was nearly isothermal. Heat generated from initiation of the electrochemical activity created hotter regions on the stack (Figure 5b) until the steady-state condition was achieved (Figure 5c). Colder air blown through the stack on shut-down then cooled the cell on the inlet side (Figure 5d). The profiles indicate that the seal thermal-stress history for thermal cycling events will depend on the actual loading cycle which should be considered in addition to the operating temperature conditions for evaluating long term mechanical response. Selected electrochemistry results for the steady-state operating condition are shown in Figure 6. Slight variations in the electrochemical results were noted between different cells in a multi-cell stack.

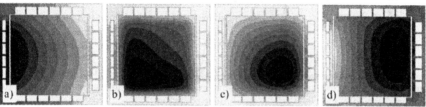

Figure 5. Series of contours representing the changing thermal profiles during a) heat-up until near isothermal condition, b) initiation of electrochemical activity, c) steady-state operation, and d) stack cool-down.

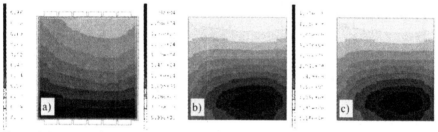

Figure 6. Contours of steady-state electrochemical results including a) hydrogen concentration, b) current density, and c) heat generation rate.

Stack Stress Analysis Results

The seal and interface responses were first evaluated using the elastic damage model. The results for the manifold and cell seals at steady-state operation are shown in Figure 7 and Figure 8, respectively, and the results for the bottom manifold and cell seals after shut-down are shown in Figure 9. The failure indicator suggests that the bottom seal exhibited complete damage during the first operational cycle that would result in unacceptable gas leakage and performance degradation. The remaining manifold seals responded better with damage only to the thin ligaments, and the cell seals were nearly undamaged. After shut-down, both the manifold and cell seals in the bottom cell exhibited significant accumulated damage leading to failure of over half the seal. Brittle seal failure during the cooling period of the stack is consistent with experimental observations during stack testing. Concentration of the damage in the bottom seals is also consistent with experiments of multi-cell stacks and is attributed to the stiff base on which the stack is mounted. This bottom cell is highly constrained which increases component stresses as compared to the upper cells which are more compliant and allowed to deform to accommodate thermal mismatch stresses. However, the extensive seal damage at the operating condition is not realistic, and this suggests the importance of including the viscoelastic response to capture high temperature relaxation for glass-ceramic materials.

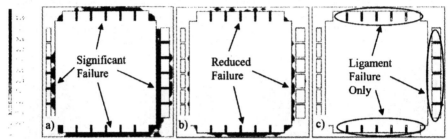

Figure 7. Contours of the failure indicator (0=undamaged, 1=failure) for the manifold seal of the a) bottom, b) middle, and c) top cells at steady-state operation.

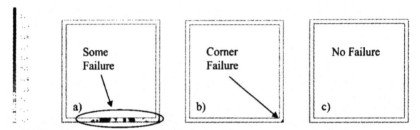

Figure 8. Contours of the failure indicator (0=undamaged, 1=failure) for the cell seal of the a) bottom, b) middle, and c) top cells at steady-state operation.

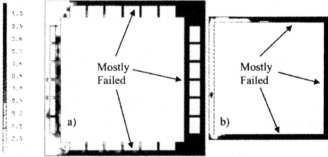

Figure 9. Contours of the failure indicator (0=undamaged, 1=failure) for the a) manifold and b) cell seals of the bottom cell after shut-down.

The seal and interface responses were also evaluated for the single-cell model using the viscoelastic damage model. At steady state operation the seals exhibited some damage (Figure 10a) but no seal failure as shown by the contour of the normalized damage variable (Figure 10b).

The damage was concentrated around the outer corners of the electrolyte seal, but the manifold seal showed no damage. This is more consistent with experimental testing than the results from the elastic damage model, as glass-ceramic seal failures at high temperatures should benefit from the known softening and stress relaxation. The results after shutdown are shown in Figure 11. The highest stresses occurred in the manifold seal, but they are acceptable and the only failures during shut-down occurred around the regions of high temperature damage in the cell seal (Figure 11a). The region of damage grew and additional elements of the cell seal failed after the second thermal cycle (Figure 11b) which could lead to leakage and performance degradation. The benefit of the viscoelastic deformation for the cell was observed as the anode principal stress decreased about 6% from a maximum of 38.4 MPa to 36.0 MPa (Figure 12). The use of the viscoelastic damage model seems to better capture the seal degradation behaviors characteristic of stack thermal cycle testing. Additional numerical experiments with this model could identify permissible heating/cooling rates under realistic operating conditions and the influence of geometric design to prevent seal damage and minimize cell component stresses.

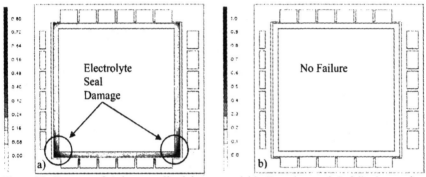

Figure 10. Contours of a) the normalized damage variable (0=undamaged, 1=fully damaged) and b) the failure indicator (0=undamaged, 1=failure) for the manifold and cell seals at steady-state operation.

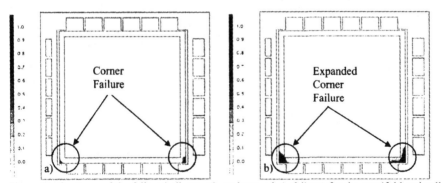

Figure 11. Contours of the failure indicator (0=undamaged, 1=failure) for the manifold and cell seals after shut-down of the a) first and b) second thermal cycles.

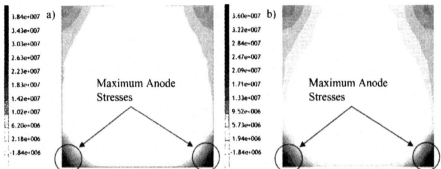

Figure 12. Contours of the maximum principal stress for the anode at operating temperatures using the a) elastic and b) viscoelastic seal damage models. The peak value was 38.4 MPa for the elastic case and 36.0 MPa for the viscoelastic case.

CONCLUSION

An analytical procedure was demonstrated to evaluate the mechanical performance of glass-ceramic seals in an SOFC stack. Testing of PNNL's G18 material was performed from room to operating temperatures to evaluate physical and mechanical properties. A viscoelastic damage model was developed based on continuum damage mechanics and thermodynamics of continuous media. This model is regarded as a modified Maxwell model in which damage is accounted for to affect the elastic modulus. The model makes use of the monotonic stress-strain curves and high temperature viscosities. It was implemented via user subroutine to the MSC MARC finite element code and applied to realistic seal geometries in a prototypical planar stack. The regions of likely seal damage during thermal cyclic loading were identified. The mechanical influence of a stack mounting base caused lower cells to exhibit greater damage which initiated during the first loading cycle and grew during shut-down. The additional compliance due to viscoelastic strains in the seal beneficially reduced anode stresses. Inclusion of the viscoelastic seal response also produced damage behavior more qualitatively characteristic of experimental testing and can be used for future analysis towards reliable glass-ceramic seal designs.

ACKNOWLEDGEMENTS

The authors would like to thank Matt Chou for his assistance in mechanical testing. This paper was funded as part of the Solid-State Energy Conversion Alliance (SECA) Core Technology Program by the U.S. Department of Energy's National Energy Technology Laboratory (NETL). Pacific Northwest National Laboratory is operated by Battelle Memorial Institute for the U.S. Department of Energy under Contract No. DE-AC05-76RL01830.

FOOTNOTES
*Manuscript authored by Battelle Memorial Institute under Contract Number DE-AC05-76RL01830 with the U.S. Department of Energy. The U.S. Government retains and the publisher, by accepting this article for publication, acknowledges that the U.S. Government retains a non-exclusive, paid-up, irrevocable, world-wide license to publish or reproduce the published form of this manuscript, or allow others to do so, for U.S. Government purposes.

REFERENCES

[1]K.S. Weil, J.E. Deibler, J.S. Hardy, D.S. Kim, G. Xia, L.A. Chick, and C.A. Coyle, "Rupture Testing as a Tool for Developing Planar Solid Oxide Fuel Cell Seals." *J. Mater. Eng. Perform.*, **13**(3), 316-326 (2004).

[2]Z. Yang, G. Xia, K.D. Meinhardt, K.S. Weil, and J.W. Stevenson, "Chemical Stability of Glass Seal Interfaces in Intermediate Temperature Solid Oxide Fuel Cells," *J. Mater. Eng. Perform.*, **13**(3), 327-34 (2004).

[3]K.D. Meinhardt, J.D. Vienna, T.R. Armstrong, and L.R. Pederson, "Glass-Ceramic Material and Method of Making," U.S. Patent 6,430,966 (2002).

[4]ASTM E494-05, "Standard Practice for Measuring Ultrasonic Velocity in Materials," ASTM International, West Conshohocken, PA (2005)

[5]ASTM C1198-01, "Standard Test Method for Dynamic Young's Modulus, Shear Modulus, and Poisson's Ratio for Advanced Ceramics by Sonic Resonance," ASTM International, West Conshohocken, PA (2005)

[6]ASTM C1211-02, "Standard Test Method for Flexural Strength of Advanced Ceramics at Elevated Temperatures," ASTM International, West Conshohocken, PA (2005).

[7]ASTM D2095-96, "Standard Test Method for Tensile Strength of Adhesives by Means of Bar and Rod Specimens," ASTM International, West Conshohocken, PA (2005).

[8]ASTM F734-95, "Standard Test Method for Shear Strength of Fusion Bonded Polycarbonate Aerospace Glazing Material," ASTM International, West Conshohocken, PA (2005).

[9]B.N. Nguyen, B.J. Koeppel, J.S. Vetrano, and M.A. Khaleel, "On the Nonlinear Behavior of a Glass-Ceramic Seal and Its Application in Planar SOFC Systems," submitted to Proceedings of FuelCell06: The 4th International Conference on Fuel Cell Science, Engineering, and Technology.

[10]B.N. Nguyen, B.J. Koeppel, S. Ahzi, M.A. Khaleel, and P. Singh, "Crack Growth in Solid Oxide Fuel Cell Materials: From Discrete to Continuum Damage Modeling," to appear in *J. Am. Ceram. Soc* (2006).

[11]G.A. Maugin, "The Thermodynamics of Plasticity and Fracture," Cambridge University Press, Cambridge (1992).

[12]M.A. Khaleel, "Finite Element SOFC Analysis with SOFC-MP and MSC.Marc/Mentat-FC," Proceedings of the Sixth Annual SECA Workshop, April 18-21, 2005, National Engineering Technology Laboratory, Morgantown, WV (2005).

[13]M.A. Khaleel, Z. Lin, P. Singh, W. Surdoval, and D. Collins, "A Finite Element Analysis Modeling Tool for Solid Oxide Fuel Cell Development: Coupled Electrochemistry, Thermal, and Flow Analysis in Marc," *J. Power Sources*, **130**(1-2), 136-148 (2004).

Mechanical Properties

FRACTURE TEST OF THIN SHEET ELECTROLYTES

Jürgen Malzbender, Rolf W. Steinbrech, Lorenz Singheiser
Forschungszentrum Jülich GmbH
Institute for Materials and Processes in Energy Systems
52425 Juelich, Germany

ABSTRACT

In electrolyte supported planar design of solid oxide fuel cells the fracture stress of the electrolyte foil determines the limit of mechanical cell integrity. However, measurement of the fracture stress of thin sheet specimens is a difficult task since high local stresses are usually obtained in bending tests where the deflection exceeds half the sheet thickness before fracture. In addition to conventional bi-axial bending tests a new method is presented to characterize the fracture stress of 100µm 6ScSZ electrolyte foils. Ring-on-ring tests are carried out with composite specimens where the thin electrolyte foil is glued to a steel substrate. The method permits to correlate the fracture stress with the stressed surface area and hence allows a prediction of failure stresses and probabilities for real size cells.

INTRODUCTION

Planar solid oxide fuel cells (SOFCs) are currently developed in two geometrical variants with different structural support and electrolyte thickness. Anode or anode substrate supported SOFC concepts allow to reduce the electrolyte thickness to films of 5-10 µm[1], whereas in the electrolyte supported variant mechanical robustness requires a much thicker membrane (100 – 200 µm). Since the cell resistance strongly depends on electrolyte thickness, the anode compared to the electrolyte supported SOFC concept has the attractive advantage of lower operation temperature at equal electrochemical performance.

However, considering mechanical aspects also some specific advantages of the electrolyte supported concept exist. The layer geometry of the SOFC composite with differences in the thermo-elastic behaviour of the individual materials generates residual stresses[2,3], which can cause problems of mechanical compatibility and integrity. The asymmetric variant with the anode support typically exhibits, if unconstrained, considerable cell curvature[4]. The electrolyte supported cell has a more or less symmetric layer structure and thus the residual stresses cause less curvature, i.e. the cells are essentially flat. Also the impact of anode re-oxidation strain on cell failure, a phenomenon well known for anode supported SOFCs[5], is considered to be less severe when a thicker electrolyte compared to the anode dominates the mechanical behaviour.

Electrolyte materials with high conductivity and high fracture stress in thin sheet geometry appear to be attractive to further promote the electrolyte supported SOFC concept. Substitution of the currently standard yttria stabilized zirconia (YSZ) electrolyte by a scandia stabilized zirconia with higher conductivity is one of the material options[6]. However, little information exists to date about the strength of thin sheet ScSZ electrolytes.

In general, the assembling, operation and residual stresses have to be balanced by the strength of a material. Although some studies of the mechanical properties of SOFC materials exist,[7,8] the measurement of the fracture stress of thin electrolyte foils is still a difficult task.[9] The present work focuses on the fracture stress characterization of ~ 110 µm thick 6ScSZ foils based on a new testing methodology for thin foils. Among the different mechanical testing methods for thin sheet specimens the ring-on ring bending is selected. Measurements are carried out with free-standing foils at room temperature and 800°C and in a novel approach with foils glued on metallic substrate. The obtained fracture stresses are statistically correlated with the stressed specimen surface area to derive fracture data relevant for larger SOFC cells.

TESTING METHODS AND ALTERNATIVE SPECIMEN DESIGN

The fracture stresses of ceramic materials typically show a considerable scatter due to the defect size distribution. This scatter can be assessed statistically via a characteristic fracture stress or modulus of rupture (MOR) and the Weibull modulus.[10] Assuming the same defect size distribution, the two statistical parameters allow a calculation of the stress for different specimen and component sizes. The probability to find a large defect increases with the specimen size, thus the MOR decreases. Depending on the defect type, the MOR could scale either with the tensile stressed surface area or with the stressed volume. Hence the characteristic fracture stress is a specific value for a discrete specimen size. For example, a calculation of the failure probability of an electrolyte in a SOFC stack requires consideration of the ratio of cell size to tested specimen size.

The ring-on-ring test (also called co-axial ring or double ring test) is conveniently used for bending plate-like brittle specimens. As an important advantage of the method defects from specimen shaping, e.g. cutting edges, do not influence the measured fracture stress. However, linear bending theory is only applicable, if the deflection of the specimen does not exceed ~ 50 % of its thickness during the test[11,12,13]. At large deflection the specimen stress has a maximum value above the loading ring, contrary to the small deflection situation of a constant tensile stress within the surface area defined by the loading ring. Thus fracture of thin sheet specimens will occur preferentially above the loading ring. Finite element analysis is often used to determine the fracture stress in such cases.[14]

Membrane tests, where vacuum or pressure rather than a mechanical load is applied for deflecting the specimen, generate in principle similar stress concentration, except that the effect is less pronounced. In addition to a relative complex experimental set up the fixation of the specimen and the sealing can lead to local changes in curvature and hence stress maxima[9,15].

The non-linear elastic behaviour should be excluded in tensile tests if preparation and gripping of the thin sheet specimens has no influence on the fracture results. However, flawless preparation of thin specimen edges is very difficult.[9]

Tensile stresses can also be induced by fixation of a thin sheet specimen to a material with different thermal expansion coefficient and subsequent heating or cooling. However, depending on the fracture stress and thermal expansion coefficients of the involved materials there might be a difference of a few 100 K necessary to cause fracture. Due to the statistical distribution of defects each specimen will fail at a different temperature and hence the characteristic fracture stresses can not be assessed from a particular temperature.

The problems associated with the determination of the fracture stress of thin sheet specimens led to the development of an alternative testing method. Modified specimens were prepared for ring-on-ring bending tests, which do not exhibit local stress maxima in the electrolyte. Essentially tests with electrolytes glued onto the surface of steel substrates and loaded under tensile bending stress were carried out. The additional substrate led to a larger total specimen thickness and reduced the stress localisation, since the composite deflection at fracture was less than half the total thickness. The ring-on-ring test with the composite specimens also allowed us to evaluate results from free-standing electrolyte specimens with respect to the stressed surface area. Otherwise this would not have been possible since the failure relevant surface area associated with large deflection can not be determined from a finite element based analysis.[14]

EXPERIMENTAL

Thin sheet electrolyte material of 6ScSZ (Nippon Shokubai) was supplied by BMW within the frame of a German SOFC project on Auxiliary Power Units (APUs). The thickness of the 50 mm × 50 mm sheets was ~ 110 μm. The foils were laser cut into smaller specimens of 24 mm × 24 mm. Part of the specimens were tested as cut, and part were glued to an INCONELL 617 substrate to obtain a composite specimen. The INCONELL 617 substrate had a thickness between 550 and 650 μm. A commercially available fast hardening adhesive (Pattex - Cyanoacrylat) was used to glue the electrolyte foils on the substrates. The free standing specimens were tested at room temperature (RT) and 800°C, the composite specimens only at RT.

All mechanical tests were carried out in ring-on-ring loading using an electro-mechanical universal testing machine (INSTRON 1362) with digital fast track control. The supporting ring had a radius of 9.5 mm, the loading ring a radius of 5 mm. The experiments were performed in load control mode with a constant rate of 50 N/min. The elastic modulus of the 6ScSZ foils was determined by depth sensing indentation (Fischerscope 100), yielding a value of 195 GPa.[16]

THEORY

The specimen stress in a ring-on-ring test with isotropic material can be calculated, as long as the deflection is less than half the specimen thickness, after[13]:

$$\sigma = \frac{3(1+\mu)}{2\pi}\left[\ln\frac{r_2}{r_1} + \frac{(1-\mu)}{(1+\mu)}\times\frac{r_2^2 - r_1^2}{2r_3^2}\right]\frac{F}{d^2} \tag{1}$$

where F is the force, d is the thickness, r_1, r_2 and r_3 are the radii of the load ring, supporting ring and (circular) specimen, respectively. Equation (1) also contains the Poisson ratio μ. In the case of quadratic specimens an equivalent average radius r_{3m} has to be calculated after[11,13]:

$$r_{3m} = \frac{(1+\sqrt{2})}{2}\times\frac{L}{2} = 0,60L \tag{2}$$

where L is the side length for quadratic specimens.

For large deflection the measured load-deflection data can be converted to stresses using calibration curves given in the literature[12,17]. Based on literature results[12,18] an analytical approximation relation, which explicitly considers the ratio of deflection to specimen thickness (d/t), can be proposed:

$$\sigma = Et^2/r_1^2\left(0.1216(d/t)^3 + 0.1643(d/t)^2 + 1.6987(d/t) + 0.02378\right) \tag{3}$$

In the case of a layered composite, like the geometry of the electrolyte foil glued on a substrate, the stress in the surface of the residual stress free electrolyte foil is[10]:

$$\sigma_{f,F} = \frac{F E_E}{4\pi\left(1-v_n^{2}\right)D^*}(t_n)\left[(1+v)\ln\left(\frac{r_2}{r_1}\right)+\frac{1-v}{2}\left(\frac{r_2^2-r_1^2}{r_3^2}\right)\right] \tag{4}$$

In Equation (4) the stiffness $D = E t^3/12\left(1-v^2\right)$ of an isotropic material was substituted by the composite stiffness D^*. Neglecting the adhesive because of the low modulus, D^* for a two layer composite (substrate-suffix S and electrolyte-suffix E) is[10]:

$$(D)^* = \frac{1}{12}\frac{d_E^4 E_E^2 + 2d_E d_S(2d_E^2 + 3d_E d_S + 2d_S^2)E_E E_S + d_S^4 E_S^2}{d_F E_F + d_S E_S} \tag{5}$$

In addition the position of the neutral axis changes in the composite to:

$$t_n = \frac{1}{2}\frac{d_E^2 E_E + 2d_E d_S E_S + d_S^2 E_S}{d_E E_E + d_S E_S} \tag{6}$$

The fracture stresses values derived from equations (3) and (4) were used for a statistical Weibull analysis based on DIN 51110 - 3 (identical to DIN ENV 843 - 5):

$$P_f = 1 - \exp\left[-\left(\frac{\sigma}{\sigma_0}\right)^m\right] \tag{7}$$

with the characteristic strength σ_0 and the Weibull parameter m. P_f is the failure probability.

RESULTS AND DISCUSSION

Typical load deflection curves obtained using ring-on-ring tests for the free standing 6ScSZ foils at RT are displayed in Fig 1. The general shape of the curves matches well, but there is considerable scatter in the fracture loads. In many cases a deflection of ~ 300 μm is reached at specimen failure, which corresponds to approximately three times the sheet thickness. Using the approximation in equation (3) and the Weibull statistics of equation (7), values of σ_0 = 780 ± 40 MPa and $m \sim 5 \pm 2$ are obtained.

The load deflection curves for free-standing foils at 800°C look similar, but failure occurs at a lower deflection (range of 1.5 to 1.9 times the thickness). Also the average fracture load is by a factor of 2.4 smaller than the value obtained at RT. In general at 800°C the force at a discrete deflection is about 10 % lower than at RT, implying a 10 % lower stiffness. Again, the experimental data can be evaluated using the approximation in equation (3). The resulting values are σ_0 = 430 ± 21 MPa and m = 6 ± 2. The Weibull modulus agrees within the limits of uncertainty with the value obtained at RT. However, the characteristic fracture stress is by a factor of ~ 1.8 lower than that at RT.

Load - deflection curves for the composite specimens with the 6ScSZ electrolyte foils glued to INCONEL 617 substrate are shown in Fig. 2. After initial setting the slope of the curves is essentially linear and predominantly governed by the mechanical behavior of the substrate.

Indeed, the two different slope regimes in Fig. 2 reflect the two different substrate thicknesses of 550 and 650 µm used in the tests. The slope of the curves is in agreement with prediction based on equation (5). Although the deformation of the substrate dominates the fracture of the 6ScSZ electrolyte foils can be clearly recognized as kink in the curves. From the associated loads a characteristic fracture stress of $\sigma_0 = 230 \pm 14$ MPa (m = 4.1 ± 1.2) is determined.

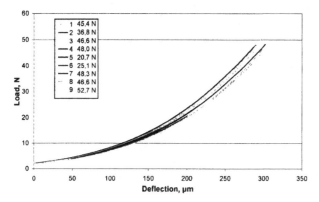

Fig. 1: Load deflection curves of free standing 6ScSZ electrolyte specimens.

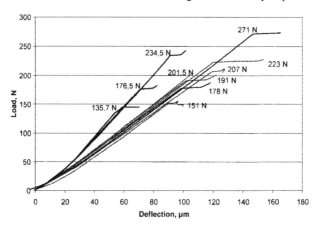

Fig. 2: Load deflection curves of 6ScSZ foils glued on metallic substrate.

A summarizing Weibull plot of the data from the three sets of experiments is shown in Fig. 3. Since the *m* values agree within the limits of uncertainty for the free-standing and glued electrolytes further statistical considerations on the relationship of characteristic fracture stress and stressed specimen size can be made.

However, it has also to be emphasized that in general the number of measurements performed in the present study is too low to obtain an accurate Weibull modulus. An increase in accuracy can only be reached by an increase in the number and/or size of tested specimens. To obtain a higher accuracy for the further evaluation of the present data we assume that all sets of the sheet specimens possess the same Weibull modulus. This assumption appears to be reasonable since all specimens were cut from the same 6ScSZ batch. Hence the fracture stresses of each set were normalized by the individual characteristic fracture stress. The normalized data of all sets were then put together and statistically analyzed. Thus a single Weibull modulus of m = 5.3 ± 1.0, now based on 27 specimens, is obtained.

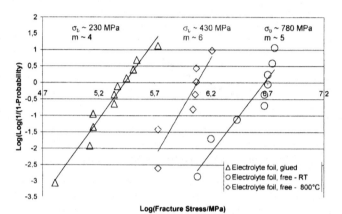

Fig. 3: Weibull plot of the fracture stresses for glued and free 6ScSZ electrolyte foils.

Assuming constant m the statistical relationship between the characteristic fracture stress $(\sigma_{0,V1}, \sigma_{0,V2})$ and the stressed surface area (A_1, A_2) of two testing geometries

$$\frac{\sigma_{0,V1}}{\sigma_{0,V2}} = \left(\frac{A_1}{A_2}\right)^{-1/m} \qquad (8)$$

can now be further elaborated with respect to the particular results obtained in the ring-on-ring tests for the free-standing and glued 6ScSZ foils.

The surface area under tension in the tests with glued foils was 78.5 mm² (area within the loading ring). Using the average Weibull modulus of $m = 5.3$ the stressed surface area of the free-standing electrolyte foil is calculated from equation (8) as $A = 0.12$ mm². The expected maximum stress at the location of the loading ring suggests that the stressed surface area correlates with the diameter of the loading ring ($D \sim 10$ mm, with $A = \pi D w$) yielding a width of the stressed concentric surface area of $w \sim 0.004$ mm.

As a first approximation this value might also be used for the stressed surface area at high temperature. However, the free-standing electrolyte foils tested at high temperatures fractured at a lower ratio of deflection to thickness, hence the stressed surface area was also larger. Using the radial stress profiles presented in[12] and assuming that the width of the stressed surface area can be associated with 90 % of the maximum stress, a ~ 20 % increase can be expected.

Based on these geometrical considerations the experimentally determined characteristic fracture stresses are plotted in Fig. 4 as a function of stressed surface area and connected by the theoretical curves of equation (8). The characteristic fracture stress of larger electrolyte sheets used in a SOFC stack can now be predicted. Accordingly the characteristic strength of a 100 mm × 100 mm cell of the 6ScSZ electrolyte material should be 90 ± 40 MPa and 50 ± 20 MPa at RT and 800°C, respectively. This decrease of ~ 44 % is similar to the decreases of ~ 61 % and ~ 33 %, respectively, reported for 3YSZ and 8YSZ electrolyte foils.[19]

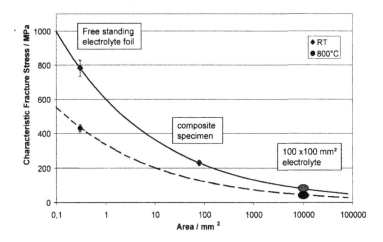

Fig. 4: Characteristic fracture stresses as a function of the stressed surface area. Cell size data of electrolyte obtained using equation (8).

A further reduction of the maximum applicable stress on 6ScSZ electrolytes needs to be taken into account for lower failure probability. If the failure of only one cell in 1000 is permitted a stress limit of 25^{+20}_{-14} MPa at RT and 14^{+11}_{-8} MPa at 800°C should not be exceeded.

CONCLUSIONS

The characteristic fracture stress and Weibull modulus have been determined for specimens from thin 6ScSZ electrolyte sheets. The results elucidate the dependence of the characteristic fracture stress on the stressed surface area and temperature. The Weibull modulus was determined to be constant with an averaging value of $m = 5.3$. Considering the large surface area of typical SOFC cells (e.g. 100 mm × 100 mm) in stacks a characteristic strength of ~ 90 MPa and ~ 50 MPa was estimated for RT and 800°C, respectively. Assuming a failure probability of 10^{-3} the maximum stress that an electrolyte foil of this size can withstand is only ~ 25 MPa at RT and ~ 14 MPa at 800°C, respectively. Stress limits corresponding to tolerable failure probabilities should be considered in SOFC stack design.

ACKNOWLEDGEMENTS

The authors are grateful to BMW for continuous support and the funding of the present work within the frame of the National German SOFC-APU Project "ENSA".

REFERENCES

[1] T.W. Napporn, X.J.-Bedard, F. Morin, and M. Meunier, "Operating Conditions of a Single-Chamber SOFC," *J. Electrochem. Soc.*, **151**, A2088–A2094 (2004).

[2] J. Malzbender, R.W. Steinbrech, and L. Singheiser, "Failure Probability of Solid Oxide Fuel Cells" *Ceram. Eng. Sci. Proc.* **26/4**, 293-297 (2005).

[3] W. Fischer, J. Malzbender, G. Blass, and R.W. Steinbrech, "Residual stresses in planar solid oxide fuel cells," *J. Power Sources*, **73**, 150-153 (2005)

[4] J. Malzbender, T. Wakui, and R.W. Steinbrech, "Deflection of Planar Solid Oxide Fuel Cells During Sealing and Cooling of Stacks," *Proc. 6th Euro. SOFC Forum*, **1**, 329 (2004).

[5] J. Malzbender, E. Wessel, R.W. Steinbrech, and L. Singheiser, "Reduction and Re-oxidation of Anodes for Solid Oxide Fuel Cells," *Solid State Ionics*, **176**, 2201-2203 (2005).

[6] D. Stöver, H.-P. Buchkremer, F. Tietz, "Material Problems of the High Temperature Fuel Cell" in German, *VDI-Berichte* Nr. 1680, (2002) 209-266

[7] J. Malzbender, R.W. Steinbrech, and L. Singheiser, "Determination of the Interfacial Fracture Energies of Cathodes and Glass/Ceramic Sealants in a Planar Solid Oxide Fuel Cell Design", *J. Mater. Res.*, **18**, 929 (2003).

[8] J. Malzbender, R.W. Steinbrech, and L. Singheiser, "Strength of Planar Cells for SOFC Application", *Proc. SOFC VIII*, 2003, 1463.

[9] K. An, H.G. Halverson, K.L. Reifsnider, S.W. Case, and M.H. McCord, "Comparison of methodologies for determination of fracture strength of 8 mol % yttria - stabilized zirconia electrolyte materials", *J. Fuel Cell Sci. Technol.*, **5**, 99-103 (2005).

[10] J. Malzbender, and R.W. Steinbrech, "Mechanical properties of coated materials and multi-layered composites determined using bending methods", *Surf. Coat. Technol.*, **176**, 165-172 (2004).

[11] R.W. Schmitt, K. Blank, and G. Schönbrunn, "Experimentelle Spannungsanalyse zum Doppelringverfahren, *Sprechsaal*, **116**, 397-405 (1983).

[12] R. Kao, N. Perrone, and W. Capps, "Large Deflection Solution of the Coaxial-Ring-Circular-Glass-Plate Flexure Problem, *J. Am. Cer. Soc.*, **54**, 566-571 (1971).

[13] Deutsche Norm, "Bestimmung der Biegefestigkeit von Glas", DIN 1288-1, 2000.

[14] A. Selcuk and A. Atkinson, "Strength and Toughness of Tape-Cast Yttria-Stabilized Zirconia", *J. Am. Cer. Soc.*, **83**, 2029-2035 (2000).

[15] D. Stolten, E. Monreal, C. Seeselberg, „Testing of Membranes via Biaxial Strength and Proof Testing Device Regarding Nonlinear Structural Behaviour", in Ceramics: Charting the Future, P. Vincenzini (Editor), Techna Srl. 1995, 2613 - 2620

[16] D. Basu, C. Funke, and R.W. Steinbrech, "Effect of heat treatment on elastic properties of separated thermal barrier coatings," *J. Mater. Res.*, **14**, 4643-4650 (1999).

[17] R. Morrel, "Biaxial Strength Testing of Ceramic Materials", *NPL Report*, 1998.

[18] R. Quinn, "Biaxial Flexure Strength Testing of Ceramic Materials," *NPL, Measurement Good Practice Guide No.12*, 1998.

[19] A. Atkinson, and A. Selçuk, "Mechanical behaviour of ceramic oxygen ion-conducting membranes," *Solid State Ionics*, **134**, 59-66 (2000).

FAILURE MODES OF THIN SUPPORTED MEMBRANES

P. V. Hendriksen, J. R. Høgsberg, A. M. Kjeldsen, B. F. Sørensen
Materials Research Department, Risø National Laboratory, Denmark
Frederiksborgvej 399, P.O. Box 49
DK-4000 Roskilde, Denmark

H. G. Pedersen
Haldor Topsøe A/S
Nymøllevej 55
DK-2800 Lyngby, Denmark

ABSTRACT
    Four different failure modes relevant to tubular supported membranes (thin dense films on a thick porous support) were analyzed. The failure modes were: 1) Structural collapse due to external pressure 2) burst of locally unsupported areas, 3) formation of surface cracks in the membrane due to TEC-mismatches, and finally 4) delamination between membrane and support due to expansion of the membrane on use. Design criteria to minimize risk of failure by the four different modes are discussed. The theoretical analysis of the two last failure modes is compared to failures observed on actual components.

INTRODUCTION
    Materials exhibiting mixed ionic/electronic conductivity (MIEC's) may be used as dense membranes for gas separation purposes, for instance for oxygen production or for supply of oxygen in production of syngas by a partial oxidation route. The achievable flux through the membrane is partly limited by the resistance to the transport of the species in the bulk, and hence to reduce the cost of the technology such membranes should be as thin as possible, and an obvious reactor design is thus an ultra thin membrane on a supporting medium. The role of the support is to ensure sufficient rigidity of the component to allow handling, and to carry most of the load if the membrane reactor is subjected to pressure differences. Units with such architecture has been pursued for both solid oxide fuel cells[1, 2] and membrane reactors for syngas production. The units may be planar or tubular. Key advantages of the former being ease of manufacture and generally compact design for a given rated output. Tubular units also have a number of advantages, key ones being: 1) Ease of manufacture by extrusion. 2) Robustness to large temperature gradients in the gas flow direction (axial direction). 3) Simplicity of sealing, and finally 4) robustness to pressure differences between the two sides. Large development programs both within the field of SOFC and in the field of membrane reactors have been based on tubular units.
    Whereas the central part of the device must be gas tight, in the case of SOFC to ensure high efficiency (and long lifetime), and in the case of the membrane to ensure selectivity, the support is porous to allow gas access to the membrane/electrolyte. Classical materials for SOFC electrolytes are Yttria stabilized Zirconia , Gd-doped Ceria or La-Sr-Gallates, and candidate materials for the membrane reactors could be perovskites belonging to the class $La_{1-x}Sr_xFe_{1-y}Co_yO_3$[3,4]. Common to both devices is thus, that they rely on ceramic com-

ponents, but yet have to operate reliably, under large pressure- as well as temperature-gradients, as the electrolyte/membrane separates a fuel from an oxidant. An analysis of possible failure modes in such devices can be of relevance both in rationalising the mechanisms behind observed failures in experimental systems, but also in formulating design criteria and material requirements to allow fail safe operation.

Stresses may build up in such bi- or multi-layered devices already in the final stages of manufacturing and/or during use. If the manufacturing involves a firing step and the layers have different TEC-values, stresses will build up during cooling. During use the component will also be exposed to various stress generating loadings. Different pressure on the two sides was already mentioned, but also if a non-homogeneous temperature distribution develops, or if one of the material changes volume (e.g. due to a chemical reaction), stresses will build up. Whether detrimental or not, depends on the magnitude compared to the material strength.

The present paper discusses mechanical failure modes of thin supported tubular membranes and the derivation criteria for fail-safe design of such. The following failure modes were analyzed, and are treated in turn in the following:

1. Collapse of support tube due to external over pressure
2. Collapse of membrane over a pore due to pressure difference over the membrane.
3. Development of surface cracks in the membrane due to TEC mismatch.
4. Buckling driven de-lamination between membrane and support due to expansion on reduction of the membrane material.

Whereas case 3 is most critical at room temperature after component manufacture, cases 1, 2 and 4 are relevant under operation. Loads associated with interactions between tube and fixture are beyond the scope of the present paper. Part of the work was carried out within an EU-funded programme aiming to develop membrane reactors (CERAM-GAS)[5] to identify rationally based material requirements and to provide guidelines for focussing the effort. Much of the analysis is based on well established theory for fracture mechanics of thin planar films[6]. However, the analysis of case 4 covers also the impacts of an external pressure on this risk failure due to buckling which is to our knowledge not available in literature.

ANALYSIS AND DISCUSSION OF FAILURE MODES
Collapse of the tube.

From a balance of plant point of view it is highly advantageous for a syngas membrane to be able to sustain an overpressure on the syngas side. Similarly, for a fuel cell and an oxygen production membrane of MIEC-type, it will be advantageous/necessary, respectively, to operate with an overpressure on the air side. This gives rise to certain design requirements. If the high pressure side is on the inside of the tube and the dense membrane lies outside the support the distribution of the load over both layers relies on the interface between the two to be sufficiently strong, whereas the interface strength is unimportant if the high pressure is on the outside. Moreover, in the former case the tube will be in tension, whereas it is in compression in the latter case. Ceramics are generally much stronger in compression than in tension and hence the most benign requirements regarding component quality, layer thicknesses and material strengths are obtained if the high pressure side is on the outside of the tube.

A tubular supported membrane of thickness, h, support thickness H and diameter $D = 2R$ (See Fig. 1 B), exposed to an external pressure q will be compressed to a degree given by:

$$\varepsilon = -\frac{qR}{E_m h + E_s H} \quad , \quad \sigma_{mem,q} = E_m \varepsilon \quad , \quad \sigma_{sup,q} = E_s \varepsilon \qquad (1, 2, 3)$$

Where σ is the radial stress and ε the radial strain., E is Young's modulus and subscripts indicate membrane or support. (Eq. 1 follows trivially from requiring force equilibrium on half the tube. Eq. 2 is Hooke's law). For a given design (h, H, R fixed) the maximum tolerable pressure will be given by the material strength ($\sigma_{mem,max}$) in compression:

$$q_{,r} \leq \frac{H}{R}\frac{E_s}{E_m}\sigma_{mem,max} \qquad (4)$$

Eq. 4. can also be used to estimate the necessary support thickness for a given strength and a required pressure. However, the tolerable pressure difference may also be limited by a tube stability criterion. If the cylinder is long (L>>R) and the membrane layer thin (h<<H), the critical pressure, q_{cr}, at which the cylinder will collapse, is given by[7]:

$$q_{cr} \leq \frac{E_s}{4(1-v_s^2)}\left(\frac{H}{R}\right)^3 \qquad (5)$$

As the tolerable pressure scales with $(H/R)^3$ from the stability criterion (Eq. 5) and only with H/R for the strength criterion (Eq. 4), it will be determined by Eq. 5 for thick walled tubes (H>H*) and by Eq. 4 for thin walled (H<H*) tubes, where:

$$H^* = \sqrt{\frac{4\sigma_{mem\;max}(1-v_2^2)}{E_m}} \qquad (6)$$

The variation in maximum tolerable pressure with tube radius is illustrated in Figure 1 for a range of different values of the support thickness.

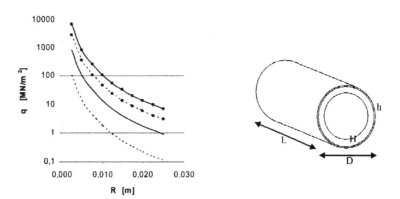

Figure 1. Calculated maximum tolerable external load (q_{cr}) on a long tube of radius R, for different wall thickness: H=0,5mm - - - , H=1mm —— , H=1,5mm • - • , H= 2 mm •—•. E_s=50 GPa, v=0,3.

Assumed values for the material constants were E_s=50 GPa, v=0,3. (The specified value is an estimate taking into account, that the support is highly porous, which has a strong impact on the modulus[8]). Considering as an example a tube of 5 cm diameter and a required external pressure of 30 atm., the necessary support thickness deduced from the stability criterion is only 1,5 mm. Hence, the stability criterion for the above values of pressure and tube diameter does not seem to impose very stringent demands on wall thickness, and practical requirements for suitable handling strength or processes requirements are likely to require larger thicknesses. It should be noted, that in practice one would have to include certain safety factors when basing design on the above criteria; in the case of the strength criterion to take into account the spread in strengths typically observed for ceramic components due to the inherent distribution in flaw sizes, and in the case of the stability criterion (Eq. 5) to take into account geometrical imperfections like small deviation from a circular cross section which will reduce the critical load.

Collapse of the membrane over a pore or a large flaw.
As discussed in the previous section the support will relieve the stresses in the membrane due to an external pressure. However, the membrane may be left locally unsupported over large pores or other types of flaws in the interface between support and membrane. As a simple model of this situation we shall consider a thin circular plate of radius=a submitted to a uniform lateral load (q) and a uniform compression (N). Hence, we neglect in the model the local curvature and treat the unsupported membrane over the flaw as a flat plate. This plate problem is treated in standard textbooks[9]. Differences in the compression forces around the periphery of the plate due to the differences in the stresses in the axial and circumferential directions in the membrane are neglected. The maximum tensile stress in the plate for the present case will occur at $r=a$ (at the plate periphery). It is given by:

$$\sigma_{mem,max} = fak \frac{6M_{max}}{h^2} - \frac{N}{h} \quad , \quad M_{max} = \frac{1}{8}qa^2 \qquad (7, 8)$$

Where N is the compressive load (per unit length) acting along the outer periphery of the plate. M_{max} is the maximum moment of a plate subjected only to the lateral load, and fak is a "correction factor" taking into account the effect of the compression on the local curvature. It is given by:

$$fak = \frac{1 - 0.473\alpha}{1 - \alpha} \quad , \text{where}: \quad \alpha = \frac{Na^2}{14.68D} \text{ and } D = \frac{E_1 h^3}{12(1 - v^2)} \qquad (9, 10, 11)$$

Here, D is the flexural rigidity of the plate. In the present case N is simply the ring stress caused by the external pressure, $\sigma_{mem,q}$ as given by Eqs. (1, 2, 3 multiplied by the membrane thickness ($N = \sigma_{mem,q} h$). If the external pressure gives rise only to small strains in the cylinder ($\varepsilon < 0.1(h/a)^2$), $\alpha = \varepsilon (a/h)^2$ will be small and fak close to unity, and Eqs. (7, 8 thus simplify to:

$$\sigma_{mem,max} = \frac{3}{4}q\left\{\frac{a}{h}\right\}^2 - \frac{E_m q R}{E_m h + E_s H} \qquad (12)$$

The maximum tensile stress in the membrane over a circular flaw, calculated according to Eq. 12, is shown in Fig. 2 for a range of geometries. The curves in Fig. 2 clearly reflect that the stress is a sum of a compression, which is proportional to qR, and a tensile term due to the bending of the unsupported section of the membrane inflicted by the external pressure. The latter is also proportional to pressure, but scales with the flaw size squared. Requiring as a design criterion that the unsupported area should always be in compression the criterion between flaw-size, membrane thickness, tube size and support thickness becomes:

$$a \leq \sqrt{\frac{4E_m R}{3E_s H}}\, h \tag{13}$$

i.e. for a given flaw size, the risk of failure by this mode can be minimized by increasing the membrane thickness or by increasing the ratio between tube size and support thickness (as this increases the compressive ring stress in the membrane). The latter is of course only a sound strategy as long as one does not violate the criterion given by Eq. 4.

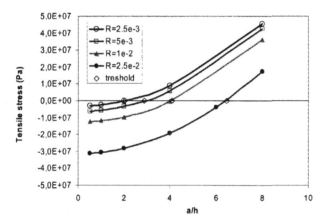

Figure 2. Maximum stress in a locally unsupported membrane plotted as a function of the flaw size for a range of tube radii (R). The pressure (q) was assumed to be 10 atm, the support thickness (H) to be 2 mm. E_s=50 GPa, h=10 μm and E_m= 125 GPa.

Also shown in Figure 2 is the $(a/h)_{crit}$ as deduced from Eq. 13 (rhombes). Evidently, the approximation fak=1 is justified for this range of parameters (The curves calculated from Eqs. 7-11 cross the axis close to the points deduced from Eq. 13). Considering as an example again a tube with R=2.5 cm, a membrane thickness of 10 μm and support thickness of 2 mm (elastic constants as stated in Fig. 1 and 2) the unsupported piece of membrane will stay in compression for flaw sizes up to a/h~6.5 or 65 μm. This is thus the tolerable size of imperfections introduced during the manufacture (measured on the length scale of the membrane thickness) before local bending of the membrane over an unsupported area introduces tensile stresses in the surface of the membrane.

Development of surface cracks due to TEC mismatches

If the difference in the thermal expansion coefficient of the membrane material and the support is too large the membrane may fail at room temperature after manufacture due to the stresses building up during cooling. Thin layers on thick supports are known in a number of technological devices and treatments of fracture mechanics in such systems can be found in literature. Evans and Hutchinson[10] have analyzed different failure modes in such systems and have shown that fail safe design criteria can be deduced from a two dimensional analysis (planar case) requiring that the energy release on failure (G) is less than the a critical material value G_c (reflecting the fracture toughness). The energy possibly released on failure is given by[4]:

$$G = \frac{h_m \sigma^2}{E_m \Omega} \leq G_c \quad , \tag{14}$$

where Ω is a cracking number (a non-dimensional constant of order unity, with small differences between different failure modes), h_m is the membrane thickness, E_m is Young's modulus of the membrane material and σ is the misfit stress in the membrane, which is simply:

$$\sigma = \frac{E_m}{(1-v)}\varepsilon \quad \text{where} \quad \varepsilon = \Delta\alpha\Delta T. \tag{15}$$

v is the Poisson's ratio (of the membrane material), and ε is the misfit strain which is proportional to the difference in TEC between membrane and support material times the temperature difference between the stress free state and the evaluated state (here manufacturing temperature and room temperature). Hence, when G_c is known one may use Eq. 14 to deduce a fail safe maximum layer thickness for a given misfit stress (given TEC-mismatch) or a maximum tolerable TEC-mismatch for a desired membrane thickness.

The cracking number depends on the failure mode. For the three most relevant failure modes the cracking numbers (assuming no elastic mismatch between support and film) are: 1) a surface crack, $\Omega \sim 0.25$, 2) "Channelling", $\Omega \sim 0.5$, and 3) debonding, $\Omega = 2$[4]. By "chanelling" is meant formation of a network of surface cracks. There are a number of important things to learn from the form of Eq. 14: I) The risk of failure for all the modes scales with layer thickness, II) The risk of failure for all the modes scales with misfit-stress squared, and III) compression is preferable to tension in the membrane as the cracking number for debonding is higher than for surface cracks, which means, that under the assumption that G_c is the same for the two modes of failure, preferably the TEC of the membrane material should be lower than that of the support. However, this may not be easy to achieve in practice. Candidate membrane materials belonging to the general class $La_{1-x}Sr_xFe_{1-y}Co_yO_3$ have TEC values[11] in the range from 12 to $20 \cdot 10^{-6}$ K^{-1}(100 - 900 $^\circ$C) strongly depending on the Sr and Co-content (increasing with both), whereas cheap strong ceramics which could be candidates for the support typically have lower TEC values. Alumina and Zirconia for instance have TEC values in the range from 8 - $11 \cdot 10^{-6}$ K^{-1} in the same temperature interval.

It follows directly from Eqs. 14, 15 that in principle large misfit stresses (large TEC-mismatches) are tolerable if one decreases layer thickness accordingly. In Table 1 design limitations in terms of allowable thicknesses for a given TEC mismatch, as well as tolerable TEC mismatch for a given membrane thickness, are assessed for a membrane relevant set of parameters. H. Lein[12] has characterised the mechanical properties of several $La_{1-x}Sr_xFe_{1-}$.

$_y$Co$_y$O$_3$-perovskites and reports E-moduli in the range from 115 GPa to 180 GP, and fracture toughnesses in the range from 1.2 to 1.8 MPa m$^{1/2}$, corresponding to critical energy release rates on the order of 20 J/m^2 ($G_c \sim K_c^2/E$).

Table 1. Tolerable TEC mismatches and allowable thickness needed to ensure membrane integrity as deduced from the criterion in Eq. 14.

Parameters	Criterion
E=125 GPa, v=0.3 ΔT=1000 Gc=20 J/m^2 h=10 μm Ω = 0.25 (surface crack)	$\Delta\alpha = \left[\Omega \dfrac{(1-v)^2}{h_m E_m} \dfrac{Gc}{(\Delta T)^2} \right]^{1/2} = 1 \cdot 10^{-6}$ K^{-1}
As above but: $\Delta\alpha$=5·10^{-6} K^{-1}	$hc = \Omega \dfrac{(1-v)^2}{E_m} \dfrac{Gc}{(\Delta\alpha\Delta T)^2} = 1.5\mu m$

The temperature difference between processing- and room-temperature may be larger than 1000 °C, but at these temperatures creep rates are very high for the analysed type of materials which may effectively relax the stresses.

There is of course some uncertainty on the above used parameter values, however the deduced numbers in the above examples show that this mode of failure inflicts very serious limitations in the number of possible material combinations as well as harsh requirements to the tube geometry (membrane thickness). The TEC mismatch between alumina and typical La$_{1-x}$Sr$_x$Fe$_{1-y}$Co$_y$O$_3$-perovskites, which is in a very conservative estimate 5·10^{-6} K^{-1}, calls for very thin layers (below 1.5 μm) which would certainly be difficult to make in the desired quality using cheap ceramic manufacturing routes. For thicknesses in the range from 10 to 20 μm, which can be readily made[13], the required degree of TEC-matching ($\sim$1·10^{-6}) strongly limits the degrees of freedom in tuning the membrane composition to optimise other relevant properties like ionic conductivity or catalytic activity.

Experimental experiences.

Within the framework of the "CERAM-GAS" project experimental bi-layer membranes with several different material combinations were manufactured. SEM micrographs of a subset of the investigated material combinations are shown in Figure 3. The TEC mismatch between membrane material decreases from upper left to lower right in the figure (see caption). Evidently, surface cracks are observed for three (A, B, C) out of the four depicted material combinations (see the zoom-in insert on 3 B). From the limited opening of the cracks, the observed pattern (Fig. 3A) as well as the observed trans-granular tracks (Fig. 3 B) it seems evident that the cracks do not originate from imperfect application of the layer, but are due to the above described failure mode related to the TEC differences. For the successful combination in the lower right corner the TEC mismatch was $\sim$0.1·10^{-6} K^{-1}. Note, that cracks are also observed in the case where the TEC differed by about 1.6·10^{-6} K^{-1}, which is close, but yet above, the formulated criterion. Many more material combinations than the four depicted in Fig.3 were investigated in the project. All successful combinations were consistent with the criterion in Table 1. (The uncertainty on the TEC measurement is ~0.1 10^{-6} K^{-1})

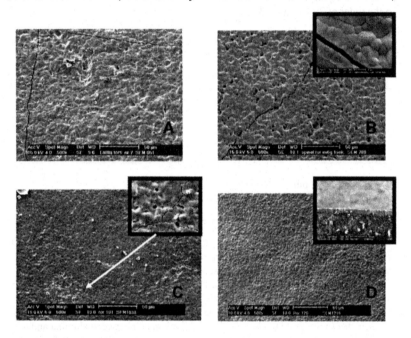

Figure 3 Top views (SEM) of different experimental supported membranes. The pictures are from different combinations of membrane material and support. A) $\Delta\alpha \sim 5 \cdot 10^{-6}$ K^{-1}, B) $\Delta\alpha \sim 4 \cdot 10^{-6}$ K^{-1}, C) $\Delta\alpha \sim 1.6 \cdot 10^{-6}$ K^{-1} and D) $\Delta\alpha \sim 0.1 \cdot 10^{-6}$ K^{-1} (insert shows a cross section).

Material expansion on reduction

It is a characteristic of materials showing high mixed conductivity (both oxide ions and electrons) that they are non-stoichiometric in oxygen, i.e. there is a considerable concentration of vacancies in the lattice. The concentration of vacancies depends on the surrounding oxygen activity. When such materials are used as membranes, where they are exposed to different oxygen activities on the two sides, a vacancy concentration profile will be established inside the material, the detailed shape of which depends on the membrane thickness as well as the rates of the oxygen exchange processes on the two surfaces[3]. In general there is a volume change associated with a change in vacancy concentration in the material – the materials expand on reduction[3,14,15,16] This is illustrated in Fig. 4, where measured expansions on reduction of a well characterized perovskite as well as for two membrane materials investigated in the CERAM GAS project are plotted versus the oxygen non-stoichiometry. Evidently, the strains associated with these stoichiometry changes are quite large (~0.3 – 0.4 %), and they are well known to lead to a number of mechanical problems for the use of such materials[4,15].

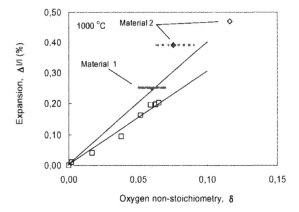

Figure 4. Measured expansions for $La_{0.8}Sr_{0.2}Cr_{0.87}Fe_{0.1}V_{0.03}O_{3-\delta}$ (squares)[16] and two materials analyzed within the CERAM-GAS project plotted versus the oxygen non stoichiometry in the materials. For two of the three measurements the oxygen activity was not measured accurately giving rise to uncertainty on the δ-values, as indicated with the horizontal bars. The upper straight line was deduced from the relation between δ and $\Delta l/l$ for $La_{0.8}Sr_{0.4}Fe_{0.8}Co_{0.2}O_{3-\delta}$[14,3]

For the present application where the membrane material is in the form of a thin layer on a thick support, they will give rise to large compressive stresses in the membrane, which may if the exceed the compressive strength of the material lead to failure. Moreover, they may result in a detachment between the membrane and the support as the stresses can be relaxed if the membrane buckles away from the support (whereby it increases its length decreasing the stress level). The situation is illustrated in Fig 5. In the following we shall analyze the implications of the buckling mode of failure (The risk of failure due to excessive compressive stresses, or other than buckling driven failures in compression[17], are not addressed further in this paper, but may indeed be relevant for this type of components).

Model of buckling driven de-lamination

Buckling driven delamination between a thick support and a thin surface film in compression has been analyzed in literature. (see the paper by Hutchinson and Suo and references therein) for the case of planar geometries. Starting from this analysis J. Høghsberg[18] has generalized the treatment to take into account the cylindrical shape as well as the effects of external overpressure. Buckling driven delamination on tubular geometries has also been treated by Storåkers et al.[19,20] and J.W. Hutchinson[21]. The situation is illustrated in Fig. 5.

Describing a segment of the membrane as a beam of constant stiffness, the differential equation describing the shape of the membrane becomes:

$$w''''+k^2(w+w_0)''-\frac{q}{EI}=0, \quad k^2=\frac{\sigma h_m}{E_m I} \qquad (16)$$

Where $w(y)$ is the vertical displacement, k is a parameter and $()'$ signifies the partial derivative with respect to y. $w_0(y)$ is a curve describing the initial shape. E_m is Young's modulus and I the moment of inertia. σ is the normal stress in the beam.

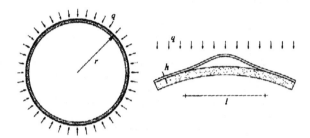

Fig. 5 Buckling of non-planar membrane subjected to external pressure (q) and additional in plane compression due to expansion of the membrane material.

In the considered case, the normal stress in the beam is the sum of the ring stresses, σ_r due to q (Eq. (1, 2, 3) and the stresses caused by the expansion of the material, σ_{exp}:

$$\sigma_t = \sigma_{exp} + \sigma_q \approx \frac{E_m}{1-\nu}\varepsilon_{exp} + \frac{qR}{H}\frac{E_m}{E_s} \tag{17}$$

We shall approximate the initial circular shape with a sinus function over the angular segment $l/2\pi R$:

$$w_0 = a_0 \sin(\frac{\pi y}{l}), \tag{18}$$

which is a fair approximation for $l/R<0,5$, as this greatly simplifies the solution of Eq.16.
Under the boundary conditions $w(0)=w'(0)=w(l)=w'(l)=0$ (clamped ends) the solution to Eq.16 can be written:

$$w(y) = \left(\frac{\pi}{lk}\alpha - \frac{l}{k}\beta\right)(C(1-\cos(ky)) - \sin(ky)) + \alpha\sin(\frac{\pi}{l}y) - \beta y(1-y), \tag{19}$$

$$C = \frac{kl\cos(kl) - 2\sin(kl) + kl}{2(\cos(kl) - 1) + kl\sin(kl)}, \quad \text{and} \quad \alpha = \frac{a_0}{\left(\pi/lk\right)^2 - 1} \quad \text{and} \quad \beta = \frac{q}{2EIk^2}. \tag{20}$$

The energy release rate, G, may be calculated from the expression:

$$G = \frac{6l}{\overline{E}h_m^3}\left(M_l^2 + \frac{h^4}{12}\Delta\sigma^2 \right),$$ (21)

where $\Delta\sigma$ is the reduction of the normal stress in the beam due to the buckling. The Moment M_l is related in to the second derivative of the displacement and the stress reduction to the elongation introduced on the buckling:

$$M_l = M(l) = -\overline{E}Iw''(l) \quad \text{and} \quad \Delta\sigma = \frac{\overline{E}_m}{2l}\int_0^l w'dy,$$ (22)

Where $\overline{E} = E/(1-\nu^2)$. Finally, note that the stress in the layer "during" buckling can be expressed:

$$\sigma = \sigma_l - \Delta\sigma$$ (23)

Equations 19 and 21 define (via Equations 17, 20, 22, and 23) a set of equations in the variables w and σ. The system may be readily solved numerically in an iterative manner and the energy release rate of the buckling calculated. A set of model results are presented in Fig. 6, illustrating the impact of parameter variations around a base case defined in the figure caption.

A number of points, relevant for materials development and design of supported membranes may readily be deduced form the results in Fig. 6:

- A certain critical membrane thickness exist in all cases, where the risk of failure is maximal. Increasing the thickness relative to this reduces the energy release rate (buckling becomes less effective for releasing the stresses) as does a decrease in thickness (less energy in the system all together). The most critical thickness is slightly thicker for the cylindrical membranes than for the planar case. The results give rather direct guidelines on how to design the membrane reactor for minimizing the risk of failure.

- The curvature of the membrane increases the risk of failure relative to the planar case i.e. G is larger for the tubular case than for the planar for a given set of parameters (Fig. 6 A).

- G_{max} (the risk of failure) increases with the lattice expansion induced strain squared, and for the base case of 0.3 %, which is not an unrealistic value (c.f. Fig. 4) G reaches a value of ~10 J/m². Whereas for tough perovskite materials one may expect higher values (see section on TEC-mismatches) Lein also reported toughness values on the order of 1 MPam$^{1/2}$, for some of the investigated perovskites, which corresponds to ~ 8 J/m² (assuming $G_c = K_c^2/E$, mode I crack opening). Moreover, for this mode of failure it is the toughness of the interface that matters, and this may well be lower than for any of the two materials of the bilayer[22]. Thus, this mode of failure may indeed be critical and put limitations to both which materials are applicable and how the reactor can be designed. Note, that relaxation by creep, which may be significant at high temperature, will alleviate the situation and thus the material requirements deduced from this mode of failure, as will taking into account tensile residual stresses in the film at the operating temperature.

- G is also sensitive to the flaw size (see Fig. 6 C). It increases with the flaw size. In the investigated range of parameters G_{max} scales with the flaw size in a parabolic manner, and hence controlling the flaw size is effective for minimizing the risk of failure. Also the most critical thickness increases with flaw size. With "small" flaws (l=200 μm) the base case of a 10 μm thick membrane exhibits a very small G value. (Note, that for this mode of failure poor local adhesion between membrane and support is "a flaw", whereas for the local burst over areas where the membrane is unsupported "a flaw" is a missing part of material in the support).
- Finally, the effect of an external over pressure is illustrated in Fig. 6 D. Evidently, a slight overpressure is very effective for decreasing G and thus the risk of failure by this mode.

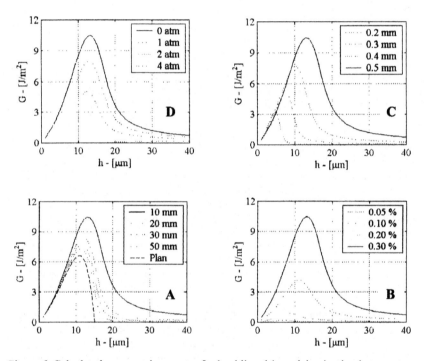

Figure 6. Calculated energy release rates for buckling driven delamination between membrane and support as a function of membrane thickness. The four figures illustrate the dependence on key parameters. A) Shows the impact of tube radius (parameter values given by the legends). B) Illustrates the impact of the membrane strain (ε_m), C) Illustrates the effects of flaw size (l), and D) the effects of external overpressure (q). The base case values of the parameters were $q = 0$ atm., ε_m=0,3%, E_m=100 Gpa, E_s =10 GPa, R =10 mm, l=0,5 mm, H=1 mm

Figure 7 shows an experimental membrane reactor after test, where the membrane layer has partly peeled off. It is likely that this piece failed due to a buckling driven delamination.

Figure 7. A section of an experimental membrane reactor that after test showed detachment between membrane and support. The reactor was prepared and tested within the framework of the CERAMGAS project.

CONCLUSION

Out of the four analysed failure modes the two ones regarding strain mismatches between membrane and support arising either from differences in TEC between the two materials or from the expansion on reduction of the membrane material were found to be the most critical. It seems likely that by proper design and careful manufacture to limit the inherent flaw size, both tube collapse and burst of local unsupported areas can be avoided for a range of technologically relevant pressures.

To avoid crack formation in the membrane on cooling after the firing step requires either extremely thin membranes or very close TEC matching between the layers. For the considered parameter values (a membrane thickness of 10 -20 μm and a TEC of the membrane exceeding that of the support) a TEC-matching of better than $1 \cdot 10^{-6} K^{-1}$ was postulated necessary on the basis of the fracture mechanical analysis. Experimental experience form manufacture of tubular supported membranes in several different material combinations, though not sufficiently numerous to establish the criterion accurately, was consistent with this threshold.

To avoid delamination by buckling due to the expansion of the membrane material imposes quite tough criteria on the dimensional stability of the membrane material and/or to the tolerable flaw size in the component. In the analysis of this case, the treatment of buckling driven delamination of a thin layer on a thick planar support was modified to describe also the case of a tubular support under external pressure. Like for the planar case two "safe" regimes exist in the limit of very thin and very thick membranes. The most critical thickness is increased relative to the planar case. Further, applying an external over pressure was found to a have great stabilizing effect towards this mode of failure. A failure observed on an experimental membrane was tentatively ascribed to this failure mode.

ACKNOWLEDGEMENTS
Financial support from EU via the CERAM GAS project contract G1RD-CT-1999-00023 is acknowledged. Dr. Luca Basini, Snamprogetti is acknowledged for providing photographic documentation of a failure observed on a tested reactor, and Dr. D. Lybye, Risø, for carrying out measurements of the expansion on reduction of the materials.

REFERENCES

[1] D. Stöver, H. P. Buchkremer and J. P. P. Huijsmans in "Hand book of Fuel Cells – Fundamentals, Technology and Applications" Ed. by W. Vielstich , H. A. Gasteiger, A. Lamm, Vol 4, Chap. 72. John Wiley & Sons, (2003).

[2] S. C. Singhal, "Advances in SOFC Technology", *Solid State Ionics* **135** 305-313 (2000).

[3] P.V. Hendriksen, P.H. Larsen, M. Mogensen, F.W. Poulsen and K. Wiik "Prospects and problems of dense oxygen permeable membranes", *Catal. Today* **56** 283 (2000).

[4] H.J.M. Bouwmeester, *Catal. Today* **82** 141-150 (2003).

[5] EU-project "CERAM GAS", contract G1RD-CT-1999-00023, See CORDIS http://ica.cordis.lu

[6] J. W. Hutchinson and Z. Suo, *Adv. Appl. Mech.* **29**, 63-191 (1992).

[7] S. P. Timoshenko, J.M. Gere, "Theory of elastic stability" McGraw-Hill, 1961.

[8] N. Ramakrishnan and V. S. Arunachalam, *J. Am. Ceram. Soc.* **76** 2745 (1993).

[9] S. Timoshenko, S. W.-Krieger, "Theory of Plates and Shells", McGraw-Hill, 1959.

[10] Evans and Hutchinson *Acta Metall. Mater* **43** (7) (1995).

[11] L. W. Tai, M. M. Nasrallah, H. U. Anderson, D. M. Sparlin and R. Sehlin, *Solid Stae Ionics*, **76**, 259 (1995).

[12] H. Lein "Mechanical properties and phase stability of oxygen permeable membranes, $La_{0.5}Sr_{0.5}Fe_{1-x}Co_xO_3$", Ph.D-thesis, IUK-thesis 114, imt-report 2005:69, Dep. of Mat. Tech. NTNU, 2005.

[13] N. Christiansen, S. Kristensen, H. Holm-Larsen, P. H. Larsen, M. Mogensen, P. V. Hendriksen and S. Linderoth, p. 168 in SOFC-IX, ed. by S. C. Singhal and J. Mizusaki, ECS Proc. Vol. **2005-07**. The Electrochemical Society (2005).

[14] J. W. Stevenson, T. R. Armstrong, L. R. Pedersen and W. J. Weber, Proc. of first int. symposium on ceramic membranes, ed. by H. U. Anderson, A.C. Khandkar, M. Lieu, The Electrochemical Soc. **PV 95-24**, p. 94 (1995).

[15] P.V. Hendriksen, J.D. Carter and M. Mogensen "Dimensional instability of doped lanthanum chromites in an oxygen pressure gradient", In: Proceedings of the fourth international symposium on solid oxide fuel cells (SOFC-IV), M. Dokiya, O. Yamamoto, H. Tagawa and S.C. Singhal (eds), Vol 95-1 p. 934 – 943 (1995).

[16] P.V. Hendriksen, J. Høgh, J.R. Hansen, P.H. Larsen, M. Solvang, M. Mogensen and F.W. Poulsen "Electrical conductivity and dimensional stability of Co-doped lanthanum chromites", Fifth International Symposium on Ionic and Mixed Conducting Ceramics, eds. T.A. Ramanarayanan, Y. Yokokawa and M. Mogensen, The Electrochemical Society, 2004.

[17] H. E. Evans, "Stress effects in high temperature oxidation of metals", *Int. Mat. Reviews*, **40**, 1, (1995)

[18] J. Riess Høgsberg. *J. Danish Ceramic Society*, **4**, (1) (ISSN 1398-3261) 2001.

[19] B. Storåkers K. F. Nielsson, "Imperfection Sensitivity at Delamination Buckling and Growth" *Int. Jour. of Solid. Struc.* **30** 1057 (1993).

[20] B. Storåkers, P. L. Larsson and C. Rohart, "On delamination Growth in Shallow Shells", *J. Appl. Mech.* **71**, 247 (2004).

[21] J. W. Hutchinson "Delamination of compressed films on curved substrates", *J. Mech and Phys. Solids* 49 1847 (2001).

[22] Sørensen, B. F., and Horsewell, A., "Crack growth along interfaces in porous ceramic layers", *J. Am. Ceram. Soc.*, Vol. 84, pp. 2051-9 (2001).

COMPARISON OF MECHANICAL PROPERTIES OF NiO/YSZ BY DIFFERENT METHODS

Dustin R. Beeaff, S. Ramousse, and Peter V. Hendriksen
Risoe National Laboratory
Frederiksborgvej 399, P.O. 49,
DK-4000 Roskilde, Denmark

ABSTRACT

Planar solid oxide fuel cells are a leading candidate for future power generation. Integral to their utility is the robustness of the component materials and their ability to tolerate the mechanical stresses expected during the lifetime, including fabrication, of the cell. Most current planar designs rely on either the anode or cathode as the structural-supporting member.

To gain a fundamental basis of data on which to compare future cell designs, a series of half-cell (electrolyte/anode) specimens was fabricated in which processes variables and additives typical of current cells were varied. These samples were tested in unreduced state in a biaxial flexure apparatus to determine the biaxial flexure strength. The samples were oriented with the electrolyte on the tensile surface. Results were compared to a concurrent study in which similar specimens were tested by a uniaxial tensile method with statistical distribution parameters used to adjust for volumetric differences to validate the biaxial flexure data.

INTRODUCTION

Flexure testing was carried out using a modified ball-on-ring biaxial flexure technique. The standard method consists of supporting a sample disc on a ring (or discreet support points equidistant from the center) and applying a load to the center of the sample, such that the specimen is flexed in two directions rather than just one [1,2]. Using the test method allows one to overcome some uniaxial testing deficiencies. The edge of the specimen is not a factor in the test as stresses decay rapidly outside the loading region towards the specimen edge. This also means that the typical sample is relatively inexpensive to produce as it requires little additional edge finishing steps during preparation as well as being suitable for thin (<5 mm) specimens. It also accounts for all flaws within a sample regardless of their orientation. Last, the test is a more rigorous approximation of actual service conditions.

Under the central loaded region of the biaxial specimen the stress is maximum and uniform. Outside this region, the stresses decay rapidly to near zero. The tangential stresses decay more rapidly than radial stresses, so outside the central region true biaxial flexure does not exist. Since the stresses at the edges of the sample are near zero, no special preparation is required for biaxial flexure specimens. Within the central region, the radial and tangential stresses are equal and the total stress is given by the equation

$$\sigma_t = \sigma_r = \sigma_f = \frac{3P(1+v)}{4\pi t^2}\left\{1+2\ln\frac{a}{b}\left[\frac{(1-v)a^2}{(1+v)R^2}\right]\left[1-\frac{b^2}{2a^2}\right]\right\} \qquad [1]$$

where P is the load, t is the thickness of the sample, a is the radius of the circle of the support points, b is the loading radius, R is the radius of the sample, and v is Poisson's ratio. The value of the loading radius is much less than the contact radius, which may be calculated using Hertzian contact theory when using a single ball to apply the load. To avoid ambiguities and uncertainties regarding the contact radius, the test was modified to use a steel hemisphere with the planar surface in contact with the specimen surface. This means the contact radius (1.8 mm) could be accurately determined.

PROCEDURE

Half-cells were produced in 130 mm x 130 mm x 0.3 mm plates. The half-cells tested consisted of "standard" production materials with YSZ electrolyte and NiO/YSZ anode support and "modified" production materials in which either the sintering conditions or anode support additives were altered. The modifications consisted of (1) higher sintering temperature, (2) finer grained (therefore, higher shrinkage on densification) starting anode support materials, (3) uncalcined oxide additions, (listed as *Additive #1*) and (4) calcined oxide additives (listed as *Additive #2*). The average thickness of the zirconia electrolyte was 11 μm. The samples were randomly chosen from a production line. From each plate, 20 circular disc specimens were cut by laser to a diameter of 2 cm. At least 10 samples of each variety were tested to obtain statistically relevant data.

The biaxial flexure apparatus consisted of a support ring of twelve steel balls forming a ring with a diameter of 1.75 cm. With more than 6 balls as a support, the stress distribution approaches that of a continuous ring, while the use of discrete balls allows for some degree of rotation of the loading points, thereby alleviating frictional contact stresses that can cause up to 15% error in strength calculations. For the samples presented here, a stress-strain curve was unavailable as the interface between the digital load cell and computer was inoperative; however, the maximum load at failure was noted and used for strength calculations. The modulus of elasticity is not directly obtainable via the biaxial method.

All samples were tested to failure with a central load applied at a rate of 1 mm·min^{-1} via a screw actuator. Samples were predominantly tested such that the zirconia electrolyte was on the tensile surface of the specimen. After fracture, the thickness of each sample was measured near the fracture origin using a digital micrometer and used to calculate strength. At the conclusion of each series of materials, the samples were screened by optical and electron microscopy for valid and invalid failures. All samples presented in this text were tested at room temperature in the unreduced state and are more applicable to stack fabrication and assembly than operation conditions, where the anode support is in the reduced state and creep properties and slow crack growth will become important.

RESULTS AND DISCUSSION

Figure1 shows the distribution of the samples tested in this experiment. The samples can be broadly categorized into two groups based on their mechanical performance. The entirety of the relatively lower strength materials includes the "standard" materials, sintered using different profiles. The relatively higher strength materials are those in which the anode support was modified.

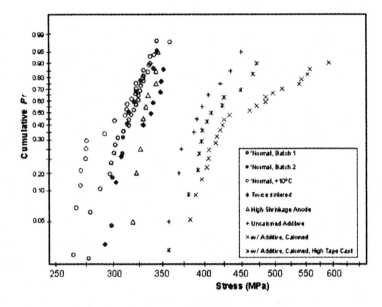

Figure 1. Weiblull plot of all samples tested thus far via biaxial flexure.

Table I gives a summary of the strengths of the materials tested. The samples can be broadly categorized into two groups based on their mechanical performance. The entirety of the relatively lower strength materials includes the "standard" materials, sintered using different profiles. The relatively higher strength materials are those in which the anode support was modified. Most notable are the high Weibull moduli seen for some samples. For the majority of these materials, an apparent bimodal distribution of strengths was observed and this is represented by the use of a slash separating the lower from the higher strength values. As will be discussed this may be due to the existence of multiple failure mechanisms or, more likely, the sensitivity of the test procedure to residual stresses present in the electrolyte at room temperature due to differential thermal expansion between the constituent materials.

Table I. Summary of Weibull statistics for tests on half-cell specimens tested in biaxial flexure. N_{total} is the total number of tests, N_{valid} is the number of valide tests. The Weibull modulus m and characteristic strength σ_o were determined by the least squares method.

Special Treatment	N_{total}	N_{valid}	m	σ_o (MPa)	R^2
none	31	30	32/12.7	290/80	0.95/0.97
twice sintered	10	10	18	105	0.92
Sintering temp +10°C	25	22	20	120	0.97
Additive #1	23	21	23/15		
Additive #2	23	20	16/5	386/100	0.94/0.96
High Shrinkage Anode	10	10	40	250	0.93

Figure 2 shows the Weibull distribution of strengths of the "standard" half-cell materials. The solid horizontal line corresponds to ln ln $(1/P_s)$ = 0, an artifact of how the Weibul plot is usually presented with units of $ln\ ln\ (1/P_s)$ as the values of the y-axis. The average biaxial strength of the standard half-cell is around 315 MPa. Sintering the half-cell twice or at a higher temperature does slightly increase the strength of the cell, but increases the variability in measured strength. It is interesting to note the existence of an inflection point, particularly noticeable for the lower-strength high-temperature sintered samples. This is typically indicative of dual failure mechanisms, although the origin is yet undetermined. It would seem from these tests that, although statistically similar, the higher sintering temperature induces a new failure mechanism.

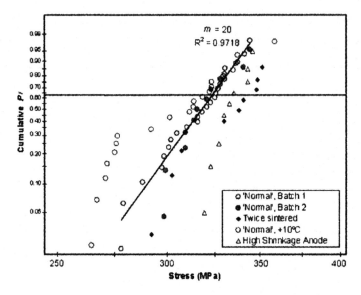

Figure 2. Weibull plot for biaxial flexure strength of half-cell specimens utilizing variable sintering conditions but no anode modifications. There is no statistically significant difference between processing temperatures.

Figure 2 shows the Weibull distribution of half-cells containing modified anodes. The modification of the anode consists of oxide additions; either calcined prior to tape casting or uncalcined. Included for reference is one of the curves from the "standard", unmodified materials sintered at the higher temperature. The overall difference resulting from the additions is quite extraordinary, giving an average strength of 400 ± 30 MPa for the uncalcined material. This represents a 30% increase in relative strength. The Weibull modulus of $m = 16$ indicates that this increase in strength does not come at a greater sacrifice in reliability. Two specimen types consisting of calcined additions were tested, the difference being the tape casting speed of the anode support material.

Interestingly, the calcined materials show a behavior similar to that seen with the standard material in which there is a high Weibull modulus ($m = 23$) low-strength portion and a low ($m = 5$) modulus upper portion.

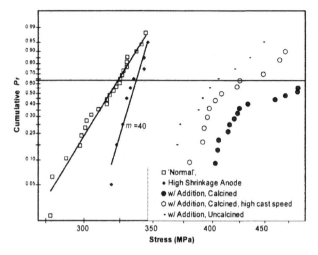

Figure 3. Weibull plot for biaxial strength of half-cell specimens containing unmodified and high shrinkage anodes (left) or modified anodes (right). Increasing the densification shrinkage shows a significant increase in specimen strength.

There are two primary results of interest. First, the modification of the anode support can result in a significant increase in the measured strength. Second, for some of the samples, there is an apparent bimodal distribution of failures consisting of a lower strength portion with a narrow distribution and a higher strength portion with a much broader distribution.

The increase in strength and apparent bimodality may be due to the imposition of residual compressive stress to the electrolyte. The well known three-term Weibull distribution is given by the equation

$$P_s = \exp\left[-\int_V \frac{\sigma - \sigma_u}{\sigma_o} dV \right] \qquad [2]$$

in which the parameter σ_u is a threshold stress level below which failure will not occur. For monolithic ceramics there always remains a remote possibility of a very large flaw existing in the specimen and this term is eliminated. In a physical sense, this implies that there is a finite probability that a sample fails without any applied stress, technically an impossibility, but for purposes of safety this term is usually neglected and the above equation becomes the more commonly used two-term Weibull distribution [3]. In the present case, however, it should be expected that a similar threshold stress be present as the tensile surface of the samples is the electrolyte, which has a compressive stress to overcome before failure can occur.

A simple way to check this is to subtract a threshold strength and replot the data relative to this value [4]. Figure 5 shows the fracture data of the half-cell with calcined anode additions (high cast speed) from Figure 2 with a σ_u correction of 380 MPa. Note that the ordinate scale is now relative to σ_u. The Weibull modulus is now $m = 1.3$, more typical of a ceramic material and the correlation factor, R^2, is 0.94 over the entire data set.

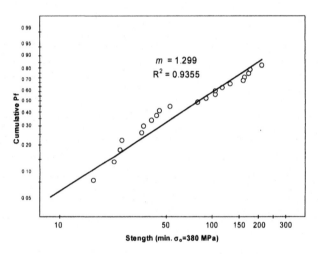

Figure 4. Weibull plot for anode modified half-cell data in Figure 2 corrected for threshold strength.

Applying a similar approach to the other materials gives the result in Table II. Again, the Weibull moduli and characteristic strength are in the σ_u domain. This simply implies that there exists some threshold the tensile surface must exceed before the sample behaves as a monolithic specimen and the flaw population distribution becomes a factor. For way of comparison, like materials to those sintered at the lower temperature with no additions were tested by Wang, in uniaxial tension [5]. Results from these tests gave a very comparable Weibull modulus of $m =$ 11.55.

Table II. Summary of Weibull statistics from Table I corrected for a threshold strength, σ_u.

Special Treatment	N_{total}	N_{valid}	σ_u (MPa)	m	σ_o (MPa)	R^2
none	31	30	200	13	90	0.97
twice sintered	10	10	18	7	105	0.92
+10°C sintering T	25	22	100	5	95	0.97
Additive #1	23	21	350	4	110	0.93
Additive #2	23	20	380	1.3	100	0.94
High Shrinkage Anode	10	10	200	40	50	0.93

The correction for a residual stress is not without precedent as there is a significant residual stress resulting from the thermal history of the samples. Using the thermal stress equation

$$\sigma_{YSZ,t} = \frac{\Delta\alpha \cdot \Delta T \cdot E_{YSZ}}{(1-v)}$$

where $\Delta\alpha$ is the difference in linear thermal expansion, E is the Young's modulus, v is the Poisson's ratio, ΔT is the temperature change from a zero-stress temperature to room temperature and σ is the resulting stress within the zirconia near the electrolyte/electrode interface. The values of thermal expansion for the individual components had been measured previously (α_{YSZ} = 10.8 ppm, $\alpha_{NiO/YSZ}$ = 12.5 ppm).

However, using a zero-stress temperature of 1200°C, this would result in a residual stress of 750 MPa. The zero-stress temperature chosen is slightly less than the sintering temperature to account for stress alleviation by material creep. This value agrees with curvature measurements of large half-cells. The use of this residual stress as a threshold value leads to an illogical interpretation of the data. Because of this, an alternative origin of the bimodal distribution must be identified and is most like related to processing flaws. Figures 3 and 4 show SEM micrographs of failed samples of a half-cell with unmodified anode support (Batch #1). Figure 3 is from a valid test, outside the central loaded region at a point where the propagating crack has bifurcated. The tensile surface is at left in the picture and the anode support is at right with the view normal to the electrolyte surface. The image distortion at the surface edge is due to electrical charging of the sample. A large flaw is seen in the electrolyte at the upper portion of the micrograph. This has been attributed to variations in slurry spray characteristics. Figure 4 is of an invalid test showing a large flaw of similar origin outside the central loading region at the point where fracture originated.

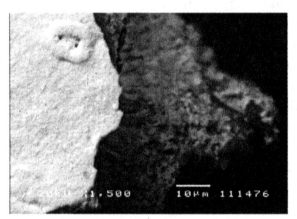

Figure 5. SEM micrograph showing surface flaw near crack bifrication point.

Figure 6. SEM micrograph showing a surface flaw at left resulting in failure outside of valid loading region.

From these observations, a better explanation may be developed if one assumes two distributions of failure stresses are likely due to a change in failure origins. The low strength values with a narrow distribution of strengths are most likely due to failure originating from large flaws in the electrolyte layer. The scale of flaws observed during processing are typically very uniform in size and are on the order of the thickness of the electrolyte. Given the relatively small area of loading (10.04 mm^2) and the low density of defects, samples would be tested without these surface flaws in the strained electrolyte section and the strength of this layer would be dominated by the residual compressive stress. At higher strains, failure most likely originates within the anode near the electrolyte interface, where the material is already under a residual tensile stress.

If this were the case, one would expect to be the failure distribution to be dominated at higher strengths by the distribution of flaws within the anode material. Wang, *et al.* has looked at uniaxial testing of monolithic anode-support materials and obtained a Weibull modulus of $m = 11.6$ for specimens in which edge effects have been minimized by rigorous sample preparation. The upper portion of the failure distribution of unmodified half-cell samples, assuming a bimodal distribution of flaw sizes, has a Weibull modulus of $m = 12.8$. This lends empirical evidence to the idea of failure origin within the anode.

Further testing will consist of orienting samples such that the anode is on the tensile surface. In this orientation, there should be no contribution of the electrolyte to the flaw distribution and the failures should be dominated by the anode material. If the observed bimodal effect is due to two exclusive flaw origins, this orientation should display only those related to the anode material. Additionally, the use of a larger loading radius should increase the probability of encountering a severe electrolyte flaw, thereby affecting the relative amount of low and high strength flaw distributions. Modeling of the effective testing volumes of the electrolyte and anode in the as tested orientation is currently underway.

CONCLUSIONS

The modification of the anode support properties, either through processing variation or by the use of additives, in such a way as to increase the densification shrinkage of the anode results in an increased strength of a half-cell when tested in biaxial flexure at room temperature. Samples were tested at room temperature, however, and this increase in strength may not to be expected at elevated temperatures. Future testing will include both reduced samples and testing at or near operating temperature (750 – 800 °C). There was a notable bimodal distribution of failures in some of the tested specimens. While the use of a threshold stress did improve the fit of the Weibull parameters, the calculated residual stress was significantly higher than the fit threshold values. Further investigation will attempt to elucidate the origin of this effect.

REFERENCES

[1]Kirstein and Woolley, *J. Res. Nat. Bur. Sta.* **71C** 1-10 (1967)

[2]De With and Wagemans, "Ball-on-Ring Revisited" *J. Am. Cer. Soc.* **72** 1538 (1989).

[3]J.B. Wachtman, in *Mechanical Properties of Ceramics*, John Wiley & Sons, Inc., New York (1996)

[4]R.B. Abernethy, in *The New Weibull Handbook*, R.B. Abernathy publ., North Palm Beach, FL (2000).

[5] F. Wang, B. Sørensen, *unpublished data*

FRACTURE TOUGHNESS AND SLOW CRACK GROWTH BEHAVIOR OF Ni-YSZ AND YSZ AS A FUNCTION OF POROSITY AND TEMPERATURE

M. Radovic, E. Lara-Curzio and G. Nelson
Metals and Ceramics, Oak Ridge National Laboratory
Oak Ridge, TN 37831-6069

ABSTRACT

In this paper we report on the fracture toughness of YSZ and Ni-YSZ and slow-crack growth behavior of Ni-YSZ at 20°C, 600°C and 800°C. Results are presented for tests carried-out in air for YSZ and in a gas mixture of $4\%H_2$ and $96\%Ar$ for Ni-YSZ containing various levels of porosity. The double-torsion test method was utilized to determine the fracture toughness from the peak load obtained during fast loading test specimens that had been precracked, while crack velocity versus stress intensity curves were obtained in the double-torsion using the load relaxation method. It was found that fracture toughness of these materials decreases with temperature and in the case of Ni-YSZ it also decreases with increasing porosity. The effect of temperature and microstructure, which was characterized by Scanning Electron Microscopy, on the fracture behavior of these materials, is discussed.

INTRODUCTION

Planar Solid Oxide Fuel Cells, SOFCs, are considered to be one of the most promising devices for power generation[1]. Fully dense yttria stabilized zirconia, YSZ, and porous Ni-YSZ cermets are the most popular materials for the manufacture of electrolytes and anodes, respectively in SOFCs. During operation at elevated temperatures (700-1000 °C), SOFC components are subjected to mechanical stresses, the magnitude and distribution of which is a complex function of several parameters including: geometry of SOFC, temperature distribution and external mechanical loads. These stresses can be classified into three groups: (a) residual stresses that result from the mismatch in thermoelastic properties and different sintering shrinkage of the constituents; (b) stresses that result from externally applied mechanical loads (e.g.- stresses applied during fuel cell assembly) and (c) thermally induced stresses that result from temperature gradients.

Fracture toughness, K_{IC}, and the susceptibility to slow crack growth, SCG, which are a measure of a material's resistance to cracking and crack propagation, respectively are key parameters to achieve structural integrity. The long-term reliability of SOFCs is dictated in great part by the fracture behavior of the materials used in their construction. Although, the fracture toughness and the susceptibility of SOFC materials to slow-crack growth behavior are relevant information for the formulation of models to predict the reliability and durability of SOFCs, very little information is available in the open literature, in particular for tests carried out in relevant environments[2-7].

In this paper, we report the results from a study of the fracture behavior of YSZ in air and Ni-YSZ cermets of different porosity in a reducing environment (gas mixture of $4\%H_2$ and $96\%Ar$). Both, the fracture toughness and slow-crack growth behavior were determined as a function of temperature at 20°C, 600°C and 800°C using the double torsion test method.

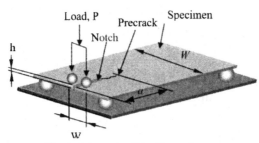

Fig. 1. Schematic of double torsion test.

The YSZ test specimens examined in this study were prepared from ZrO_2 stabilized with 8 mol% Y_2O_3 (TOSOH Corp.[*], Grove City, OH) powder while Ni-based materials were prepared using NiO (J.T.Baker[*], Phillipsburg, NJ) powders and the same YSZ powders used for the preparation of YSZ test specimens. The powder mixture used for the preparation of Ni-based materials contained 75mol% of NiO and 25mol% of Y_2O_3 stabilized ZrO_2 (YSZ). Different amount of organic pore former (rice starch, ICN Biomedicals[*], Inc Irvine, CA) were added to the NiO-YSZ powder mixture to obtain test specimens with different porosity. Green test specimens were prepared by tape casting NiO-YSZ and YSZ slurries in ≈ 250 μm thick single layers. Four green tapes were subsequently laminated to make 1 mm thick test specimens. Rectangular test specimens for double torsion testing were cut from tapes previously sintered at 1450°C in air for 2 hours. NiO-YSZ test specimens for double torsion testing were exposed to a gas mixture of 4%H_2 and 96%Ar at 1000°C for 30 minutes, to reduce NiO completely into metallic Ni. The relative porosity of the test specimens before and after reduction was determined

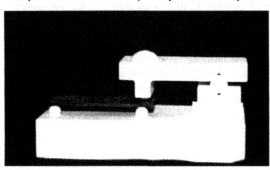

Fig. 2. Alumna fixture for double torsion.

by alcohol immersion according to ASTM standard C20-00. The weight of the Ni-based test specimens before and after reduction in hydrogen were measured and used to calculate the weigh loss for every reduced test specimen. Calculated weight loss was compared to the theoretical weight loss to ensure that all test specimens were fully reduced.

The fracture toughness, K_{IC}, and slow-crack growth behavior of the YSZ and Ni-based materials were determined using the double torsion method[7-11]. The basic geometry of rectangular double torsion test specimens (20 mm in width, W, 40 mm in length, L, and ≈ 1 mm thick, t) and loading configuration used in this study are shown on Fig. 1. Initial notches 1 mm in width and 12.5 mm in length were cut into one side of the test specimen using a circular diamond blade. The notch tip was machined such that the thickness of the specimen at the notch tip tapered from very thin to the full thickness. The tests were carried-out using a fixture fabricated from alumina with alumina supporting and loading balls (See Figure 2). The fore-and-aft and side-to-side alignment of the base and upper fixtures is adjusted by modifying the position of two alumina pivot rollers. Once the specimen and fixture are aligned and setup, the support fixture

Figure 3. Furnace and environmental chamber for high temperature double torsion testing.

for the top portion plays no role during loading. The tests were carried out in air or 4%H$_2$ and 96%Ar using the environmental testing chamber shown in Figure 3.

Prior to testing, all notched test specimens were pre-cracked by loading the test specimen at a rate of 0.01 mm/min in the double torsion fixture. The reduction in the thickness at the notch tip facilitates the formation of a sharp pre-crack at relatively low loads, well below that required to cause fast fracture of the test specimen, as illustrated in Figure 4. For valid measurements, it is necessary to ensure that the cumulative length of notch and precrack is in the constant K_I regime, which is described as 0.55W<(notch length + precrack length)<L-0.65W^*. This condition was satisfied for all tests that are reported in this paper.

Fracture toughness was determined from the maximum load, P_f at which pre-cracked test specimens failed during fast loading at a displacement rate of 2 mm/min using the following relationship[8-9]:

$$K_{IC} = P_f W_m \left[\frac{3(1+\nu)}{W h^3 \xi} \right]^{1/2} \tag{1}$$

where W_m, W, and h are defined in Figure 1a, ν is Poisson's ratio and ξ thickness correction factor given as: $\xi=1-1.26(h/W)+2.4(h/W)\exp(-2\pi W/2h)^8$.

The slow crack growth behavior of Ni-YSZ materials was determined using the load relaxation version of the double torsion test configuration[10,11]. According to this test procedure,

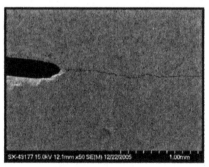

SX-43177 15.0kV 12.1mm x50 SE(M) 12/22/2005   1.00mm

Figure 4. Scanning electron micrograph of pre-crack at the notch tip in Ni-YSZ test specimen.

a pre-cracked double-torsion test specimen is fast loaded at a constant displacement rate of 2 mm/min up to a load equivalent to 95% of the load associated with the fracture toughness of the material at a particular temperature. At that point, the crosshead displacement of the mechanical testing machine is arrested and the load is monitored and recorded as a function of time. Under those conditions, cracks will propagate in a stable manner, resulting in the relaxation of the load. Prior to testing, a series of tests were carried out to ensure that the relaxation of the fixture and testing machine was negligible for the applied load range.

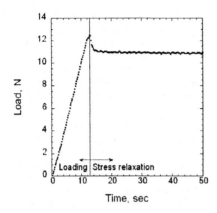

Figure 5. Typical stress vs. time curve obtained by double torsion testing using the stress relaxation technique for 40% porous Ni-YSZ test specimen at room temperature.

Figure 5 shows a plot of load versus time for a test of a 40% porous Ni-YSZ test specimen. The crack growth rate, da/dt was calculated directly from the rate of load relaxation, dP/dt using following relation[10,11]:

$$\frac{da}{d\upsilon} = \frac{a_i P_i}{P^2}\frac{dP}{dt} = \frac{a_f P_f}{P^2}\frac{dP}{dt} \qquad (2)$$

Where P is load, t is time, a_i and a_f are initial and final crack lengths, respectively, and P_i and P_f are initial and final loads, respectively. At the same time, the stress intensity K_I at each loa. ing load relaxation testing can be calculated using Equation 1 but with the applied load P, instead of the critical load P_f. The crack velocity was calculated using both, initial and final crack lengths and loads to ensure validity of the assumption of linear relationship between test specimen compliance and the length of the precrack, a, that was used to derive Equation 2. In all examined cases, the differences between crack velocities that were calculated using initial, and final crack lengths and loads were negligible.

The microstructure of, and cracks in, YSZ and Ni-based test specimens were examined by Field Emission Scanning Electron Microscope (FESEM) Hitachi S4700*.

RESULTS AND DISCUSSION

Fracture toughness of YSZ and Ni-YSZ

The fracture toughness of the YSZ test specimens as a function of temperature is summarized in Table I and Figure 6. The value of fracture toughness at ambient temperature is comparable to the previously reported value of 1.61 MPam$^{1/2}$ for tape-cast 8mol% YSZ[2-3]. It was f   at fracture toughness initially decreases with temperature, and then it increases slightly between 600 and 800°C.

Table I. Fracture Toughness of YSZ

T, °C	#	Fracture Toughness*, K_{IC}, MPam$^{1/2}$
25	5	1.65±0.03
600	5	1.24±0.27
800	5	1.51±0.13

T-Temperature, # - number of tested specimens
*Average value ± one standard deviation

The fracture toughness results for 26% ?   40% porous Ni-YSZ test specimens are also summarized in Table II and plotted in f i.   o as a function of temperature. It was found that the fracture toughness of Ni-YSZ decreased both with temperature and porosity from 4.5 MPam$^{0.5}$ for 26% porous test specimens at 22°C to 1.6 MPam$^{0.5}$ for 40% porous test specimens at 800°C.

Table II. Fracture Toughness of Ni-YSZ.

Porosity	# of test specimens	Fracture Toughness*, MPam$^{1/2}$		
		22°C	600°C	800°C
26%	5	3.49±0.33	2.68±0.16	2.51±0.19
34%	5	2.82±0.29	2.15±0.43	1.73±0.21
40%	5	2.56±0.04	1.61±0.17	1.63±0.32

*Average value ± Standard Deviation

Although Ni-YSZ test specimens were highly porous, their fracture toughness is still significantly higher than that of fully dense YSZ test specimens (Figure 6), highlighting the role of the ductile, metallic Ni phase for the increased fracture resistance of Ni-YSZ cermets. However, the decrease in fracture toughness of Ni-YSZ cermets, for all examined porosities, is more pronounced with increasing temperature than for the case of YSZ. The beneficial effect of the Ni phase on the fracture resistance of Ni-YSZ diminishes quite rapidly with temperature, as illustrated in Figure 6. For example, the fracture toughness of 40% porous Ni-YSZ cermet at room temperature is higher than that of YSZ by more than 50%, but their magnitudes are comparable at 800°C.

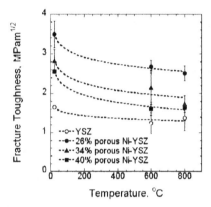

Figure 6. Fracture toughness of YSZ and Ni-YSZ test specimens with different porosity as a function of temperature.

The room temperature fracture toughness results obtained for Ni-YSZ cermets of different porosity that are reported in this paper are in good agreement with previously published results for similar materials[6]. Furthermore, the arguments regarding the role that the Ni phase plays in contributing to the fracture toughness of Ni-YSZ cermets is also in good agreement with previous results[6] that show that the fracture toughness of NiO-YSZ increases when it is reduced in hydrogen and the brittle NiO phase is transformed into the ductile, metallic Ni phase, in spite of the increase in porosity due to the volumetric shrinkage associated with the conversion of NiO into Ni.

Susceptibility of Ni-YSZ to Slow Crack Growth

Figures 7 and 8 show typical log-log plots of crack velocity versus stress intensity factor for Ni-YSZ test specimens of different porosity. Most of the da/dt-K curves exhibited two regions: a threshold region and a Region I. In some cases, like in the case of the curve for 40-anode in Figure 7, Region II is also present. The threshold region is characterized by a threshold stress intensity, K_{th} which is the value of stress intensity below which the crack will not grow. However, crack velocity, *da/dt*, is extremely sensitive to K_I within the Region I (slow crack growth region) and can be related to it by following:

$$\frac{da}{dt} = AK_I^n \tag{3}$$

where A is constant and n is exponent. n and K_{th} were determined from the series of slow crack growth curves, such as those shown in Figures 7 and 8, for Ni-YSZ test specimens with different porosity at 22°C, 600°C, and 800°C. The values of n and K_{th} are summarized in Table III and plotted as a function of temperature in Figures 9 and 10.

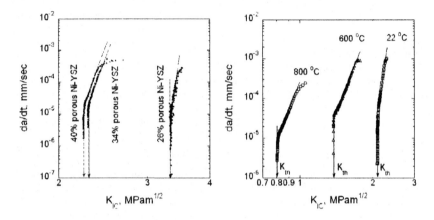

Figure 7. Typical room temperature slow-crack growth curves for Ni-YSZ test specimens with different porosity.

Figure 8. Typical slow crack growth curves for 40% porous Ni-YSZ materials as a function of temperature.

Table III. Fracture Toughness of Ni-YSZ material.

Porosity	# of test specimens	Temperature					
		22°C		600°C		800°C	
		$K_{th},$ MPam$^{1/2}$	n	$K_{th},$ MPam$^{1/2}$	n	$K_{th},$ MPam$^{1/2}$	n
26%	4	3.3±0.4	18.9±1.5	1.9±0.2	11.8±2.2	1.5±0.1	9.5±2.0
34%	4	2.3±0.1	19.7±2.4	1.7±0.2	12.6±1.6	1.3±0.2	10.9±2.1
40%	4	2.1±0.1	20.0±2.7	1.3±0.2	12.1±1.6	1.0±0.2	10.5±1.8

*Average value ± Standard Deviation

These results show that the threshold stress decreases with porosity of the Ni-YSZ cermets and temperature. However, the exponent n slightly increases with porosity and decreases with temperature.

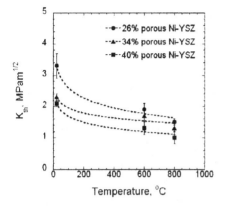 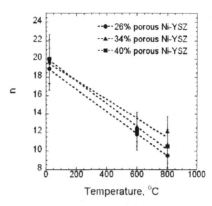

Figure 9. Threshold K_{ic} vs. temperature for Ni-YSZ test specimens of different porosity.

Figure 10. Exponent n from Equation (3) vs. temperature for Ni-YSZ test specimens of different porosity.

Figure 11. Scanning electron micrograph of a crack in 40% porous Ni-YSZ double torsion test specimens after slow-crack growth testing at room temperature.

The cracks that were formed during slow-crack growth testing were examined by scanning electron microscopy to identify the characteristics of crack propagation. It was found that cracks propagate transgranularly in zirconia grains or intergranularly between Ni-YSZ, YSZ-YSZ or Ni-Ni grain boundaries. Figure 11 illustrates transgranular fracture of YSZ grains. The concentration of secondary cracks increased with the porosity of the test specimens.

The fractographic analysis also confirmed the role that the Ni phase plays in the fracture behavior of Ni-YSZ cermets. While there was no evidence of transgranular fracture of Ni grains a large number of highly deformed Ni ligaments were observed in the vicinity of crack tip. Figure 12 shows typical Ni ligaments close to the crack tip. The plastic deformation of the Ni grains at the crack tip is probably responsible for the tougher behavior of Ni-YSZ cermets compared to YSZ. However, it was also found that the concentration of Ni ligaments decreased with test temperature. Therefore, it appears that crack bridging by plastic deformation of Ni grains is a less effective toughening mechanism at elevated temperatures. This observation is consistent with the decrease in fracture resistance of Ni-YSZ cermet with temperature (Figures 6 and 9).

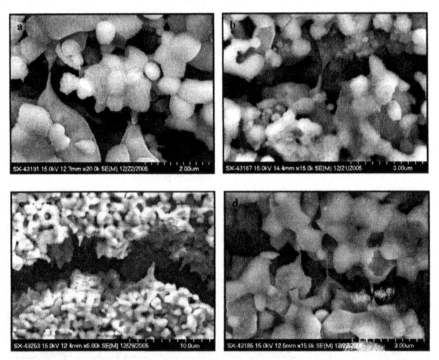

Figure 12. Scanning electron micrograph illustrating Ni ligaments in the wake of a crack after slow crack growth tests for (a) and (b) 0% porous Ni-YSZ at room temperature, (c) 40% porous Ni-YSZ and 600°C and (d) 40% porous Ni-YSZ at room temperature.

SUMMARY

The fracture toughness and slow-crack growth behavior of tape cast YSZ and Ni-YSZ were characterized using the double torsion test method at temperatures between 22 and 800°C. It was found that the fracture toughness of both YSZ and Ni-YSZ decreases with temperature and also with porosity for the case of Ni-YSZ test specimens. The fracture toughness of Ni-YSZ was found to be significantly higher than that of fully dense YSZ test specimens, regardless of the level of porosity of the Ni-YSZ cermets examined. This indicates that the ductile metallic Ni phase plays an important role in the toughness of Ni-YSZ. Fractographic analyses of test specimens after double torsion testing revealed the existence of highly deformed Ni ligaments in the wake of the crack.

The slow crack growth behavior of Ni-YSZ cermet in reducing environment was fully characterized as a function of temperature and porosity. It was found that the threshold stress intensity factor decreases with temperature and porosity of the test specimens. On the other hand, the slow-crack growth exponent decreases with increasing temperature, but decreases with porosity.

ACKNOWLEDGMENTS

This research work was sponsored by the US Department of Energy, Office of Fossil Energy, SECA Core Technology Program at ORNL under Contract DE-AC05-00OR22725 with UT-Battelle, LLC. The authors are grateful for the support of NETL program managers Wayne Surdoval and Travis Shultz. The authors are indebted to Beth Armstrong, Claudia Walls and Rosa Trejo of ORNL for help with specimen's preparation and to Chris Coffer for help with pre-cracking the test specimens.

REFERENCES
[1]Minh N. Q. and Takahashi T., "Science and Technology of Ceramic Fuel Cells", Elsevier, Amsterdam (1995)

[2]Atkinson A and Selcuk A., "Mechanical Behavior of Ceramic Oxygen Ion-conducting Membranes", *Solid State Ionics* 134, 59-66 (2000)

[3]Selcuk A. and Atkinson A., "Strength and Toughness of Tape-Cast Yttria-Stabilized Zirconia", *J. Am. Ceram.* Soc. 83, 2029-35 (2000)

[4]Kumar A.N. and Soresen B.F., "Fracture Energy and Crack Growth in Surface Treated Yttria Stabilized Zirconia of SOFC Applications", Mat. Sci. Eng. A333, 380-398 (2002)

[5]Lowrie FL and Rawlings RD, "Room and High Temperatures Failure Mechanisms in Solid Oxide Fuel Cell Electrolytes", *J. Euro. Ceram. Soc.* 20, 751-776 (2000)

[6]Radovic M. and Lara-Curzio E., "Mechanical Properties of Tape Cast Nickel-based Anode Materials for Solid Oxide Fuel Cells Before and After Reduction in Hydrogen", *Acta. Mater.* 52 5747-5756 (2004)

[7]Fuller E. R. Jr, "An Evaluation of Double-Torsion Testing – Analysis" in *Fracture Mechanics Applied to Brittle Materials*, ASTM Special Technical Publication No 678 Edited by Freiman S. W., ASTM, Philadelphia (1997)

[8]Pletka B.J, Fuller E. R. Jr and Koepke B. G. "An Evaluation of Double-Torsion Testing – Analysis", ibid 7

[9]Tait R. B., Fry P. R. and Garrett G. G., "Review and Evaluation of the Double –Torsion Techniques for Fracture Toughness and Fatigue Testing of Brittle Materials", *Exp. Mech.* 14, 14-(1987)

[10]Evans A.G. and Wiederhorn S.M., "Crack Propagation and Failure Prediction in Silicone Nitride at Elevated Temperatures", J. Mater. Sci. 9, 270-278 (1974)

[11]Evans A.G. "A Method for Evaluating the Time-dependant Failure Characteristics of Brittle Materials and its Application to Polycrystalline Alumina", *J. Mater. Sic.* 7, 1137-1146 (1972)

---

[*] Certain commercial equipment, instruments, or materials are identified in this paper in order to specify the experimental procedure adequately. Such identification is not intended to imply recommendation or endorsement by Oak Ridge National Laboratory, nor is it intended to imply that the materials or equipment identified are necessarily the best available for the purpose

EFFECT OF THERMAL CYCLING AND THERMAL AGING ON THE MECHANICAL PROPERTIES OF, AND RESIDUAL STRESSES IN, NI-YSZ/YSZ BI-LAYERS

E. Lara-Curzio, M. Radovic, R. M. Trejo, C. Cofer, T. R. Watkins and K. L. More,
Metals and Ceramics, Oak Ridge National Laboratory
Oak Ridge, TN 37831-6069

ABSTRACT

In this paper we report the effect of thermal cycling, between 20 °C and 800 °C in a gas mixture of 4% H_2 and 96%Ar, on the physical and mechanical properties of Ni-YSZ/YSZ bi-layers. It was found that the porosity and Young's modulus of the bi-layers did not change significantly after thermal cycling. However, the characteristic biaxial strength, as determined by the ring-on-ring test method, was found to decrease by as much as 15% after 1250 thermal cycles. Similar trends were observed from the evaluation of test specimens that had been thermally aged in a similar environment at 800 °C for 625 hours. It was also found that the magnitude of the compressive residual stress in the YSZ layer changes significantly during thermal cycling and aging. Potential mechanisms responsible for this behavior are discussed.

INTRODUCTION

The durability and reliability of Solid Oxide Full Cells, SOFCs, depend not only on their electrochemical performance, but also on the ability of its components to withstand mechanical stresses that arise during processing and service[1]. Specifically, the mechanical reliability and durability of SOFC is determined by the stress distribution to which its components are subjected and by their distribution of strengths. Furthermore, the evolution of these two distributions with time will dictate the durability of these systems with time.

During normal operation some SOFCs might experience temperatures as high as 1000 °C and thermal cycling[2]. Large-amplitude, low-frequency thermal cycles will occur during start-ups and shut-downs, while small-amplitude, high-frequency thermal cycles will occur during transients that are intrinsic to the operation and control processes of the SOFC. Under such operational conditions, the long-term reliability (durability) of SOFCs will be dictated in great measure by the resistance of the SOFC materials and components to degradation mechanisms that will become activated during thermal cycling and thermal aging, such as creep deformation, thermal fatigue and slow-crack growth.

These processes will not only affect the distribution of strengths of the material with time, but they will also affect the distribution of stresses to which these are subjected, including residual stresses. Residual stresses result when joined materials with dissimilar thermoelastic properties are subjected to temperatures different from that at which they were fabricated. This is particularly true for SOFCs because the electrolyte, electrodes and interconnects will be in physical contact during SOFC operation over a wide range of temperatures.

The main objective of this study is to characterize the changes in the physical and mechanical properties and state of residual stress of SOFC materials when subjected to thermal aging and thermal cycling. To quantify these effects, bi-layer test specimens that consist of YSZ and Ni-YSZ layers were subjected to thermal cycling and thermal aging. The changes in the test specimens' porosity, elastic moduli and distribution of biaxial strengths were monitored as a function of number of thermal cycles and aging time. Changes in the test specimens' curvature and residual stresses in the YSZ layer were also evaluated.

EXPERIMENTAL PROCEDURE

The Ni-based layers were prepared from a powder mixture of 75mol% NiO and ZrO_2 stabilized with 8mol% Y_2O_3 (YSZ). Different amounts of organic pore former (0, 15 or 30 vol% of rice starch) were added to the powder mixture to obtain test specimens with different levels of porosity. Green test specimens were prepared by tape casting Ni-YSZ slurries into 250-μm thick layers, which were subsequently laminated. Discs with nominal diameter of 25.4 mm for biaxial testing and determination of elastic properties were hot-knifed from the laminated green tapes. A 15-μm thick layer of YSZ was screen-printed over the NiO-YSZ layer to obtain bi-layers. Bi-layer discs were sintered at 1400°C in air for 2 hours. All test specimens were subsequently reduced in a gas mixture of 4%H_2 and 96%Ar at 1000°C for 30 minutes. The weight of the test specimens was determined before and after reduction to confirm that NiO had been completely reduced to metallic Ni and their relative porosity was determined by alcohol immersion[3]. After reduction the thickness of the test specimens varied from 0.89 ± 0.12 mm for 40-bi-layers to 1.02

Table I. Nomenclature of the examined test specimens

Test specimen name	Amount of pore former in NiO-YSZ slurry	Porosity of Ni-YSZ layer after sintering and reduction in H_2	Symbol used in plots	
			thermal cycling	thermal aging
26-bi-layer	0%	26%	■	□
34-bi-layer	15%	34%	●	○
40-bi-layer	30%	40%	◆	◇

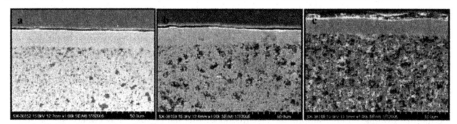

Fig. 1. SEM micrographs of polished cross-sections of (a) 26-bi-layer, (b) 34-bi-layer and (c) 40-bi-layer specimens.

± 0.10 mm for 26-bi-layers. Table I lists the materials that were evaluated, their porosity and the nomenclature that is used throughout this paper to identify them. Figure 1 shows scanning electron micrographs of polished cross-sections of representative test specimens of these materials. The porous structure of Ni-YSZ is evident in these micrographs. Also evident in these micrographs is the different pore size distribution in 26-bi-layer test specimens, which were prepared without the addition of pore formers. In general, pores in this material were smaller than those in other bi-layers.

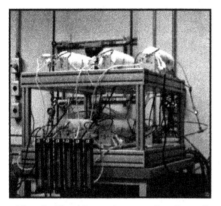

Fig. 2. Experimental setup for thermal cycling and aging.

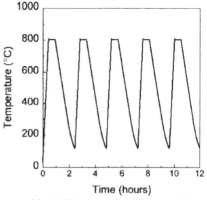

Fig. 3. The thermal cycling schedule

Bi-layer test specimens were subjected to 1, 25, 250 or 1250 thermal cycles between 100°C and 800°C under a constant flow rate (100 cc/min) of a gas mixture of 4%H₂ and 96%Ar using the experimental set-up shown in Figure 2. The set-up consists of six tubular furnaces with quartz retorts and a gas distribution system. The test specimens were placed at the center of the retort and furnace using a wire frame. Figure 3 shows the thermal cycling schedule used for those tests. The thermal cycling schedule consisted of heating at a constant rate of 30°C/minute up to 800°C, followed by a 30-minute long dwell at 800°C and cooling down to 100°C at an approximate rate of 10°C/minute. To distinguish the effects of thermal exposure at 800°C from thermal cycling on the physical and mechanical properties of these materials, thermal aging tests were carried-out at 800°C for up to 625 hours, which is the cumulative time that test specimens experienced at 800°C after 1250 cycles. The changes in porosity, elastic modulus and distribution of biaxial strengths were determined as a function of number of thermal cycles or aging time using total of 12 bi-layer test specimens per each testing condition.

To assess the effect of thermal cycling on residual stresses, measurements were obtained at 20°C using a Scintag PTS rotating anode goniometer using Cu radiation and parallel beam optics after a prescribed number of thermal cycles or aging time. At the same time, the curvature of the test specimens was determined in two mutually orthogonal directions using a laser profilometer (Rodenstock RM 600-S).

RESULTS AND DISCUSSION

Effect of thermal cycling and aging on porosity

The result of porosity measurements for test specimens after different numbers of thermal cycles and aging are listed in Tables II and III, respectively, while the relative changes in porosity are plotted as a function of the number of thermal cycles and aging time in Figure 4. In all cases, except for 26-bi-layers for which porosity changes by as much as 10%, the porosity of these materials didn't change significantly as a result of thermal aging for 625 hours or thermal cycling for up to 1250 thermal cycles.

Table II. Porosity, elastic moduli, characteristic biaxial strength and Weibull modulus as a function of aging time.

Material	# of test specimens	Aging time, h	Porosity%	Young's Modulus (GPa)	Shear Modulus (GPa)	Characteristic biaxial strength, (MPa)	Weibull modulus
40-Bi-layer	12	0	39.4±0.9	59.8±3.1	26.8±1.5	106.5(118.9/94.6)	5.7(8.6/3.5)
	12	625	38.8±1.0	59.0±2.6	28.0±3.8	87.11(94.25/80.1)	7.9(11.5/5.0)
34-Bi-layer	12	0	35.5±1.4	68.9±4.9	26.8±1.5	95.91(110.0/83.6)	5.3(7.9/3.5)
	12	625	35.0±0.7	68.9±2.9	28.0±3.8	90.72(98.8/82.3)	9.5(1 ; 0/6.4)
26-Bi-layer	12	0	27.4±1.6	90.0±3.2	34.8±1.2	127.0(136.6/117.5)	8.0(12.6/5.4)
	12	625	26.7±0.7	85.3±4.6	33.4±2.0	136.6(151.2/123.4)	6.2(9.5/4.1)

Average ± one standard deviation
Average (upper 95% confidence bound, lower 95% confidence bound)

Table III. Porosity, elastic moduli, characteristic biaxial strength and Weibull modulus as a function of number of thermal cycles.

Material	# of test specimens	# of cycles	Porosity%	Young's Modulus GPa	Shear Modulus GPa	Characteristic Strength MPa	Weibull Modulus
40-Bi-layer	12	1	39.5±0.5	57.9±2.5	22.4±1.3	90.7(98.2/83.3)	7.9(11.9/4.8)
	12	25	40.2±1.0	58.0±1.9	22.0±0.9	85.8(99.4/74.1)	4.1(6.2/2.6)
	12	250	40.2±1.0	58.7±4.3	22.6±1.6	79.9(92.0/69.4)	4.2(6.6/2.8)
	12	1250	39.5±1.0	57.2±4.3	21.9±1.3	81.8(89.9/74.1)	6.7 (9.9/4.1)
34-Bi-layer	12	1	33.9±0.5	72.3±2.2	27.5±1.3	126.0(137.0/115.4)	7.5(11.0/4.7)
	12	25	34.1±0.5	72.4±8.4	27.6±3.5	126.6(142.4/112.6)	5.1(7.5/3.5)
	12	250	34.6±0.6	70.9±3.4	26.4±2.9	108.18(120/8,96.1)	5.7(8.4/3.5)
	12	1250	34.9±0.7	68.5±5.4	26.6±0.9	108.3 (117.9/98.9)	7.3(10.9/4.5)
26-Bi-layer	12	1	26.6±0.3	90.1±2.3	35.0±0.8	151.2(167.1/136.0)	6.3(9.3/3.8)
	12	25	26.2±0.4	91.2±1.9	35.3±0.7	166.1(182.3/151.3)	6.4(9.9/4.2)
	12	250	26.4±0.5	92.6±2.4	35.5±2.1	162.3(177.7/147.2)	6.9(10.1/4.3)
	12	1250	27.0±0.7	90.5±1.8	34.9±0.8	161.5(178.7/144.9)	6.9(9.3/3.8)

Average ± Standard deviation
Average (upper 95% confidence bound, lower 95% confidence bound)

This suggests that additional sintering of YSZ and/or Ni-YSZ is unlikely to occur at 800°C. However, the reasons for the significant change in porosity during thermal cycling of 26-

blilayers are still unclear at this time, although the nature of the initial porosity in these materials is much more different than that of the other materials analyzed in this work.

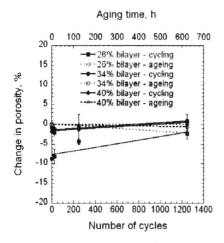

Fig.4. Changes in porosity with thermal cycling and aging.

Effect of thermal cycling and thermal aging on elastic properties

The elastic properties, namely Young's, E, and shear, G, moduli, of bi-layers, were determined by impulse excitation[4-6] using the commercially available Buz-o-sonic software program (BuzMac Software, Glendale, WI). Disc-shaped test specimens supported by a foam material on its nodal lines were excited by a light mechanical impulse. A microphone, located in the vicinity of the test specimen is used to transmit sound vibrations to the signal-processing unit. The fundamental resonance frequencies, in both torsional and flexural mode were identified, which in turn was used to calculate values of Young's and Shear moduli for a test specimen of known mass and dimensions[4-6].

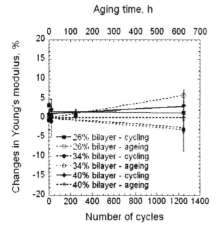

Fig.5. Changes in Young's modulus with thermal cycling and aging

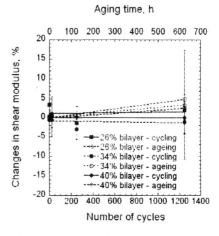

Fig.6. Changes in shear modulus with thermal cycling and aging

Tables II and III, as well as Figures 5 and 6, summarize the effect of thermal cycling and thermal aging on the elastic properties of bi-layers. As in the case of porosity, elastic Young's and shear moduli didn't change significantly as a result of thermal aging or thermal cycling. This suggests that microstructural changes, such as microcracking, do not occur during thermal cycling, because such changes would cause a decrease in the stiffness of the test specimens.

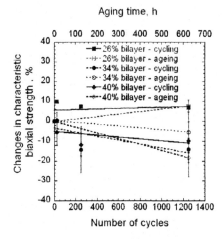

Fig.7. Changes in characteristic biaxial strength with thermal cycling and aging

Effect of thermal cycling and thermal aging on biaxial strength

Given the nature of stresses to which SOFC materials will be subjected during service, the equibiaxial flexural strength is an appropriate measure to quantify the mechanical strength of these materials. The equibiaxial flexural strength of Ni-YSZ/YSZ bi-layers was determined before and after thermal cycling and thermal aging by the ring-on-ring test method [7] and the results were analyzed using Weibull statistics. In the ring-on-ring configuration the disk-shaped test specimens with diameter (D) of ≈25 mm were spaced concentrically between a loading ring with diameter (D_l) of 5.5 mm and a supporting ring with diameter (D_s) of 20 mm[8]. The test fixture was articulated by using three rods and a semi-sphere. The loading and unloading rings were machined from Delrin® plastic to minimize friction and contact stresses between the ring and the test specimen. The load was applied to the test specimens through the loading ring at a constant cross head displacement rate of 1 mm/min using an electromechanical testing machine. In these tests, the Ni-YSZ layer was always subjected to the maximum tensile stress. The equibiaxial strength was calculated using the following equation for known breaking load, F, test specimen dimension and Poisson's ratio, v of the material:

$$\sigma_f = \frac{3 \cdot F}{2\pi h^2}\left[(1-v)\frac{D_s^2 - D_l^2}{2D^2} + (1+v)\ln\frac{D_s}{D_l}\right]$$

where h, D, D_s, and D_l are sample thickness, sample diameter, diameter of supporting ring and, diameter of loading ring respectively.

The Weibull analysis of the strength results, namely characteristic strength and Weibull modulus, are summarized in Tables II and III. Figure 7 presents plots of the changes in characteristic strength for Ni-YSZ/YSZ bi-layers as a function of number of cycles and aging time. The characteristic biaxial strength of 34-bi-layers and 40-bi-layers was found to decrease with the number of thermal cycles or aging time by as much as 20%. However, the biaxial strength of 26-bi-layers followed an opposite trend and increased by as much as 5% with number

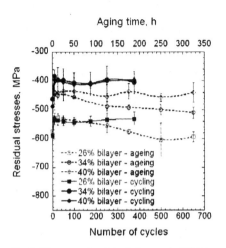

Fig.8. Changes of residual stresses in YSZ layer with thermal cycling and aging

of thermal cycles. These results are important because they demonstrate that the strength of Ni-YSZ materials can be affected by thermal cycling and thermal aging.

Effect of thermal cycling and aging on residual stresses and curvature of the test specimens

The magnitude of residual stresses in the YSZ layer of bi-layer test specimens was determined by X-ray diffraction at room temperature using a Scintag PTS goniometer with rotating anode generator after different numbers of thermal cycles and different aging times. Figure 8 shows a plot of the residual stresses in the YSZ layer as a function of time/number of thermal cycles. Each data point in Figure 8 represents the average value obtained from measurements on three bi-layer

test specimens. It was found that the YSZ layer was invariably in residual compression and that the magnitude of the compressive stress decreased with increasing porosity of the Ni-YSZ layer.

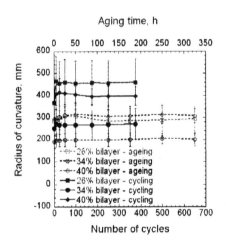

Fig.9. Changes of radius of curvature in YSZ layer with thermal cycling and aging

The compressive residual stress in the YSZ layer was found to relax. i.e. to become less compressive during the first 10 cycles or corresponding 5h of aging in all test specimens. Figure 8. After the first 10 cycles, the magnitude of the compressive residual stresses doesn't change significantly in the case of thermal cycling. However, compressive residual stresses decrease slightly with time in the case of thermal aging after the first 5 h of aging. It is worth noting that initial increases in compressive stresses are more pronounced in the case of thermal cycling than thermal aging. These results were found to be very reproducible for all test specimens.

The decrease in the magnitude of residual stresses with thermal cycling could be explained if Ni-YSZ experienced cracking or if the porosity in this layer increased as a result of thermal cycling or thermal aging. However, the changes observed in the porosity and elastic constants of these materials after thermal cycling cannot support this argument. Another possible explanation for the stress relaxation in the YSZ layer could be creep of the Ni-YSZ layer, which would be subjected to tensile residual stresses, albeit small, as required by equilibrium. This seems to be a very likely mechanism because the propensity of NiO-YSZ to undergo creep deformation at 800°C was demonstrated recently through stress-relaxation testing[1].

While the decrease in the magnitude of the compressive residual stress in the YSZ layer can be explained by the creep deformation of the Ni-YSZ layer, this argument cannot explain the opposite trend observed after long term thermal aging. Preliminary quantitative image analysis of the microstructure of these materials has revealed the slight coarsening of small Ni grains, which could explain these trends, but additional work is in progress to understand these results.

Figure 9 shows the changes in the radius of curvature of the same bi-layer test specimens that were used for residual stresses measurements. Values for the radius of curvature, which was measured on the YSZ layer in two mutually orthogonal directions, and for the average values for the three bi-layer test specimens of the same porosity are presented. The positive value of the radius of curvature means that the YSZ side of the bi-layer test specimen is convex. As in the case of residual stresses, the change in the radius of curvature is very pronounced at the initial stage of thermal cycling or thermal aging. The increase in radius of curvature at this stage indicates that the test specimens become more flat. This would be in good agreement with the observed relaxation of residual stresses during initial thermal cycling or aging. This is indirect evidence that creep of the Ni-YSZ layer could be responsible for initial relaxation of the residual stresses. However, after initial ≈25 cycles or ≈12.5 h of thermal aging, the radius of curvature does not change significantly either for thermal cycling or thermal aging. This could likely be the result of creep deformation reaching a saturation stage point at which no further creep deformation occurs.

SUMMARY

Bi-layer test specimens of YSZ and Ni-YSZ were subjected to thermal cycling between 22°C and 800°C and thermal aging at 800°C in a gas mixture of 4% H_2 and 96%Ar. It was found that the porosity and elastic moduli of the YSZ/Ni-YSZ bi-layers do not change significantly with the number of thermal cycles or thermal aging time. However, it was found that the characteristic strength decreases by as much as 20%.

Residual stresses in the YSZ layer were found to be compressive and to undergo relaxation during the initial stage of thermal cycling and aging. These results, along with the flattening of the test specimens, suggest that creep deformation of Ni-YSZ could be the dominant mechanism in the initial stage of thermal aging and cycling at 800°C. It was also found that after an initial stage, residual stresses do not change significantly with the number of thermal cycles, but that they increase and become more compressive with prolonged thermal aging.

ACKNOWLEDGMENTS

This research work was sponsored by the US Department of Energy, Office of Fossil Energy, SECA Core Technology Program at ORNL under Contract DE-AC05-00OR22725 with UT-Battelle, LLC. The authors are grateful for the support of NETL program managers Wayne Surdoval and Travis Shultz. The authors are grateful to Beth Armstrong, Claudia Walls and for help with specimen's preparation.

REFERENCES

[1]E. Lara-Curzio, M. Radovic, R. Trejo. "Reliability and Durability of Materials and Components for Solid Oxide Fuel Cells", FY 2005 Annual Report, Office of Fossil Energy Fuel Cell Program, (2005), http://204.154.137.14/technologies/coal_and_power_systems/ distributed_generation/seca/refshelf.html

[2]Minh N. Q. and Takahashi T., "Science and Technology of Ceramic Fuel Cells", Elsevier, Amsterdam (1995)

[3]ASTM standard C20

[4]ASTM standard C1259

[5]Radovic M. and Lara-Curzio E., "Change of Elastic Properties of Nickel-based Anodes for Solid Oxide Fuel Cells as a Function of the Fraction of Reduced NiO". J. Am. Ceram. Soc., 87, 2242-2246 (2004)

[6]Radovic M., Lara-Curzio E., Rieser L., "Comparison of Different Experimental Techniques for Determination of Elastic Properties of Solids", Mater. Sci. Eng. 368, 56-70 (2004)

[7]ASTM standard C1499

[8]Radovic M. and Lara-Curzio E., "Mechanical Properties of Tape Cast Nickel-based Anode Materials for Solid Oxide Fuel Cells Before and After Reduction in Hydrogen", Acta. Mater. 52 5747-5756 (2004)

THREE-DIMENSIONAL NUMERICAL SIMULATION TOOLS FOR FRACTURE ANALYSIS IN PLANAR SOLID OXIDE FUEL CELLS (SOFCs)

Janine Johnson and Jianmin Qu
School of Mechanical Engineering
Georgia Institute of Technology
Atlanta, GA, 30332-0405

ABSTRACT

A major step in future development of solid oxide fuel cells (SOFCs) is comprehension of the relationship between critical electrochemical and thermomechanical processes and the structural failure of the fuel cells. The reported research makes use of several finite element modeling tools to gain an overall understanding of fracture in a planar cell model. In our analysis, the ANSYS finite element (FEM) software is used to create a simplified cell structure with thermally induced stresses, which is then used to determine areas of high stresses in the anode-electrolyte-cathode (PEN) layers of an anode-supported planar SOFC. Refined fracture models are analyzed using the Fracture Mechanical Analyzer (FMA) code developed at Georgia Tech, which is a post-processing program capable of calculating fracture parameters in conjunction with finite element programs. The FMA code enables prediction of both crack growth and direction of three dimensional curvilinear cracks in a PEN structure under combined thermal and mechanical loading conditions. Examples of flaws in the anode, electrolyte, and at the anode-electrolyte interface will be given to demonstrate the robustness of the FMA software and to study possible failure modes of the PEN.

INTRODUCTION

While fuel cells are undergoing renewed development in industry, due to their compact configuration and high power density, a common industry complaint remains the unreliable lifespan of current planar configurations. This is due to electrochemical and thermomechanical reactions and strict structural requirements, making long term operation of planar solid oxide fuel cells (SOFCs) extremely difficult. To address this, numerical models have successfully incorporated electrochemistry and structural stresses, but detailed fracture analyses have yet to be performed. [1, 2] The recent shift of SOFCs to anode-supported structures, in which a thin film electrolyte is sintered onto the anode support, has further hampered structural modeling due to the large aspect ratio between the anode/electrolyte layers and cell length. In a fracture analysis there is also difficulty in modeling fracture at areas of interest, i.e. the PEN (anode-electrolyte-cathode layer) region or at the seal interfaces. Yet due to thermal mismatch between these layers and the cell operating conditions, significant thermal stresses are created within the cell structure, which may eventually lead to failure, making fracture analysis a critical part of thermomechanical modeling of SOFCs. Specifically, simulation tools are needed to obtain fracture mechanics parameters such as the stress intensity factors (SIFs), and to understand the influence of thermal gradients on crack behavior.

To meet this need, a computer program called Fracture Mechanical Analyzer (FMA) was developed to calculate the SIFs of 3D cracks, including interfacial cracks in the PEN structure subjected to combined mechanical and thermal loadings. [3] The FMA program, written in MatLab language, is essentially an "add-on" to any commercial finite element software. It computes the energy release rate and the individual SIFs based on the crack-tip displacement fields computed from any commercial finite element software. To illustrate its usage and capabilities, the FMA program is used here in conjunction with the finite element software ANSYS to study various crack geometries occurring in the PEN layer. Initially a simplified model of an anode supported PEN is created and studied under both constant and linear temperatures. The results of this analysis led to the creation of three different fracture models; a vertical straight crack in the electrolyte, vertical penny crack in the anode, and an interfacial penny crack. Fracture parameters are calculated for each of these situations, and finally predictions are made about failure possibilities in the planar PEN structure.

FRACTURE THEORY

In linear elastic fracture mechanics (LEFM), three stress intensity factors (SIFs) are used to describe the intensity of the singular stresses occurring at the crack tip and eventually to predict crack growth.[4] Each parameter can be connected to a particular crack deformation and loading mode. For instance a tensile load perpendicular to the crack plane will result in opening behavior generating a Mode I SIF, or symbolically K_I. SIFs resulting from shear stresses are designated mode II which is in-plane shear (K_{II}) and mode III which is out-of-plane shear (K_{III}). The combination of these three stress intensity factors can then be used to describe the stress fields occurring for a crack for any boundary and loading conditions. While analytical solutions exist for simple problems it is extremely difficult to calculate parameters for varying and/or more complex loading configurations, therefore the stress intensity factors often need to be calculated numerically for each set of given boundary conditions and applied loads.

Study of cracks occurring at the interface requires the introduction of several material parameters and a different SIF value as discussed in detail by Hutchinson and Sou.[5] For instance, the mode I and mode II SIFs are coupled due to the material mismatch at the interface, i.e. tensile loads which result in a pure Mode I condition for a homogenous crack will create a Mode II SIF when applied to a bimaterial crack. This is brought about by the different constitutive responses of the two materials, which can be characterized by the Dunder's parameters (α, β) in equation (1).

$$\alpha = \frac{\mu_1(\kappa_2+1) - \mu_2(\kappa_1+1)}{\mu_1(\kappa_2+1) + \mu_2(\kappa_1+1)}$$
$$\beta = \frac{\mu_1(\kappa_2-1) - \mu_2(\kappa_1-1)}{\mu_1(\kappa_2+1) + \mu_2(\kappa_1+1)}$$

(1)

where κ_i is equal to $3-4\nu_i$ for plane strain and $\kappa_i = (3-\nu_i)/(1+\nu_i)$ for plane stress and i designates the material number. The symbol ν is the Poisson's ratio and μ is the shear modulus.

The Dunder's parameters are used to define the bimaterial constant (ε) which is a function of the Dunder's parameters and ultimately the material properties of the two materials and is defined in equation (2). It can be seen from the formula that as β approaches zero that the bimaterial constant will also approach zero resulting in zero coupling between the Mode I and Mode II SIFs, or more simply a homogenous material.

$$\varepsilon = \frac{1}{2\pi} \ln\left(\frac{1-\beta}{1+\beta}\right) \tag{2}$$

The coupling of K_I and K_{II} requires the use of a complex SIF (**K**) and the magnitude of K_{II} is determined by the bimaterial constant (ε). Definition of the bimaterial constant allows for a complex SIF to be used in interfacial fracture mechanics.

$$\mathbf{K} = K_I + iK_{II} \equiv (\text{applied stress}) \times FL^{1/2 - i\varepsilon} \tag{3}$$

where F is a complex valued function and L is some characteristic length such as the crack length.[5]

Knowledge of the SIFs can be used to determine if the crack will grow and in traditional linear elastic fracture mechanics theories, a necessary condition for fracture is

$$\frac{1}{E^*} \frac{\mathbf{K}\overline{\mathbf{K}}}{\cosh^2(\pi\varepsilon)} + \frac{K_{III}^2}{2\mu^*} = G_{Ic} \tag{4}$$

where G_{Ic} is the fracture toughness of the interface, which is a intrinsic material property that needs to be measured experimentally, and the asterisk refers to the calculated effective material properties shown below

$$E^* = \frac{E_1' E_2'}{E_1' + E_2'} \quad , \quad \mu^* = \frac{\mu_1 \mu_2}{\mu_1 + \mu_2} \text{, where} \tag{5}$$

$$E' = \begin{cases} E & \text{for plane stress} \\ \dfrac{E}{1-v^2} & \text{for plane strain} \end{cases} \tag{6}$$

In the above equations, E_i is the Young's modulus and μ_i is the shear modulus. Their subscripts indicate the material below ($i = 1$) and above ($i = 2$) the interface crack.

THERMAL INTERACTION INTEGRAL

The SIFs can be calculated numerically using the thermal interaction integral method. A detailed derivation of the method can be found in the referenced work by Johnson.[3] Here only the key equations involved are briefly discussed. The interaction integral that accounts for the non-uniform temperature distribution can be written as

$$I = -\int_V \left(P_{jk} q_{j,k} + P_{kj,j} q_k\right) dV \text{, where} \tag{7}$$

$$P_{jk} = \sigma_{mn} \varepsilon_{mn}^{aux} \delta_{jk} - \sigma_{ik} u_{i,j} - \sigma_{ik}^{aux} u_{i,j} \tag{8}$$

$$P_{kj,j} = \sigma_{ij} \varepsilon_{ij,k}^{aux} - \sigma_{ij} u_{j,ik}^{aux} - \sigma_{ij,j}^{aux} u_{j,k} - \alpha \sigma_{ii}^{aux} u_{i,j} \tag{9}$$

In the above, σ_{ij} and ε_{ij} are the stress and strain tensors, respectively, α is the coefficient of thermal expansion (CTE) and δ_{ij} is the Kronecker delta. The summation convention applies to the subscripts and a comma denotes the spatial derivative. The quantities with a superscript aux are called the auxiliary fields which are the known analytical solutions to a semi-infinite interfacial crack with a straight front in a bimaterial of infinite extent. In equation (7) integration is carried out over a volume V surrounding the crack front. The function q_i is an artifact constructed to facilitate the formulation and must be a differentiable vector function within the volume V, but must be zero on the boundary V. It was shown by Shih et. al. that the interaction integral is path-independent, i.e., the value of I is the same for any volume V encompassing the crack front.[6] Therefore, once the stress and displacement fields near the crack front are known, I can be evaluated from for a volume of choice that is convenient for numerical integration.

On the other hand, for a very small volume around the crack front, the field quantities in (7) can be replaced by their corresponding asymptotic singular fields which leads to

$$I(s) = \frac{2}{E^* \cosh^2(\pi\varepsilon)}\left[K_I K_I^{aux} + K_{II} K_{II}^{aux}\right] + \frac{1}{\mu^*} K_{III} K_{III}^{aux} \tag{10}$$

where the variable s denotes the location along the crack front.[7] Since I is known from the interaction integral, the above equation can be simplified to solve for KI, which can be calculated from (11) by choosing an auxiliary field that has $K_I^{aux} = 1$ and $K_{II}^{aux} = K_{III}^{aux} = 0$, i.e.,

$$K_I = \frac{I(s)}{2} E^* \cosh^2(\pi\varepsilon) \tag{11}$$

The other two SIFs can be similarly obtained by selecting the appropriate auxiliary fields.

NUMERICAL PROCEDURE

The 3d structural analysis was separated into two steps: a global cell model and a local fracture model. The global–local modeling technique extrapolated the boundary displacements and temperature fields from a location in the full scale model to the smaller local fracture model. This fracture model would be able to incorporate a higher density of nodes at the crack edge over that of a full scale model featuring a crack. The global–local modeling is acceptable because the occurrence of flaws within the stack will influence only those regions immediately surrounding the crack. Fracture parameters were then determined by using the FMA program which is based on equations (7) through (10).

GLOBAL ANALYSIS
Numerical Model

The global cell model created for this study focused on the impact of the thermal mismatch of the PEN layers. No mechanical constraints were placed on the model and periodic boundary conditions were applied such that the PEN layer behaved approximately like a beam in pure bending. Three loading conditions were applied to the model to simulate the different operating conditions and are listed in Table I. The conditions are characterized as a total temperature change from a stress-free state, which is defined as the temperature in the PEN layer in which all three layers of the PEN layer are stress free. For example, due to micro-cracking after sintering, a PEN layer could be stress free at 500° C, making an operating temperature of 700°C result in a 200° temperature difference from a stress free state.

Table I: Loading Conditions in Global PEN model

Case	Loading Type	Operating Condition
I	Uniform Temperature Increase	represents temperature increase during operation from stress free state
II	Uniform Temperature Decrease	represents cooling to room temperature from stress free state
III	Linear Temperature Increase in Direction of Fuel Flow	represents temperature change due to electrochemical reactions

The layer thicknesses in the model are normalized by the anode height and the electrolyte and cathode heights can be described as percentages of the anode height. Different models are created with the electrolyte height varying from 3, 5, 7, 10 and 15% of the anode height. The cathode height is held constant for each model at 15%. The total length of the model is ten times greater than the anode, although symmetry allows the temperature constant model to be half modeled. The width of the model is five times greater than the anode and symmetry conditions are used to half model this width. An illustration of the PEN model is in Figure 1 and it should be noted that the origin of the global coordinates is at the bottom front left corner. It is from this origin that the future fracture models are located. Each model is meshed using 10 node tetrahedral elements.

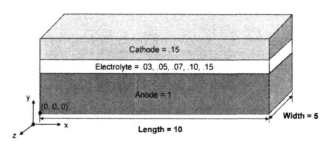

Figure 1: Model configuration for PEN layer.

Table II is a list of typical material properties of anode supported cells. The literature values for the mode I fracture toughness (K_{lc}) of the electrolyte and anode are 1.65 MPa•m$^{1/2}$ and 1.04 MPa•m$^{1/2}$, respectively.[8]

Table II: Material properties of PEN materials

Material	Modulus-E(GPa)	Poisson's Ratio-v	CTE-α (10-6/°C)
Electrolyte (YSZ)[a]	200	0.3	10.56
Anode (Ni+YSZ) [a]	96	0.3	12.22
Cathode (LSM) [b]	96	0.3	10.56

a. Values take from Yakabe, Baba et al.[9]
b. Approximate values matched to anode modulus and electrolyte CTE respectively.

Numerical Results

Figure 2 plots the normalized stress versus the PEN height away from the edge singularities for positive temperatures and multiple electrolyte heights. The stresses plotted are in-plane normal stresses (σ_x) and it is this stress that will have the main impact on the future fracture models. The shear stresses in the model are insignificant for these loading conditions, except at the edges.[10]

All stresses are normalized by a thermal stress (σ_e) which is the electrolyte stress calculated from a one dimensional (1d) analysis of the PEN layer shown, see equation (12). This stress is the maximum stress seen in the electrolyte for a 1d model with an electrolyte height of .07, and while not exact, use of this equation allows good approximation of comparable stresses at other temperatures, without the need for additional numerical analyses.

$$\sigma_e = \left(\frac{(\alpha_e - \alpha_a)h_a E_a + (\alpha_e - \alpha_c)h_c E_c}{h_a E_a + h_e E_e + h_c E_c} \right) \Delta T E_e \tag{12}$$

In equation (12) the subscripts a, e, and c refer to the anode, electrolyte and cathode material properties respectively. ΔT is the temperature difference from the stress free state.

Further examination of Figure 2 and consideration of basic mechanical principles allow the following conclusions to be made, which can be used to guide the fracture analysis of the next section.

- The curves shown in Figure 2 envelope the stresses for electrolyte heights between 3% and 15% of the anode height.
- The stress will reverse signs when the temperature changes sign, for instance in the case of a negative temperature the electrolyte will be in compression as in case II.
- Since normalized and a linear elastic problem this curve is the same for both the constant temperature (case I) and a linear temperature (case II).
- For the boundary conditions used, the stresses are independent of the length and width of the cell, i.e. these stresses will occur for any size cell.

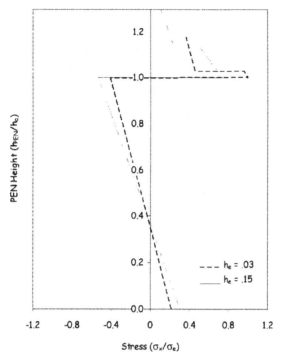

Figure 2: In-plane stress (σ_x) for positive temperatures versus PEN height

FRACTURE ANALYSIS
Fracture Models

From study of the stress behavior in the global model, three different cases were picked for fracture analysis and are listed in Table III. First since the electrolyte was a high tensile area during heating, vertical cracks in the electrolyte are studied. However, during cooling the electrolyte is now in tension halting growth and it is the anode that needs to be studied for vertical cracks. Lastly since the growths of delaminations are a possibility that would reduce power output an interfacial delamination is studied between the electrolyte and anode for variable temperature loading. For each case a standard small scale fracture model is located with respect to the coordinates of the global model in Figure 1. Boundary conditions and temperatures are then extracted from the global model and applied to the local fracture model.

Table III. Fracture models and relevance to cell operating conditions

Case	Fracture Model	Operating Conditions and Potential Failure
i	Vertical straight crack in electrolyte	Case I—initial defects in electrolyte could grow resulting in leakage at seals or interface.
ii	Vertical penny crack in anode	Case II—during cooling crack could grow towards electrolyte leading to leakage.
iii	Penny crack at anode-electrolyte interface	Case III—temperature gradients lead to increased delamination size resulting in reduced power output

Case i: Straight vertical crack in electrolyte

Figure 3 shows the location of the vertical crack within the local fracture model. The flaw is the entire height of the electrolyte. Due to symmetry only half the crack is model and the half length is designated a. The bottom cracktip is located at point $P(2.5, 1.0, -a)$ from the global origin designated in Figure 1. The overall dimensions of the model are set by the points $Q_1(2.3, 0.9, 0)$ and $Q_2(2.7, 1.1, -0.2)$. A positive constant temperature of 200°C is applied to the global model and extracted to the fracture model. Several cases are analyzed for an increasing crack size with a constant element height ($e = .01$) and a constant electrolyte height of 7% the anode height. The entire model is meshed using 20 node brick elements.

Case ii: Vertical penny crack in the anode

Next a penny shaped crack was placed in the tensile region of the anode during cooling. To conduct the fracture mechanics analysis, a cylindrical region containing the crack is meshed into 20-node brick elements as shown in Figure 4. Because of the symmetry, only one half of the crack needed to be modeled. Element size at the crack tip is controlled by a ratio with respect to the crack radius (a), such that $e = a/18$. The location of the model is controlled by the points designated in Figure 4 such that; $P(2.5, 0.6, 0)$, $Q_1(2.1, 0.2, 0)$, and $Q_2(2.9, 1.0, 0)$.

Case iii: Delamination at anode-electrolyte interface

The last model examined is that of a penny shaped delamination at the interface of the anode and electrolyte. The cylinder model used in Case ii is modified to create a crack with a radius of 1.5 mm at the interface. Once again the layers are considered perfectly bonded except at the crack interface. The center of the crack is located at $P(5, 1, 0)$ with the edge points located at Q_1 (2.5, 0.8, 0) and $Q_2(7.5, 1.22, 0)$. The element size is set to a ratio of $e = a/18$.

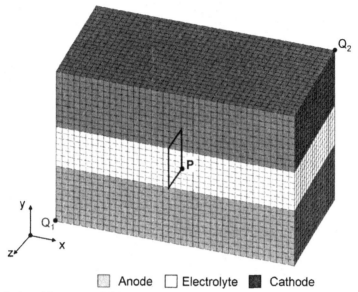

Anode ▢ Electrolyte ▮ Cathode

Figure 3: Local fracture model for case i (vertical crack in electrolyte) with $h_e = .07$.

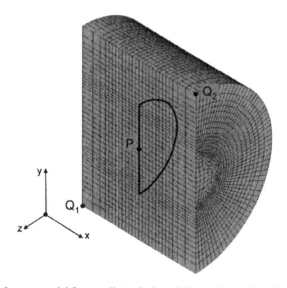

Figure 4: Local fracture model for case ii (vertical crack in anode); crack radius is $a = .1$

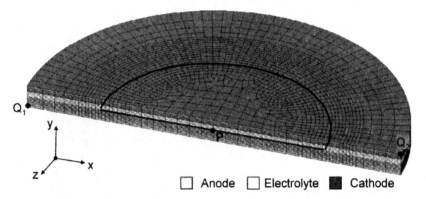

Figure 5: Local fracture model for case iii (delamination at anode-electrolyte interface)

Numerical Results
Case i: Straight vertical crack in electrolyte
     The vertical crack in the electrolyte is studied under a constant temperature increase of 200°C and it was determined that the only significant SIF was K_I. Figure 6 is a plot of the normalized K_I values with respect to the PEN height. The SIF values are normalized by $\sigma_e \sqrt{\pi a_n}$, where the standard crack length (a_n) is equal to 0.1 mm. At the interface of the electrolyte and anode a slightly higher K_I occurs over that of the electrolyte-cathode interface corresponding to the lower stress at this interface, reference Figure 2. However, the K_I values are at a maximum in the center since this is the area where the crack is least constrained by the geometry of the model. The actual maximum calculated K_I value is 0.34 MPa•m$^{1/2}$ at $a = 0.1$ mm which is still significantly less than literature fracture toughness values.

Case ii: Vertical penny crack in the anode
     The vertical crack in the anode also underwent a pure Mode I condition under a negative temperature change, but with added complexity due to the circular nature of the flaw. For cracks in the anode the values were normalized by a K_I value calculated from the 1d stress in the anode as shown in equation (13).

$$\sigma_a = \frac{h_c E_e (\alpha_c - \alpha_a) + h_e E_c (\alpha_c - \alpha_a)}{h_a E_a + h_c E_c + h_e E_e} \Delta T E_a \qquad (13)$$

     The area most susceptible to crack growth is at the crack edge nearest the electrolyte, which is the location of the highest stress on the crack edge. Also due to the large variation of the stress across the anode the maximum K_I value was determined for increasing crack sizes.

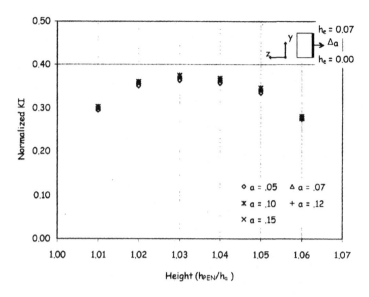

Figure 6: Mode I SIF $\left(K_I / \sigma_e \sqrt{\pi a_n} \right)$ for crack in electrolyte at positive constant temperature.

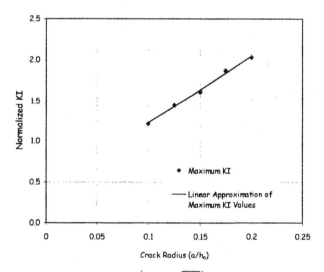

Figure 7: Maximum normalized mode I SIF $\left(K_I / \sigma_u \sqrt{\pi a_n} \right)$ values for increasing crack radius

Figure 7 compares the maximum stress along the crack front for different crack sizes for the electrolyte height of .07. It can be seen that as the crack radius increases the edge is closer to the electrolyte and correspondingly the K_I is higher. It should also be noted that there is a linear relationship between the maximum K_I and the crack radius. For the 200°C temperature decrease the maximum actual K_I value is 0.26 MPa·mm$^{1/2}$ which means that for this temperature difference a flaw up to 0.2 mm radius will not increase in size.

Case iii: Delamination at anode-electrolyte interface
        The final fracture model studied was the interfacial circular flaw. Since the global model results showed that there were no significant stresses generated in the y-direction and that all shear stresses were minimal the study shifted from study of constant temperature changes to linear changes. However, even for temperature changes up to 100°C/mm no K_I existed and the values for K_{II} and K_{III} were extremely small. The total energy release rate, calculated from equation (4), never exceeded 11.10 (10^{-3}) N/m, an extremely small value making crack growth unlikely, even for a flaw size with a 1.5 mm radius.

DISCUSSION
        The creation of a simplified PEN model showed the basic behavior of a planar fuel cell due to thermal mismatch without mechanical constraints.   Study of this model allowed several generalizations to be made about the PEN behavior such that the two curves shown in Figure 2 can be used to estimate the stresses of the PEN for multiple loading conditions and PEN sizes.
        The three different fracture models examined were placed in locations where either failure would be catastrophic and/or the stresses from the global model appeared especially high. In the first case of the electrolyte in the PEN layer, it was found that the model geometry constrained significant crack growth within the electrolyte itself and that the Mode I SIF was virtually independent of length. Eventually modification of the FMA program will be needed to study the possibility of crack growth in the direction of the cathode or anode materials.
        The second case, the vertical crack in the anode is the most likely source of failure, although the models used here never did exceed the fracture toughness of the material. It was this model that showed the most variation along the cracktip and was also least constrained by the geometry of the model. If the vertical crack was to grow it would also grow towards the electrolyte leading to the possibility of leakage, but this will only occur when the fuel cell is cooled since the anode is primarily tensile only for a negative temperature change.
        Finally, the study of an interfacial crack eliminates the possibility for delamination growth except for extremely large crack radiuses and only for unrealistic and very steep temperature gradients. This knowledge is significant since delamination between the anode and electrolyte could lead to decreased power output of the cell.   It was also this model that eliminates the possibility of the shear SIFs K_{II} and K_{III} of being a possible source of failure. None of the fracture models studied had K_{II} or K_{III} at significant numerical values.

SUMMARY AND CONCLUSIONS
        This was an introductory work into the study of fracture analysis in anode supported cells. The FMA program was used in conjunction with the ANSYS commercial software to demonstrate a method to study structural failure in the PEN despite issues with the aspect ratios

of the material layers and difficulties in performing a complex fracture analysis. After studying three different fracture models it was found that thermal mismatch and modest temperature gradients will likely not be a leading cause of crack growth in the PEN layer. If crack growth was to become a significant factor additional factors such as mechanical constraints from hard seals would probably be the dominant driving force. For the global model studied here if failure was to occur it would most likely occur in the anode as crack growth towards the electrolyte during cooling.

This study did suggest several useful possibilities for future analysis. Analysis of the PEN layer shows that temperature sign has a significant impact on fracture behavior, which leads to the question of the impact of thermal fatigue on the model, especially in the anode which will cycle between tension and compression. The relative insensitivity of the delamination to temperature change also suggests that future work would be most useful if focused on the effects of mechanical constraints on the fuel cell and not strictly on the complex thermal gradients brought about from electrochemistry.

ACKNOWLEDGMENT

This work was supported by SECA/DOE under Contract No. DE-AC26-02NT41571. The authors acknowledge the stimulating discussions with T. Shultz (Program Manager, NETL), M. Khaleel and B. Koeppel (PNNL).

REFERENCES

[1]	Khaleel, M. A., Z. Lin, P. Singh, W. Surdoval and D. Collin, A finite element analysis modeling tool for solid oxid fuel cell development: coupled electrochemisty, thermal, and flow analysis in MARC. Journal of Power Sources, 2004. 130: p. 136-148.

[2]	Yakabe, H., T. Ogiwara, M. Hishinuma and I. Yasuda, 3-D model calculation for planar SOFC. Journal of Power Sources, 2001. 102: p. 144-154.

[3]	Johnson, J., Fracture failure of solid oxide fuel cells, M.S Thesis, Mechanical Engineering. 2004, Georgia Institute of Technology: Atlanta. p. 114.

[4]	Anderson, T. L., Fracture Mechanics: Fundamentals and Applications. 2 ed., ed. Vol. 1995: CRC Press.

[5]	Hutchinson, J. W. and Z. Suo, Mixed Mode Cracking in Layered Materials. Advances in Applied Mechanics, 1991. 29: p. 64-187.

[6]	Shih, C. F., B. Moran and T. Nakamura, Energy release rate along a three-dimensional crack front in a thermally stressed body. International Journal of Fracture, 1986. 30(2): p. 79-102.

[7]	Nakamura, T., Three-dimensional stress fields of elastic interface cracks. Transactions of the ASME. Journal of Applied Mechanics, 1991. 58(4): p. 939-46.

[8]	Radovic, M., E. Lara-Curzio, B. Armstrong and C. Walls. Effect of thickness and porosity on the mechanical properties of planar components for solid oxide fuel cells at ambient and elevated temperatures. in. 2003. Cocoa Beach, FL, United States: American Ceramic Society.

[9]	Yakabe, H., Y. Baba and I. Yasuda. X-Ray Stress Measurements for Anode-Supported Planar SOFC. in. 2001. Electrochemical Society.

[10]	Chiu, C.-C. and Y. Liou, Residual stresses and stress-induce cracks in coated components. Thin Solid Films, 1995. 268: p. 91-97.

Modeling

ELECTROCHEMISTRY AND ON-CELL REFORMATION MODELING FOR SOLID OXIDE FUEL CELL STACKS*

K.P. Recknagle, D.T. Jarboe, K.I. Johnson, V. Korolev, M.A. Khaleel, P. Singh
Pacific Northwest National Laboratory, Richland, WA. 99352

ABSTRACT

Providing adequate and efficient cooling schemes for solid-oxide-fuel-cell (SOFC) stacks continues to be a challenge coincident with the development of larger, more powerful stacks. The endothermic steam-methane reformation reaction can provide cooling and improved system efficiency when performed directly on the electrochemically active anode. Rapid kinetics of the endothermic reaction typically causes a localized temperature depression on the anode near the fuel inlet. It is desirable to extend the endothermic effect over more of the cell area and mitigate the associated differences in temperature on the cell to alleviate subsequent thermal stresses. In this study, modeling tools validated for the prediction of fuel use, on-cell methane reforming, and the distribution of temperature within SOFC stacks are employed to provide direction for modifying the catalytic activity of anode materials to control the methane conversion rate. Improvements in thermal management that can be achieved through on-cell reforming is predicted and discussed. Two operating scenarios are considered, one in which the methane fuel is fully pre-reformed and another in which a substantial percentage of the methane is reformed on-cell. For the latter, a range of catalytic activity is considered, and the predicted thermal effects on the cell are presented. Simulations of the cell electrochemical and thermal performance with and without on-cell reforming, including structural analyses, show a substantial decrease in thermal stresses for an on-cell reforming case with slowed methane conversion rate.

INTRODUCTION

For high power density operations and many high power applications of SOFC stacks, the current development trend is toward the low-cost fabrication of larger active area cells and stacks that also incorporate advanced high-performance electrodes and current collector materials. For cell stacks such as these, removal of excess heat becomes increasingly important for long term stable and reliable operation. Of the various heat-removal methods available for use in SOFC stacks, including radiative and excess cathode air cooling, on-cell reformation (OCR) of methane could perhaps be the most beneficial. The endothermic steam-methane reformation reaction (Eq. 1) occurs on the conventional nickel base cell anode at high temperatures to form hydrogen and carbon monoxide. The hydrogen and carbon dioxide gas products can be consumed electrochemically in the oxidation reactions (Eq. 2 and 3) and are also subject to the rapid water-gas shift reaction (Eq. 4). Typical SOFC operating temperatures in the 750–800°C range provide sufficient heat, and Ni-YSZ cermet anode material, typical to SOFCs, is a suitable catalyst for the steam-methane reforming reaction. Thus OCR of methane can be used in SOFC systems to remove excess heat generated on the anode by the electrochemical oxidation of hydrogen. When OCR is used, the demand on an external fuel reformer and the amount of heat it requires are decreased proportionally to the fraction of fuel reformed on-cell. Thus the size of the air blower, external reformer, and possibly a heat exchanger can all be decreased, yielding improved system efficiency and lower cost.

$$CH_4 + H_2O \Rightarrow 3H_2 + CO \tag{1}$$

$$H_2 + \frac{1}{2}O_2 \rightarrow H_2O + 2e^- \tag{2}$$

$$CO + \frac{1}{2}O_2 \rightarrow CO_2 + 2e^- \tag{3}$$

$$CO + H_2O \leftrightarrow CO_2 + H_2 \tag{4}$$

On-cell reformation allows for the coupling of endothermic reformation reaction with the exothermic electrochemical oxidation of reformed fuel species. While the thermal coupling is an attractive means of removing excess heat, the kinetics of the reformation reaction on the conventional cell anode is normally quite rapid. and the increase in thermal stresses that can develop from the resultant endotherm presents a challenge associated with OCR operation that requires attention. By slowing the reformation rate and smoothing the endothermic effect over a larger portion of the cell, the temperature difference on the cell (and the associated thermal stresses) can be decreased.

At Pacific Northwest National Laboratory (PNNL), modifications of conventional anode material has been developed and tested under the U.S. Department of Energy-sponsored Solid State Energy Conversion Alliance (SECA) program to decrease the catalytic activity for reformation and to slow methane conversion[1]. The degree to which the conversion rate should be reduced for optimal thermal and electrochemical performance within planar SOFC stacks, however, is not well documented. The objective of this paper is to document the numerical study that was performed to determine reduction in cell temperature difference and subsequent thermal stresses by manipulating reformation activity on the cell. It was assumed in this work that the catalytic activity of the anode can be reduced for methane reformation without a reduction in the electro-catalytic activity of the anode for the oxidation of H_2 and CO.

NUMERICAL MODELING

Methodology

A computational modeling tool for simulating the multi-physics of SOFC operation was used in this study. The PNNL developed SOFC-MP solves the finite-element equations for mass transport. energy, and electrochemistry required to predict the fluid flow, temperature, species, and current density distributions in a three-dimensional SOFC geometry[2,3]. A methodology complementary to SOFC-MP was used for some of the cases presented in this writing. A description of this methodology is provided elsewhere[4]. The electrochemistry model used was described by Chick et al.[5], calibrated[6,7] for application to planar stack simulations, and updated to provide an improved anode concentration polarization model[8]. The capabilities of these tools have also been expanded to incorporate steam-methane reformation for simulating on-cell reforming[9], and have been updated with a rate expression derived experimentally at PNNL[1]:

$$(-r_{CH_4})(mol / gm_{cat} / s) = (2.188E8)e^{-\frac{E_{act}}{RT}} C_{CH_4} C_{CO_2}^{-0.0134}$$ (4)

In this expression, the temperature, T, is in Kelvin, R is the universal gas constant (8.314 J/mol-K), the activation energy, E_{act}, is 94,950 J/mol, and the concentrations, C_i, are in units of mol/cc. The conversion rate is first order with CH_4, and CO_2 has a slight hindering effect. King's testing included a broad range of steam-to-carbon ratios, and no effect of H_2O concentration was observed. The pre-exponential of the rate expression carries the units of moles per gram bulk catalyst per second because the testing was performed using anode powder in a plug reactor. The pre-exponential could also be considered in terms of an area of exposed nickel per gram of bulk catalyst (i.e., cm^2/gm_{cat}). Thus the conversion rate could be slowed by some process that effectively masks some of the exposed nickel area. Alternatively, the methane conversion could be slowed by an effective increase in the activation energy due to some competing reaction or other mechanism. Throughout this study, slowed methane conversion was simulated by artificially increasing the catalyst (Ni-YSZ anode) activation energy by various amounts. The impact of these modifications on the cell's temperature distribution is presented.

Two fuel compositions were considered, one completely reformed gas stream with no methane (pre-reformed fuel) and another in which a substantial fraction of the methane conversion was performed on-cell. In the stoichiometric relations (1, 2, and 3), one mole of CH_4 is converted to three moles of H_2 and one mole of CO. In the shift reaction, one mole of CO is converted to one mole of H_2. In the oxidation reaction, H_2 is the fuel for generating electrical current. From these considerations, the fuel content of gas mixtures that contain CO and CH_4 (on a mole/sec basis) is

$$(mol/s)H_2 + (mol/s)CO + (4mol/s)CH_4$$ (5)

The molar fuel composition for the OCR cases was 13% H_2, 59% H_2O, 18% CH_4, and 10% N_2. The molar composition for the pre-reformed fueling case was 55% H_2, 24% H_2O, 8% CO, 6% CO_2, and 7% N_2. The fuel content in each case was 4.7E-04 mol/sec.

Approach

The subject of this work is a generic cross-flow planar SOFC design with 10.4 x 10.6 cm (110.24 cm^2) active cell area and internal manifolds. The basic geometry and flow orientations are shown schematically in Figure 1. A three-dimensional computational model containing 57,200 elements was constructed to represent a single repeating cell unit assumed to be situated at the mid-level of a large, multiple-cell stack. It was assumed the stack was operating within an insulated enclosure with an air gap between the stack and enclosure walls. At mid-level in a large stack, the end effects of the stack are not important; thus, cyclic boundaries were used at

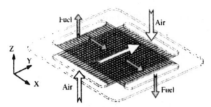

Figure 1. Schematic View of a Typical Cross-Flow Planar SOFC Interconnection Plate.

the top and bottom surfaces of the model to represent this condition. Wall boundaries at the stack perimeter accounted for natural convection of air surrounding the stack and radiation across the gap to the container walls. Inflow boundaries were set at constant mass flow with adjustable temperatures to enable control of average cell temperature to 1023K (750°C).

For this study, the electrochemical performance of the 110.24 cm^2 cell operating on the baseline pre-reformed fuel was taken to be 0.53 A/cm^2 (58.4 A) at 0.692 V and 64% fuel utilization with average cell temperature of 750°C. Anode, electrolyte, and cathode thicknesses were 600, 10, and 50 microns, respectively. The anode and cathode porosity was 30%, and the tortuosity was 2.5 for both electrodes. The materials and their properties used in the model for the steady state flow/electrochemical/thermal solution are summarized in Table 1.

Table 1. Material Properties Used in the Simulations

Material	Density kg/m^3	Thermal Conductivity W/m-K	Specific Heat, J/kg-K
Gases (air, fuel)	Ideal gas: $\rho = \rho(T)$	Multicomponent mass weighted	Multicomponent: mass weighted
PEN	4300	3	6.23 (steady)
Interconnect / separator	7700	26	8
Seal (glass)	2800	1.05	6

RESULTS

Uniform Manipulation of Activation Energy

The endothermic effect associated with the rapid kinetics of steam-methane reformation on an SOFC anode can cause an undesirable temperature depression where the fuel first contacts the cell. Thus, the desired effect of activity manipulation is to slow the reformation process and smooth the thermal gradients. As discussed above, the conversion rate (in the simulations) was slowed by artificially increasing the catalyst (Ni-YSZ anode) activation energy. To analyze the relative effects of conversion rate on the temperature distributions and power, the fuel and oxidant gas flow rates were held constant while the cell voltage and air/fuel inlet temperatures were automatically adjusted during the simulations to achieve the desired average current density and cell temperature respectively. Cases were completed to establish the baseline of expected performance related to uniform reformation activity manipulation. The baseline cases (cases 1A–5A) consisted of a simulation using the uniform standard activation energy of 94,950 J/mol

(Std-E_{act}), and cases in which Std-E_{act} was increased uniformly by 10, 17, 20, and 22% to slow the methane conversion. Increasing the Std-E_{act} by roughly 20% slows the conversion rate by as much as two orders of magnitude.

Figure 2 shows the effect of methane reformation rate on cell temperature difference and power. As the activation energy was increased from the Std-E_{act} (Case 1A) and methane conversion was slowed, the cell temperature difference (ΔT) became smaller. Coincident with the decreasing reformation rate was a decrease in the electrical power generated. The power decrease was due to decreased Nernst potential; thus, a decreased cell voltage was required to maintain the desired cell current. The decreased Nernst potential was driven by increased water vapor pressure, which also elevated the O_2 pressure in the fuel mixture. The curve in Figure 2 shows that substantial improvements in ΔT can be achieved by increasing the Std-E_{act} by up to 20% with relatively small loss of power. Increases to the Std-E_{act} of more than 20% were less beneficial for decreasing ΔT and had relatively large decreases in cell power as the limit of full conversion of methane was surpassed.

Also in Figure 2 is a data point for the performance of a fully pre-reformed fueling case. This case had an intermediate temperature difference relative to the OCR cases and slightly higher power. The pre-reformed case had lower water partial pressure than the OCR cases and hence slightly increased Nernst potential, cell voltage, and subsequent gross power. This is not to say that the net power of the pre-reformed case would be relatively larger. Knowledge or assumptions of the overall system would be required to make that assessment, and such system considerations are beyond the scope of this work.

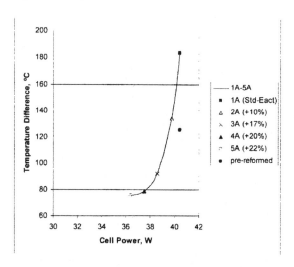

Figure 2.   Influence of Activation Energy (methane conversion rate) on
Cell Temperature Difference and Power

Temperature Distributions and Thermal Stresses

The predicted temperature distributions (at the cell level in the stack) for the pre-reformed fuel case and Cases 1A (Std-E_{act}) and 4A (Std-E_{act} +20%) are shown in Figures 3a, b, and c, respectively. For all cases, in the orientation shown, the cathode air flowed from top to bottom and fuel flowed from right to left.

As indicated by the magnitude of the temperature depression of Case 1A, shown in blue along the right edge of the cell in Figure 3b, the endothermic reformation of methane was essentially completed within roughly one-fourth of the cell length, as measured from the fuel inlet at right. Clearly, the speed of this reaction created a severe temperature depression near the fuel inlet and a large temperature difference on the cell. The temperature distribution for the case with activation energy increased by 20% (Case 4A) is shown in Figure 3c. In this case the endothermic reaction was extended over more of the cell area, the temperature depression became less pronounced, and the ΔT became much smaller. In fact, the ΔT was smaller than in the case with fully pre-reformed fuel (Figure 3a). The effect of these temperature distributions on the thermal stresses is summarized in Table 2. The maximum principal stress in the anode for Case 1A (with standard reformation activity) was 30% larger than that of the pre-reformed fuel case, while the maximum stress in the seal for the latter case was slightly larger. The larger anode stress in Case 1A was driven by the larger ΔT while the larger seal stress in the pre-reformed case was driven by the temperature distribution and the location of the maximum stress. In the pre-reformed case the seal stress was concentrated at the lower corner near the fuel inlet where the thermal gradients were largest. In Case 1A the seal stress was higher on average, but distributed along the fuel inlet side. In Case 4A, with slowed reforming, the maximum stresses were substantially less in both the anode and the seal due to smaller ΔT and improved temperature distribution relative to the other cases.

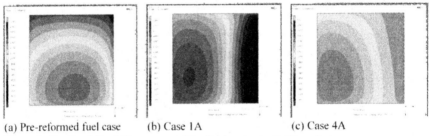

(a) Pre-reformed fuel case   (b) Case 1A   (c) Case 4A

Figure 3. Predicted Stack Temperature Distributions at the Cell Level

Table 2. Summary of Temperature and Maximum Principal Stress

Case	Temperature, °C Min, Max	ΔT	Sl_{max}. MPa Anode	Sl_{max}. MPa Seal
Pre-reformed	671, 796	125	27.6	11.3
1A (Std-E_{act})	636, 817	181	40.0	10.9
4A (Std-E_{act} +20%)	712, 787	75	11.8	5.9

Nonuniform Manipulation of Activation Energy

Taking advantage of the flexibility provided by numerical modeling, several attempts were made to further decrease ΔT without significant power decrease by thoughtfully specifying non-uniform distributions of the activation energy over the anode. Thus nonuniform modifications in activation energy were contrived and simulated to counter undesirable features that were identi-fied in the distributions of fuel and current density of a case with uniform activation energy attempting to minimize the thermal gradients. Several nonuniform cases (1B-11B) were attempted; these 2-dimensinal distributions of activity on the anode are represented graphically in Figure 4. As in Figure 3, the orientation of the active cell areas shown in Figure 4 is such that cathode air flowed from top to bottom and fuel flowed from right to left. In Cases 1B–5B the activation energy was higher near the fuel inlet at right to slow the methane conversion and was decreased across the cell to the left. The distributions in Cases 6B–11B used lower activation energy for fast conversion along the upper and lower cell edges.

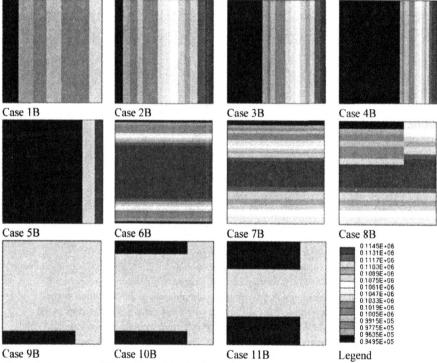

Figure 4. Summary of Nonuniform Distributions of Activation Energy Cases Tested.

The resulting temperature differences for Cases 1B–11B are compared with those of the uniform cases (1A–5A from the curve in Figure 2) in Figure 5. A performance improvement would correspond to a data point below and/or to the right of the curve. In general, the results from the nonuniform activation energy cases were very near the curve. Those farthest from the curve were found above it, indicating that the nonuniform modification was less effective than a uniform modification. A few cases were below the curve but were nonetheless very close to it. Consequently, these results indicated that nonuniform manipulations of the activation energy distribution were not significantly more effective than uniform modifications. Considering the added challenges when attempting to implement such nonuniform manipulations, there seemed little reason to pursue such efforts, at least for the planar geometry selected.

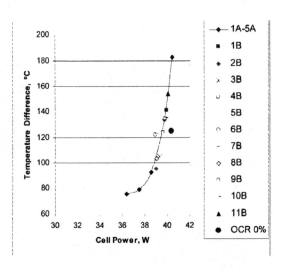

Figure 5. Comparison of Temperature Difference and Power Results from Nonuniform Activation Energy Cases with Uniform Activation Energy Cases.

CONCLUSION

Numerical simulations of OCR-SOFCs were performed to study the role of activation energy of the reformation reaction and how it could be manipulated to decrease the temperature difference while maintaining high power. Uniform manipulation of the activation energy revealed a trend in which the temperature difference decreased quickly for increases in the standard activation energy of 20% and less, while the subsequent power decrease remained small. Consequently, the temperature difference was significantly reduced without paying a sizable power penalty by increasing the activation energy. The immediate benefit of a smaller temperature difference is decreased thermal stresses within the stack components, resulting in higher mechanical reliability. A stress analysis was performed that demonstrated decreased maximum principal stresses in primary stack components using of OCR and a decreased reformation rate.

Nonuniform manipulations of the activation energy distribution were shown to improve performance very much like uniform manipulations. Due to the added complexity associated with implementing a nonuniform distribution of activation energy in an actual SOFC, this methodology is not encouraged for the planar cell used in this study. Improvements and benefits may be realized in other cell topologies due to differences in cell size and fuel flows.

ACKNOWLEDGEMENTS

Technical work described in this paper was performed as part of the U.S. Department of Energy Solid-State Energy Conversion Alliance (SECA) Core Technology Program. Pacific Northwest National Laboratory is operated by Battelle for the U.S. Department of Energy under Contract DE-AC05-76RL01830.

FOOTNOTES

*This manuscript has been authored by Pacific Northwest National Laboratory under Contract DE-AC05-76RL01830 with the U.S. Department of Energy. The United States Government retains and the publisher, by accepting the article for publication, acknowledges that the United States Government retains a non-exclusive, paid-up, irrevocable, world-wide license to publish or reproduce the published form of this manuscript, or allow others to do so, for United States Government purposes.

REFERENCES

[1]D.L. King, Y. Wang, Y. Chin, Y. Lin, H. Roh, R. Romiarek. "Controlling Activity and Stability of Ni-YSZ Catalysts for On-Anode Reforming." Presented at SECA Core Technology Program Review Meeting, Tampa, Florida. (2005).

[2]M.A. Khaleel. "Finite Element SOFC Analysis with SOFC-MP and MSC.Marc/Mentat-FC." *Proceedings of the Sixth Annual SECA Workshop.* National Engineering Technology Laboratory, Morgantown, WV.
http://www.netl.doe.gov/publications/proceedings/05/SECA_Workshop/SECAWorkshop05.html

[3]M. A. Khaleel, Z. Lin, P. Singh, W. Surdoval, D. Collins. "A Finite Element Analysis Modeling Tool for Solid Oxide Fuel Cell Development: Coupled Electrochemistry, Thermal, and Flow Analysis in Marc." *J. Power Sources,* 130 [1-2] pp. 136-148 (2004).

[4]K.P. Recknagle, RE Williford, LA Chick, DR Rector, and MA Khaleel. "Three-dimensional thermo-fluid electrochemical modeling of planar SOFC stacks." *J. Power Sources,* Vol. 113, pp. 109-114 (2003).

[5]L.A. Chick, J.W. Stevenson, K.D. Meinhardt, S.P. Simner, J.E. Jaffe, R.E. Williford. "Modeling and Performance of Anode-Supported SOFC." *2000 Fuel Cell Seminar – Abstracts,* pp. 619-622 (2000).

[6]L.A. Chick, R.E. Williford, J.W. Stevenson, C.F. Windisch, Jr, S.P. Simner. "Experimentally-Calibrated, Spreadsheet-Based SOFC Unit-Cell Performance Model." *2002 Fuel Cell Seminar - Abstracts* (2002).

[7]K. Keegan, M. Khaleel, L. Chick, K. Recknagle, S. Simner, J. Deibler. "Analysis of a Planar Solid Oxide Fuel Cell-Based Automotive Auxiliary Power Unit." *Society of Automotive Engineers, Congress 2002 Proceedings,* 2002-01-0413 (2002).

[8]R.E. Williford, L.A. Chick, G.D. Maupin, S.P. Simner, J.W. Stevenson. "Diffusion Limitations in the Porous Anodes of SOFCs." *J. Electrochemical Soc.*, 150 (8) A1067-A1072 (2003).

[9]K.P. Recknagle, P. Singh, L.A. Chick, M.A. Khaleel. "Modeling of SOFC Stacks with On-Cell Steam-Methane Reformation at PNNL." PNNL-SA-43248, *Proceedings of the Fuel Cell 2004 Seminar,* San Antonio, TX (2004).

MODELING OF HEAT/MASS TRANSPORT AND ELECTROCHEMISTRY OF A SOLID OXIDE FUEL CELL

Yan Ji, J. N. Chung, Kun Yuan
Department of Mechanical and Aerospace Engineering,
University of Florida, FL 32611-6300, USA
Email: jnchung@ufl.edu, Tel.: 352-3929607

ABATRACT

A three-dimensional thermo-fluid-electrochemical model is developed to simulate the heat/mass transport process in a solid oxide fuel cell. A network circuit is applied to simulate the electrical potential, ohmic losses and activation polarization. Governing equations of mass, momentum and energy conservation are simultaneously solved. A parametric study examines the effects of channel dimensions, rib width and electrolyte thickness on the temperature, species concentration, local current density and power density. Results demonstrate that decreasing the height of flow channels can lower the average solid temperature and improve cell efficiency. However, this improvement is limited for the smallest channel. The cell with a thicker rib width and a thinner electrolyte layer has higher efficiency and lower average temperature. Numerical simulation will be expected to help optimize the design of a solid oxide fuel cell.

INTRODUCTION

The SOFC is considered to be the most desirable fuel cell type for generating electricity from hydrocarbon fuels. The reason is that the SOFC is simple, highly efficient, tolerant to impurities, and can at least partially internally reform hydrocarbon fuels. In general, the SOFC can offer combined heat and power (CHP) at an electrical efficiency between 40%-50%, and an overall LHV efficiency between 80%-90%. There are three kinds of mechanisms causing the terminal voltage loss at a given operation condition [1]: (1) Ohmic polarization, which is from the ohmic resistances of all components and contact interfaces; (2) Activation polarizations at the anode and cathode, which are related to charge transfer processes and depends on the nature of electrode-electrolyte interfaces; (3) Concentration polarization, which is associated with the transport of gas species and thus depends on the channel dimensions, electrode thickness, and the nature of electrode microstructure, i.e., porosity and tortuosity. All losses should be minimized in order to improve the efficiency of a SOFC. In an electrolyte-supported SOFC, the typical thickness of electrolyte layer is larger than $100\,\mu m$ with thin electrode layers (20-$50\,\mu m$). Thus, the main contribution to ohmic polarization is from electrolyte and the concentration polarizations in electrode layers are much less than electrode-supported cells. In addition, the temperatures in a SOFC must be kept at a higher level ($>1000K$) to sufficiently activate the electrochemical reaction and reduce the resistance in electrolyte, which avoidably causes larger thermal stresses. In order to minimize the overall losses, lower the thermal stress and improve the electrolyte-supported SOFC efficiency, key parameters governing the ohmic polarization effect and heat/mass transfer efficiency are needed to be optimized. Cell's geometric parameters, such as channel dimension, electrode/electrolyte thickness and so on, are among those key parameters. The principle objective of present work is to examine the effects of geometry on the performance of an electrolyte-supported SOFC and optimize those parameters.

As a matter of fact, heat transfer, species diffusion, kinetics of chemical reactions and charge transfer in SOFCs' energy transport process are interdependent and very complicated.

Therefore, a mathematical model is a good tool to analyze the phenomena inside a SOFC and predict its performance. Many models of different complexity have been proposed for this purpose [2-7]. All efforts were expected to achieve a more accurate description of physical-electro-chemical processes and physical structure/microstructure. For example, Yakabe et al. [2] reported an investigation of thermal stress due to mismatch in thermal expansion coefficients among the cell components. In view of the microstructure, Kim et al. [5] studied polarization effects on an anode-supported cell performance in terms of a charge-transfer resistance. Li et al. [6] provided a coupled continuum-level electrochemical, flow and thermal model by approximately dealing with the heat/mass transport in electrodes. In the present study, a 3-D thermo-fluid electrochemical model is developed to simulate the heat/mass transport and electrochemical process. Then, the geometric effects on the SOFC performance are investigated. The geometry of interconnector and phenomenally microstructures of electrodes were considered. In addition, chemical reactions, activation overpotentials, heat generation in different locations were included in the model. Temperature distribution, chemical reaction rates, concentrations of species, local current density and output voltage were obtained and analyzed.

MATHEMATICAL MODELING

The stack of a solid oxide fuel cell is made up of many repeating unit cells, and most of them work under similar operation conditions. To reduce calculation work, we take one unit cell as the computational domain as illustrated in Fig.1. Symmetry conditions are applied at the boundaries in the z-direction. So only half of gas-flow channels need to be modeled. The single cell consists of a porous layer of anode, a porous layer of cathode, a dense layer of electrolyte, and interconnectors. The interconnectors support the structure and also serve as the current collectors. Air and fuel streams flow along the x-direction (co-flow arrangement).

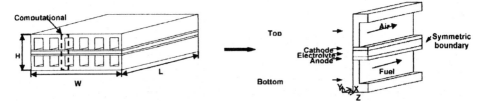

Fig.1.Schematic of a co-flow planar SOFC and the computational domain.

O_2 gas from the air stream diffuses through the cathode and accepts electrons from the external circuit at the cathode/electrolyte interface. Then, the oxygen ions pass through the electrolyte and react with the H_2 at the anode/electrolyte interface. The reaction product H_2O diffuses back into fuel stream. If the fuel is natural gas (CH_4, CO_2, CO H_2 and H_2O), the reforming and shift reactions take place through the anode layer as follows:

Reforming: $CH_4 + H_2O \leftrightarrow 3H_2 + CO + 206kJ/mol$

Water-gas shift: $CO + H_2O \leftrightarrow H_2 + CO_2 - 41kJ/mol$.

There are several assumptions that should be pointed out: (1) Transport and gradients in all three-directions are considered. (2) Energy transfer due to species diffusion in channel flow and porous layers are neglected. (3) Because the single cell model represents a repeating cell unit in the center of a larger stack, cyclic boundary conditions for current are imposed at the boundaries of model domain. Walls at the periphery of the single cell are assumed to be adiabatic.

Thermal-fluid model

In this model, the entire computational domain consists of three sub-domains: porous (electrodes), solid (interconnectors and electrolyte) and pure fluid regions (channel flow). Previous experiments have demonstrated that the fluid flow in the porous media is a low speed mixture gas flow and the Reynolds number is much less than unity. Therefore, this flow can be treated as laminar and incompressible. The properties of porous media are homogeneous and in local thermodynamic equilibrium with the fluid. In the porous region, the flow is governed by the Brinkman-Forchheimer extended Darcy model, which takes into account the effects of flow inertia as well as friction caused by macroscopic shear stresses. While in the fluid region, flow is governed by the Navier-Stokes equations. The equations of continuity, momentum, energy and species for both pure fluid and porous domains are:

$$\nabla \cdot (\rho_f \vec{V}) = 0 \tag{1}$$

$$\frac{\rho_f}{\varepsilon} \nabla \cdot (\vec{V}\vec{V}) = -\frac{\nabla p}{\varepsilon} + \frac{\mu_f}{\varepsilon} \nabla^2 \vec{V} - \eta(\frac{\mu_f}{K}\vec{V} + \frac{\rho_f F_1}{\sqrt{K}}\frac{\varepsilon}{}|\vec{V}| \cdot \vec{V}) \tag{2}$$

$$\nabla \cdot (\varepsilon \rho_f c_f \vec{V} T) = \nabla \cdot (k_{eff} \nabla T) + Q_{g,reaction} + Q_{g,ohmic} \tag{3}$$

$$\nabla \cdot (\varepsilon \rho_f Y_i \vec{V}) = -\nabla \cdot \vec{J}_i + S_i \tag{4}$$

$$\sum_{i=1}^{n} Y_i = 1 \tag{5}$$

$$\rho_f = \frac{P_{op}}{R_g T (\sum_i \frac{Y_i}{M_i})} \tag{6}$$

where parameter η is set to unity for flow in porous medium ($0 < \varepsilon < 1$) and to zero in regions without porous material ($\varepsilon = 1$). V and Y_i represent velocity vector and local mass fraction of each species, respectively. ε is porosity and K is permeability of the porous medium. F_1, Forchheimer coefficient, is calculated by $F_1 = 1.8/(180\varepsilon^5)^{0.5}$. The effective thermal conductivity k_{eff} is equal to $\varepsilon k_f + (1-\varepsilon)k_s$. Equation (3) is only valid for solid region. Ohmic heat source $Q_{g,ohmic}$ in the solid or porous part is a function of local current and resistance. Another source term, $Q_{g,reaction.}$ denotes the heat produced by reforming and shift reactions which take place in any point of anode layer. Thus, once the information on the electrical field is obtained, the source terms can be exactly determined. S_i is the net rate of production of species i by chemical reaction as discussed specifically later. $\vec{J}_i$ is the diffusion mass flux vector derived from the Stefan-Maxwell relations [7]:

$$\frac{J_i}{D_{k,i}} + \sum_{\substack{j=1 \\ j \neq i}}^{n} (\frac{X_i \vec{J}_j}{D_{ij,e}} - \frac{X_j \vec{J}_i}{D_{ij,e}}) = \frac{P}{R_g T} \nabla X_i \tag{7}$$

where X_i is the mole fraction of species i and $D_{ij,e}$ is the effective binary diffusion coefficient of a mixture of species i and j and equal to φD_{ij}. φ is the ratio of porosity to tortuosity. $D_{k,i}$ is the effective Knudsen diffusion coefficient and calculated according to the kinetic theory of gas:

$$D_{k,i} = \varphi \frac{2}{3} \sqrt{\frac{8 R_g T}{\pi M_i}} \bar{r} \tag{8}$$

$\bar{r}$ is the mean pore radius. It should be noted that velocity and temperature in the porous region are volume-averaged quantities. The density, viscosity, specific heat, and thermal conductivity are functions of temperature and mass fraction of gas mixture. In a SOFC, the shift reaction takes place at high temperatures and can be assumed to react very quickly and almost in equilibrium [8]. The reforming kinetics is assumed to be quasi homogeneous. The forward reaction rate (k_s^+, k_r^+) and backward reaction rate (k_s^-, k_r^-) for the two reactions at three different temperatures are reported by Lehmert et al. [8]. Therefore, the reaction rates of shift and reforming reactions are:

$$R_{s,H_2} = k_s^+ (R_g T)^2 C_{CO} C_{H_2O} - k_s^- (R_g T)^2 C_{CO_2} C_{H_2} \tag{9}$$

$$R_{r,CH_4} = k_r^+ (R_g T)^2 C_{CH_4} C_{H_2O} - k_r^- (R_g T)^4 C_{CO} C_{H_2}^3 \tag{10}$$

where C_i ($i = CH_4, CO, CO_2, H_2O$) is the molar concentration of component i. The net rate of production for each species S_i can be stated as the following:

$$S_{CH_4} = -M_{CH_4} R_{r,CH_4} \tag{11}$$

$$S_{CO} = M_{CO} R_{r,CH_4} - M_{CO} R_{s,H_2} \tag{12}$$

$$S_{H_2O} = -M_{H_2O} R_{r,CH_4} - M_{H_2O} R_{s,H_2} \tag{13}$$

$$S_{H_2} = 3 M_{H_2} R_{r,CH_4} + M_{H_2} R_{s,H_2} \tag{14}$$

$$S_{CO_2} = M_{CO_2} R_{s,H_2} . \tag{15}$$

When the reacted molar numbers of methane and hydrogen are determined, the heat generated from chemical reactions is expressed by:

$$Q_{g,reaction} = \Delta H_{reforming} \cdot R_{r,CH_4} + \Delta H_{shift} \cdot R_{s,H_2} \tag{16}$$

In addition, the thermodynamic heat generation from electrochemical reaction at the anode/electrolyte interface is:

$$Q_{electro} = (\Delta H - \Delta G) \cdot m_{H_2}'' \tag{17}$$

here ΔG is the chemical potential. m_{H_2}'' is molar consumption rate of hydrogen at the interface.

Electrochemical model

The electrochemical model predicts local electrical potential and current density, which responds to changes in geometry, local temperature and gas compositions. Numerical simulation for a three-dimensional electrical field is directly analogous to the calculation of heat transfer using a finite volume method. A separate anode/electrolyte/cathode unit is modeled and the discretization of equivalent electric circuit is shown in Fig. 2. Within an electrolyte layer, the current primarily flows from the anode to the cathode. Electrostatic potential within all computational elements must satisfy the Laplace equation:

$$\nabla \cdot (\sigma \nabla \phi) = 0 \tag{18}$$

where σ is the electrical conductivity and ϕ is the electrical potential. The electric field potential calculation combines the following contributions: (1) ohmic losses in all the conducting materials, including the electrolyte, electrodes, and current collectors. (2) The contact resistance at interfaces, and interconnector resistance. (3) Activation overpotentials η_{act} in the anode and cathode layers, which is described by the well-known Butler-Volmer expression [6]:

$$i = i_0 \left\{ \exp(\frac{\beta n_e F \eta_{act}}{R_g T}) - \exp[-(1-\beta)\frac{n_e F \eta_{act}}{R_g T})] \right\} \tag{19}$$

where β is the transfer coefficient and i_0 is the exchange current density. R_g and F are universal gas constant and Faraday constant, respectively. If β is chosen as 0.5, activation overpotential for anode and cathode can be written as follows:

$$\eta_{act,an} = \frac{2R_g T}{n_e F} \sinh^{-1}(\frac{i}{2i_{0a}}) \tag{20}$$

$$\eta_{act,ca} = \frac{2R_g T}{n_e F} \sinh^{-1}(\frac{i}{2i_{0c}}) \tag{21}$$

The concentration overpotentials at electrodes are calculated as:

$$\eta_{con} = \frac{R_g T}{4F} \ln(\frac{p_{O_2}}{p_{O_2}'}) \tag{22}$$

where p_{O_2} and p_{O_2}' are the partial pressures of oxygen at electrode/channel interface and on the reaction site, respectively. The distribution of local current density must be calculated repeatedly and converged to satisfy the same output voltage (V_{out}) between the upper and lower surfaces of interconnectors.

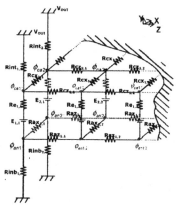

Fig. 2. Schematic of equivalent electric circuit

Once the electrical potential at each node is known, the local current density can be calculated:

$$i(x,z) = \frac{E - \eta_{act}(i) - \phi_{ca} - \phi_{an}}{Re \cdot A_{area}} \tag{23}$$

with Nernst potential:

$$E = E_0 + \frac{R_g T}{2F} \ln(\frac{P_{H_2} P_{O_2}^{0.5}}{P_{H_2O}}) \tag{24}$$

where Re is the electrolyte resistance and A_{area} is the active area. ϕ_{ca} and ϕ_{an} are electrical potentials at cathode/air and anode/fuel interfaces, respectively. p_{H_2}, p_{H_2O} and p_{O_2} are partial pressure of hydrogen, water and oxygen at interfaces between electrodes and electrolyte, respectively. $E_0 = 1.2723 - 2.7645 \times 10^4 T$ is a function of temperature at reaction locations [9]. The thermo-fluid model and electrochemical model are coupled by the boundary conditions at interfaces:

$$-m_{O_2}^{''} = -M_{O_2} D_{O_2,air} \frac{\partial C_{O_2}}{\partial y} + M_{O_2} C_{O_2} v_{int,air} \tag{25}$$

$$-m_{H_2O}^{''} = -M_{H_2O} D_{H_2O,fuel} \frac{\partial C_{H_2O}}{\partial y} + M_{H_2O} C_{H_2O} v_{int,fuel} \tag{26}$$

$$\frac{m_{H_2O}^{''}}{M_{H_2O}} = -\frac{m_{H_2}^{''}}{M_{H_2}} \tag{27}$$

Here $m_{O_2}^{''}$ and $m_{H_2}^{''}$ are molar production or consumption rate of oxygen and water at the cathode/electrolyte and anode/electrolyte interfaces, respectively. $D_{H_2O,fuel}$ and $D_{O_2,air}$ are diffusivities of water and oxygen in air or fuel channel flows, respectively. $v_{int,air}$ and $v_{int,fuel}$ are interfacial velocities and defined as:

$$v_{int,air} = \frac{\sum_{i=O_2,N_2} m_i^{''}}{\rho_{air}} \tag{28}$$

$$v_{int,fuel} = \frac{\sum_{i=H_2,H_2O,etc} m_i^{''}}{\rho_{fuel}} \tag{29}$$

with ρ_{air} and ρ_{fuel} are densities of air and fuel mixture.

PARAMETERS

Air (79% N_2 and 21% O_2) is delivered to the air channel. Fuel (29.58%CO, 13.72%CO_2, 5.5%H_2, 33.05%H_2O, and 18.15% CH_4) is delivered to the fuel channel. The average current density is set to 4000A/m^2. The SIMPLEC method is applied to solve the discretized equations of momentum, energy, concentration and electrical potential. Some input parameters are extracted from literatures [9-11] and listed in Table 1. For common materials used for electrodes and electrolyte, the physical properties of solids strongly depend on temperature, and thus significantly affect the performance of a SOFC. For example, the resistivity of electrolyte at 1200K could be one or two orders of magnitude smaller than that at 900K. This temperature effect is included in the present model. In order to study the geometry effects, we vary the dimensions of channel height, rib width and electrolyte thickness as shown in Table 2. In actual calculations for each case, only one geometrical parameter is varied while keeping Reynolds number and other input parameters fixed.

Table 1 Model input parameters

Parameters and conditions	Value	Parameters and conditions	Value
Fuel inlet temperature (K)	1000	Interconnect resistivity ($\Omega \cdot m$)	2.2×10^{-3}
Air inlet temperature (K)	1000	Contact resistance ($\Omega \cdot m^2$)	1×10^{-6}
Inlet pressure (Pa)	1.01×10^5	Density of electrodes/electrolyte (kg/m^3)	4400
Fuel inlet velocity (m/s)	0.6 (case1)	Density of interconnector (kg/m^3)	5700
Air inlet velocity (m/s)	2.5 (case1)	Thermal conductivity for anode/cathode ($W/m \cdot K$)	12
Cell length (m)	0.05	Thermal conductivity for interconnector ($W/m \cdot K$)	11
Cell width (m)	1.6×10^{-3}	Thermal conductivity for electrolyte ($W/m \cdot K$)	2.7
Anode thickness (m)	5×10^{-5}	Porosity (%)	50
Cathode thickness (m)	5×10^{-5}	Tortuosity	3
Electric resistivity of electrolyte ($\Omega \cdot m$)	$2.94 \times 10^{-5} \times \exp(-10350/T)$	Anode exchange current density (A/m^2)	6300
Ionic resistivity for cathode ($\Omega \cdot m$)	$8.114 \times 10^{-5} \times \exp(500/T)$	Cathode exchange current density (A/m^2)	3000
Ionic resistivity for anode ($\Omega \cdot m$)	$2.98 \times 10^{-5} \times \exp(-1392/T)$	Average pore radius (m)	0.5×10^{-6}

Table 2 Simulation cases

Case No.	Channel height (m)	Rib width (m)	Electrolyte thickness (m)
1	1×10^{-3}	0.4×10^{-3}	150×10^{-6}
2	5×10^{-3}	0.4×10^{-3}	150×10^{-6}
3	2.5×10^{-3}	0.4×10^{-3}	150×10^{-6}
4	0.5×10^{-3}	0.4×10^{-3}	150×10^{-6}
5	0.2×10^{-3}	0.4×10^{-3}	150×10^{-6}
6	1×10^{-3}	0.2×10^{-3}	150×10^{-6}
7	1×10^{-3}	0.4×10^{-3}	100×10^{-6}
8	1×10^{-3}	0.4×10^{-3}	50×10^{-6}

RESULTS AND DISCUSSION

Simulation results for standard case (case 1)

The performance of a SOFC highly depends on the temperature. Electrolyte-supported cells are always designed to operate at >1000K to avoid larger ohmic resistance in the electrolyte. Therefore, the inlet fuel and air temperatures are chosen as 1000K in the present model. Fig. 3 shows the temperature contour on the z=0.001m plane for case 1. Along the flow direction, the temperatures increase monotonically from inlet to outlet due to the contribution of Joule heating and chemical reaction. The maximum temperature is around 1192K near the fuel outlet. On the air side, the temperature rise is not so large as that in the fuel side due to high mass rates. Fig. 4 exhibits the average heat generations in all solid (or porous) components. Clearly, the ohmic heating in electrolyte mainly contributes to the increase in temperature. The heat generation in anode is lower than that in cathode since the smaller activation polarization and resistivity. The heat released from electrochemical reaction on the electrolyte/anode interface is also larger. However, further calculations indicate that the heat releases from shift and reforming reactions are relatively smaller, which only result in the temperature rise in the anode layer about 10~15K. Therefore, for electrolyte-supported cells, minimization of the ohimc resistance and suppressing temperature increase without significantly decreasing the cell performance are key issues, which requires an optimization of parameters governing the ohmic polarization effect and heat transfer efficiency.

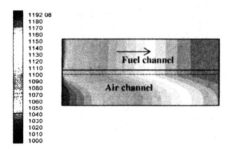

Fig.3 Temperature contours on the z=0.001m plane (arrow denotes flow direction)

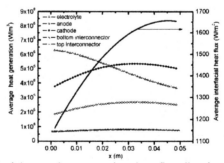

Fig.4 Average heat generation along flow direction

Fig. 5 gives the molar fraction profiles for different gas components at an average current density of 4000A/m². There is a highly non-uniform distribution for oxygen concentration in the z-direction as shown in Fig. 5(a). The molar fraction of oxygen under the rib is lower than that in other areas. This is because oxygen can not diffuse well underneath the interconnector. The molar fraction of H₂ significantly decreases along the flow direction (Fig. 5b). It should be noted that the rib does not cause the remarkable concentration gradients for H₂ in the z-direction since the diffusion of H₂ is about two or three times faster than those of other species. The concentration gradients in the y-direction, normal to the electrolyte/anode interface, produce concentration loss, which will be discussed later. Similarly, molar fraction of CH₄ decreases from fuel inlet to outlet due to reforming reaction (Fig. 5c). However, this decrease is slight, which means the reacted hydrogen in the electrochemical reaction is mainly from the supplied fuel, not from the reforming reaction. Fig. 6(a)-(c) show the calculated overpotential distributions. The activation and concentration polarization at the cathode are relatively smaller. Especially, the concentration polarization is on order of 10^{-4} and almost negligible in this case. However, this polarization is higher near the rib since oxygen can not diffuse well. In experiments, it is usually observed that the concentration overpotential may be several times larger. This is because this overpotential is associated with the detailed nonhomogenous microstructures of electrodes.

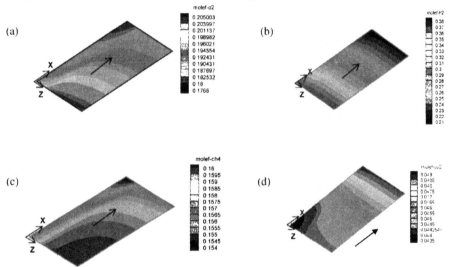

Fig.5 Concentration contours: (a) oxygen at cathode/electrolyte interface, (b) hydrogen at anode/electrolyte interface, (c) methane at anode/electrolyte interface, (d) carbon-dioxide (arrow denotes flow direction).

In the present study, the miscrostructures are assumed to be homogeneous. Similar tendency exists for the anode. The maximum loss is due to the electrolyte and electrodes ohmic polarization as shown in Fig. 6(c), which is in the range of 0.13V-0.27V. In addition, the ohmic overpotential in interconnectors is also substantial and high up to ~0.1V. The output voltage is 0.526V for this case. Fig. 6(d) demonstrated that the local current density profile has a peak value roughly near the middle area. The local current density near the symmetric line is relatively lower because of the longest current path. At the region under the rib, the current density is higher and very smooth. Therefore, the right one to be focused on for improvement should be the ohmic polarization in all components of a cell. In the following section, the effect of channel height, rib width and electrolyte thickness on the SOFC performance will be examined and analyzed.

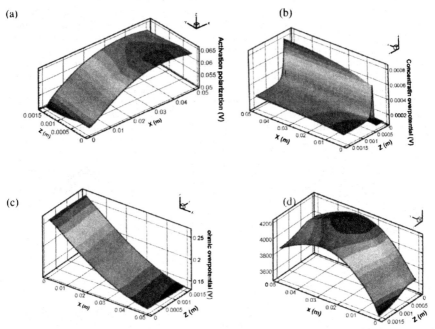

Fig.6 Overpotential and current density distributions: (a) activation polarization at cathode, (b) concentration polarization at cathode, (c) ohmic polarization in electrodes & electrolyte, (d) current density (unit: A/m²).

Influence of channel height

When the channel height is reduced, the current path will be shorter and heat/mass transport rates will increase. This leads to some variations in temperature and concentration distributions and in power efficiency in a SOFC. The average temperature profile for the top interconnector is shown in Fig. 7. With the decreasing of channel height, the average solid temperature level steadily decreases. However, the temperature slopes for smaller channels are steeper except for the region near inlet. This means that although the larger channels have higher temperature levels, the more gradual temperature increases will alleviate the risk of severe thermal stress and thus help ensure the structural integrity of the cell components. Fig. 8(a) and Fig. 8(b) give the calculated mass transfer coefficients for all cases. As expected, the channels with smaller heights greatly improve the mass transfer coefficients. Thus, the concentration polarization related to the non-uniform distribution of the species existing in the flow channels will be significantly reduced in cells with smaller channels. Similarly, the increase in heat transfer rate on air side will help improve the cooling effect and remove more heat.

However, the efficiency improvement of a SOFC is always our final interest. For a cell with smaller channel height, on one hand, low temperature increases the ohmic loss of solid part, which is a function of temperature. On the other hand, shorter current path and lower concentration loss may partially or totally counteract this ohmic polarization increase. Fig. 8(c) illustrates the comparison of output voltages. The output voltage for smaller channels (case 4) increased by 86.87% compared with largest channels (case 2). The power density (total power/cell weight) will be greatly improved due to an increase of output voltage and area/volume ratio However, when the channel height is further reduced to 0.2mm, the output voltage would not be improved, but decreased slightly. Therefore, in the present study, the smallest channels do not improve the efficiency of cell and the SOFC with a 0.5mm height shows the best performance.

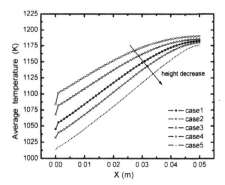

Fig.7 Comparision of average top interconnector temperature

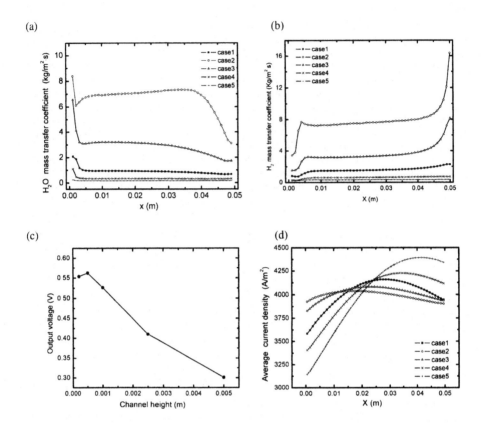

Fig.8 Effects of channel height on cell performance: (a) mass transfer coefficient of H_2O, (b) mass transfer coefficient of H_2, (c) output voltage, (d) local current density

In addition, with the decreasing of channel dimensions, the peak value of local current density shifts to the downstream of the flow as indicated in Fig. 8(d).Consistent with the average temperature distributions, the current density profiles flatten with increasing channel heights. For the smallest channels, peak current density occurs near the gas outlet and current density distribution is highly non-uniform.

Influence of rib width and electrolyte thickness

If the width of a rib is decreased with a fixed cell width, the contact resistances between the rib and anode or cathode will increase. On the other hand, the active reaction area will increase and concentration overpotential in the porous electrode underneath the interconnector rib will decrease. To obtain the maximum cell performance, the rib width should be optimized. Fig. 9 exhibits oxygen and methane molar concentration distributions at interfaces for case 6. It is clear that the concentration gradients significantly decrease by comparing with those with thicker ribs (case 1) as shown in Fig. 5(a) and (b). Calculation shows that the overall efficiency of a cell decreases since the output voltage drops from 0.5263V to 0.438V.

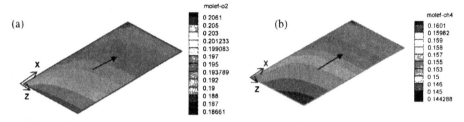

Fig.9. Concentration distributions for case 6 (with thin rib): (a) oxygen at cathode/electrolyte interface, (b) methane at cathode/electrolyte interface

From Fig. 10(a), it can be seen that the current density distribution is more uniform and the average innterconnector temperature rise is suppressed by about 50K for case 6 as compared with case 1, which will alleviate thermal stress. Therefore, for the present two cases, the smaller ratio of the channel width to the rib width, i.e., 3:1, is beneficial to reduce the ohmic losses at the interface and improve performance. Just as mentioned before, the ohmic polarization in the electrolyte layer is a major contributor to the voltage loss. Fig. 10(b) and Fig. 10(c) show the performance improvement when reducing the electrolyte thickness. The solid temperatures decrease and the ohmic polarizations in the electrolyte and electrodes are greatly reduced. Terminal voltages are increased to 0.5582V with $100\,\mu m$ of electrolyte thickness, and 0.61V with $50\,\mu m$ thickness, respectively.

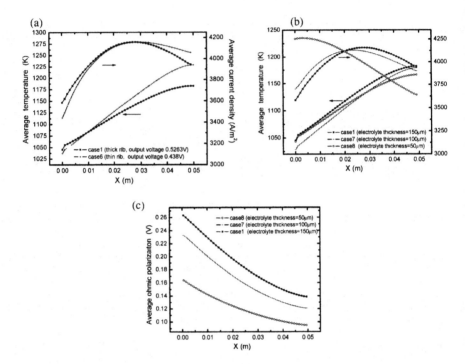

Fig.10. Effect of rib width and electrolyte thickness: (a) different rib width, (b) different electrolyte thickness, (c) ohmic polarizations

CONCLUSION

A 3D thermo-fluid/electrochemical model is developed to evaluate the heat/mass transfer and electrochemical performance for an electrolyte-supported SOFC. Based on this model, the effects of geometry on the distributions of temperature, local concentration, local current density and output voltage are further investigated. Simulation results indicate that when the height of flow channels is decreased, the average solid temperature is reduced accordingly that improves the cell efficiency due to higher heat/mass transfer coefficient between the channel wall and flow stream, and a shorter current path. However, the smallest channels do not further improve the efficiency. The cell with a thicker rib width and a thinner electrolyte layer has higher efficiency and lower average temperature. Actually, the present study has demonstrated that there is a trade-off among factors influencing the performance of a SOFC. Therefore, our future work is to find better approaches or other critical parameters to optimize the design of a solid oxide fuel cell.

ACKNOLEGEMENTS

This research was partially supported by the Andrew H. Hines, Jr./Florida Progress Endowment Fund.

REFERENCES

[1] S. C. Singhal, K. Kendall, *High temperature solid oxide fuel cells*, Elsevier, New York, 2003.

[2] H. Yakabe, T. Ogiwara, M. Hishinuma, I. Yasuda, "3-D model calculation for planar SOFC," *J. Power Sources*, **102**, pp. 144-154 (2001).

[3] H. Yakabe, K. Sakurai, "3D simulation on the current path in planar SOFCs," *Solid State Ionics*, **174**, pp. 295-302 (2004).

[4] J. J. Hwang, C. K.Chen, D. Y. Lai, "Detailed characteristic comparision betweenplanar and MOLB-type SOFCs," *J. Power Sources*, **143**, pp. 75-83 (2005).

[5] J. W. Kim, A. V. Virkar, K. Z. Fung, K. Mehta, S. C. Singhal, "Polarization effects in intermediate temperature, anode-supported solid oxide fuel cells," *J. Electrochem. Soc.*, **146**, pp.69-78 (1999).

[6] P. W. Li, L. Schaefer, M. K. Chyu, "A numerical model coupling the heat and gas species' transport processes in a tubular SOFC," *ASME J. of Heat Transfer*, **126**, pp.219-229 (2004).

[7] A. F. Mills, *Mass transfer*, Prentice Hall (2001).

[8] W. Lehnert, J. Meusinger, F. Thom, "Modeling of gas transport phenomena in SOFC anodes," *J. Power Sources*, **87**, pp. 57-63 (2000).

[9] S. H. Chan, K. A. Khor, Z. T. Xia, "A complete polarization model of a solid oxide fuel cell and its sensitivity to the change of cell component thickness," *J. Power Sources*, **93**, pp. 130-140 (2001).

[10] N. F. Bessette, W. J. Wefer, J. Winnick, "A mathematical model of a solid oxide fuel cell," *J. Electrochem. Soc.*, **142**, 3792-3780 (1995).

[11] M. Iwata, T. Hikosaka, M. Morita, T. Iwanari, et. al, "Performance analysis of planar-type unit SOFC considering current and temperature distributions," *Solid State Ionics*, **132**, pp. 297-308 (2000).

Author Index